CHEMICAL INDUSTRY
with special reference to the U.K.
General Editors: J. DAVIDSON PRATT AND T. F. WEST

A History of the Modern British Chemical Industry

FIG. 1. ONE OF THE FIRST BRITISH CHEMICAL FACTORIES
A general view of James Muspratt's works in Vauxhall Road, Liverpool, *ca.* 1830, from a contemporary engraving in the possession of Liverpool Public Libraries. The wide-based chimney was, before Gossage, in 1836, discovered water film absorption of hydrochloric acid, the only means available of dispersing the great volumes of acid from the saltcake pots. (The windmill in the background is that, still extant, on Bidston Hill on the opposite bank of the Mersey.)

A History of the Modern British Chemical Industry

BY

D. W. F. HARDIE

AND

J. DAVIDSON PRATT

PERGAMON PRESS

OXFORD · LONDON · EDINBURGH · NEW YORK

TORONTO · SYDNEY · PARIS · BRAUNSCHWEIG

PERGAMON PRESS LTD.
Headington Hill Hall, Oxford
4 & 5 Fitzroy Square, London W.1
PERGAMON PRESS (SCOTLAND) LTD.
2 & 3 Teviot Place, Edinburgh 1
PERGAMON PRESS INC.
Maxwell House, Fairview Park, Elmsford, New York 10523
PERGAMON OF CANADA LTD.
207 Queen's Quay West, Toronto 1
PERGAMON PRESS (AUST.) PTY. LTD.
19a Boundary Street, Rushcutters Bay, N.S.W. 2011, Australia
PERGAMON PRESS S.A.R.L.
24 rue des Écoles, Paris 5^{e}
VIEWEG & SOHN GmbH
Burgplatz 1, Braunschweig

First edition 1966
Reprinted with corrections 1969
Library of Congress Catalog Card No. 66-16455

Printed in Great Britain by Bell and Bain Limited, Glasgow

08 011686 8 (flexicover)
08 011687 6 (hard cover)

Contents

EDITORS' PREFACE vii

FOREWORD ix

PART A. A HISTORY OF THE MODERN BRITISH CHEMICAL INDUSTRY 1

Introduction 3

Chap. I. Origins and Rise of the Industry 15

Chap. II. Transition to the Modern Industry 81

Chap. III. The Modern Industry 116
- Heavy Inorganic Chemicals 116
- Dyes and Organic Pigments 156
- Biologically-active Chemicals 166
- New Materials 187
- Heavy Organic Chemicals 225

Chap. IV. The Economic Importance of the Chemical Industry 256

PART B. COMPANIES OF IMPORTANCE IN THE BRITISH CHEMICAL INDUSTRY 267

PART C. TRADE ASSOCIATIONS 330

INDEXES 345
- Part A 347
- Part B 361
- Part C 377

CHART OF THE PRINCIPAL PRODUCTS OF THE BRITISH CHEMICAL INDUSTRY *facing page* 380

Editors' Preface

WE WERE asked by Sir Robert Robinson, O.M., P.P.R.S., to organize the preparation of a series of monographs as teaching manuals for senior students on the Chemical Industry, having special reference to the United Kingdom, to be published by Pergamon Press as part of the Commonwealth and International Library of Science, Technology and Engineering, of which Sir Robert is Chairman of the Honorary Editorial Advisory Board. Apart from the proviso that they were not intended to be reference books or dictionaries, the authors were free to develop their subject in the manner which appeared to them to be most appropriate.

The first problem was to define the chemical industry. Any manufacture in which a chemical change takes place in the material treated might well be classed as " chemical ". This definition was obviously too broad as it would include, for example, the production of coal gas and the extraction of metals from their ores; these are not generally regarded as part of the chemical industry. We have used a more restricted but still a very wide definition, following broadly the example set in the special report (now out of print), prepared in 1949 by The Association of British Chemical Manufacturers at the request of the Board of Trade. Within this scope, there will be included monographs on subjects such as coal carbonization products, heavy chemicals, dyestuffs, agricultural chemicals, fine chemicals, medicinal products, explosives, surface active agents, paints and pigments, plastics and man-made fibres. The chart at the end of the book sets out some of the ramifications and intra-connections of this diversified group of chemical manufactures.

A list of monographs now available and under preparation is appended.

As our task progressed, it became apparent that there was a great need for a monograph of a semi-popular nature, which would provide a brief up-to-date history of this scientific and progressive industry, and of the tremendous developments which have taken place, particularly since the two Great Wars. This first volume is intended to serve as an introduction to the whole series.

We wish to acknowledge our indebtedness to Sir Robert Robinson for his wise guidance and to express our sincere appreciation for the encouragement and help which we have received from so many individuals and organizations in the industry, particularly The Association of British Chemical Manufacturers.

The lino-cut used for the covers of this series of monographs was designed and cut by Miss N. J. Somerville West, to whom our thanks are due.

J. DAVIDSON PRATT
T. F. WEST
Editors

Foreword

WHILE this book is designed primarily as an historical introduction to a series of monographs, to be published by the Pergamon Press, on specialized aspects of chemical industry, it is also intended to supply a gap in the techno-historical literature. Only two histories* dealing exclusively with the British chemical industry have been published in this century and both of these appeared before World War II and, therefore, before the characteristically modern developments of the industry became fully established.

A technological history, besides being an explanatory account of the chronological sequence of events, should have regard to the particular mode of development of the branch of technology studied. The authors see the mode of development of chemical industry mainly in the changing web of its processes, and regard its organizational and economic developments as of secondary significance and as substantially determined by the processes employed. A thorough-going treatment, as this is, of the history of the British chemical industry from this standpoint is believed to be novel. Recognition by its few historians of this the principal matter-transforming industry as the archetype of process industries has long been delayed, although it is this very status which makes the chemical industry a peculiarly interesting one for techno-historical study. Demand for the individual products is largely determined by the general development of technology,

---

* *A History of the British Chemical Industry*, Stephen Miall, Ernest Benn Ltd, London, 1931; *British Chemical Industry, its Rise and Development*, Sir Gilbert T. Morgan and David D. Pratt, Edward Arnold & Co., London, 1938.

but process development obeys its own laws: thus, the same chemicals, over the years, are often made by a series of processes, succeeding one another as their economic and technical advantages are discerned. It is in the process field that the chemical industry is changed by the advance of applied science and engineering, and it is there that invention and research have their impact upon it. For example, a few short years ago, ethylene, the most important " building brick " in the plastics industry, was made by a single process, today it is produced by a constellation of closely-competing methods.

A general understanding of the nature of the chemical industry is most readily acquired through acquaintance with its techno-historical development. Knowledge of chemistry alone, or merely of the economic or organizational history of the industry will not suffice. Prestige histories, prepared for individual chemical concerns, often, as might be expected, give success stories of the onward-and-upward type and tend to obscure the historical realities. Some excellent histories of sections of the modern British chemical industry have been published by chemical companies, for example, the history of Unilever by Wilson* and of Albright & Wilson by Threlfall,† but these do not reach the general reader.

The British contribution to chemical industry, which it is a principal object of this book to isolate clearly, has not been inconsiderable: the new materials " Polythene " and " Terylene " and five out of the seven important developments in the dyestuffs field in recent years were of purely British origin. It is, however, shown that a large number of the processes at present operated in British chemical factories came from abroad, a circumstance for which there are in most instances good historical reasons.

The techno-historical section, which is the first and main part of this book, deals with its subject in three chapters; Chapter I is concerned with the industry in the nineteenth century, Chapter II

---

* *The History of Unilever*, C. Wilson, Cassell & Co., London, 1954.

† 100 *Years of Phosphorus Making*, R. E. Threlfall, Albright, & Wilson Ltd, Oldbury, 1951.

with the transition to the modern phase, and Chapter III with the history of the twentieth century developments. The second chapter contains probably the first attempt systematically to examine the whole complex of factors which marked the transition from the old, largely inorganic, chemical industry to that which emerged in this century. The history of the modern industry is treated under heavy inorganic chemicals, dyes and pigments, biologically-active chemicals, new materials (i.e., plastics and the like), and heavy organic chemicals. The other two parts of the book contain brief historical summaries of the organization of the main firms in the industry and of its important trade associations; these are primarily non-technical.

In preparing the technical history it has been assumed that the reader is acquainted only with the fundamental notions of chemistry. Where some understanding of the scientific basis of a development beyond that level is necessary, this has been supplied in as non-technical language as possible. It is to be hoped that not only will this work serve its primary purpose of introducing the other volumes on chemical technology, but also have some value as a contribution to the history of an industry which is the point of application of one of the great positive sciences to human affairs and as such be of interest to those in search of landmarks in the no-man's land between what are now fashionably known as the Two Cultures.

PART A

A History of the Modern British Chemical Industry

"*Technology concerns itself with the creative acts of man. To enthuse about Gothic churches and Tudor town halls and not even to glance at Viscount aeroplanes and stressed concrete bridges cannot be justified on the grounds that only the former are creative arts of man. It is simply an unwillingness to face the present.*"*

"*It is more important to know the properties of chlorine than the improprieties of Claudius!*"†

"*The whole fabric of modern civilisation becomes everyday more interwoven with the endless ramifications of applied chemistry.*"‡

---

* Sir Eric Ashby (in a speech to the Scottish Institute of Education at Dunblane, 1956).

† Lord Cherwell.

‡ Leo Hendrik Baekeland (in an address to the Society of Chemical Industry, 1938).

Introduction

Historical writing on the chemical industry has not, in this country, been notably characterized by attempts to arrive at a general view of its development similar to those discerned in the fields of political, constitutional and economic history. The task of synthesizing chemical industrial history from the wealth of fragments dealing with its many departments has still largely to be performed; this is particularly so in the case of the industry in its twentieth century development. Part at least of the explanation which may be advanced for the present state of chemical industrial history is that it has been studied *backwards*: present-day branches of the industry have been individually traced back too often in isolation from their interconnections with other branches of technology and within the limits of the activities of particular industrial concerns engaged in them. While this is not a wholly invalid approach, its dangers are that it leads to modern conceptions being read into past stages of the industry and to the production of prestige history, a form of sophisticated advertising now so voluminously produced by science-based industries—such as the chemical industry—in order to explain themselves to the educated public, or to enhance their image.

In the present work an historical account is given of the modern chemical industry as it has evolved in the United Kingdom. A particular object of the work has been to separate the purely British contribution to the inventive and other aspects of the industry. To make the account comprehensible, the developments of the earlier industry have been outlined and the transition from that to the modern stage analysed. It is assumed that the reader has a sufficient grounding in the fundamentals of chemistry to appreciate the technical content of the exposition in its essentials. In the belief that a brief examination of certain general aspects of the chemical industry will facilitate understanding of the historical account, the remainder of this *Introduction* has been devoted to that purpose.

1. DEFINITIONS OF CHEMICAL INDUSTRY

Chemical industry is the archetypal matter-transforming industry; it is also a principal exemplar of process technology. The function of a matter-transforming industry is to produce chemical or physical changes in the nature of substances; in effect it is a kind of *external metabolism*, whereby man converts the substance of his environment into technologically useful (" digestible ") forms. Definition of the field of chemical technology is a subject with a curious and instructive history. Currently accepted definitions generally approximate to that set out in 1949 by the Association of British Chemical Manufacturers in their *Report on the Chemical Industry*. The taxonomy* of modern chemical industry would form an absorbing topic for a technological history thesis; it is one of some philosophical and practical interest, indeed much of the history of the industry is implicit in its definitional complexities. All definitions that have been attempted obscure the actual scope of chemical technology. In recognition of the violence done to actualities by proposed definitions, the phrase " chemical and *allied industries* " is in general use. The allied industries include most of the matter-transforming industries which produce material mixtures rather than substantially pure substances, e.g., coal gas, soap, and glass manufacture. All definitions agree on exclusion of the ferrous metal industries, even from the " allied " category, although production of some metals, e.g., titanium, aluminium, and perhaps nickel, is commonly regarded as lying within, or at least near to the frontiers of chemical industry. Sodium manufacture and the

* I.e., the science or principle of classification.

production of the other alkali metals, the alkaline earth and rare earth metals are generally accepted as chemical industrial operations.

The Oxford English Dictionary defines a process as " a continuous and regular action or succession of actions, taking place or carried out in a definite manner ". All chemical industrial processes comprise at least the two stages of (1) bringing the reactive substances into contact under suitable conditions of mixing, pressure and temperature, often in the presence of catalysts (the conversion stage), and (2) separating as effectively as possible the product or products from the mixture resulting from the reaction. At the centre of a chemical process is the conversion stage at which the matter-transforming reaction takes place. This is the stage which determines the way in which the process as a whole is carried out, the raw materials required, materials of construction of the reaction vessels, thermal, pressure and other conditions, wastes to be disposed of, or co-products to be recovered. While the increasing variety of its products parallels the increase of its importance in the general technological field, *the fundamental history of chemical industry is that of the evolution of its processes*, and its eras may be discerned in the general changes in their nature. The individual histories of chemical firms modify and mark the scale and tempo of the evolving operations, but the dynamic determinant is the rate of invention of new techniques and processes. The logic of chemical technology, the interrelation of its processes, the laws of fixed chemical proportions, which determine the relative amounts of co-products and of by-products have predominantly woven the evolving pattern of the chemical industry from its earliest to its latest phase. The development and refinement of processes by chemical engineering and the application of physical science form the bases of modern chemical technology. In the modern chemical industry there are often whole constellations of processes grouped round the manufacture of a single chemical commodity, each having its particular, often temporary, generally economic, advantage. It is an important feature of modern chemical technology that processes

B

are superseding one another at a steadily increasing rate. In West Germany, for example, about half of the products of the chemical industry are being made by processes having only a seven- to ten-year life period. Chemicals seldom carry any indication of the process by which they were produced; unlike other artefacts, their nature is determined by fixed quantitative natural laws.

2. WHAT IS A CHEMICAL?

Just as gas meters had to be devised in order that coal gas could become a commercial commodity by the establishment of a price–quantity relationship, chemical analysis had to advance to a certain empirical level before any important trade in chemicals could evolve—the consumer had to know how much of the technologically effective substance he was buying. In the early years of the synthetic alkali industry soap boilers had to purchase filthy messes, containing often only a minor quantity of actual sodium carbonate. As early as 1756, Francis Home estimated the alkali content of wood ashes (i.e., potash) by titration with acid, and William Lewis, in 1767, advanced this procedure by standardizing the acid against a potassium carbonate solution of known strength, using litmus paper as indicator. By 1806 volumetric methods were firmly established in the chemical industry. The notion of a " chemical " as a commodity of specified purity, containing only impurities which do not interfere with its technological effectiveness, arose at a very early stage in the history of chemical industry. A " chemical ", as opposed to technical products such as tar, pitch, crude petroleum, paraffin wax, is a technologically useful substance containing, except for minor amounts of impurities quantitatively and qualitatively known, a single chemical species.* Most of the principal

* Synthetic materials such as polyvinyl chloride, polymethyl methacrylate, strictly fall outside this definition, since they contain mixtures of closely-related molecular species. They are products of their monomers, however, and these are undoubtedly " chemicals " within the definition. Writers generally refer to these high molecular polymers as " plastics ", " polymers ", or " polymeric substances ".

industrial chemicals contain over 95% of the " chemical ", and some more than 99%. Many common commercial chemicals may contain between ten and twenty impurities, some of these in amounts detectable only by means of the spectroscope. Chemicals, not specifically purified for use in the manufacture of foodstuffs or pharmaceuticals or in chemical analysis (AR chemicals) are generally referred to as being of " technical " or " commercial " grade. BP chemicals are substances complying in purity with the requirements of the *British Pharmacopoeia*, which carry the force of law in the case of all chemicals sold and used as pharmaceuticals. Definitions of " heavy " and other chemical groups are given in the main text.

During the two centuries or so in which scientific chemistry has been practised probably some three-quarters of a million chemical compounds have been synthesized, many of them already found in nature, but many more entirely unknown and purely artefactual.* The rate of multiplication of new forms of matter greatly accelerated from about 1850 onwards, when the highly fruitful field of organic chemistry began to be systematically cultivated. So great has the proliferation of organic compounds become in modern chemistry that chemists are faced with the need to devise new systems of nomenclature in order unambiguously to identify and classify the new products of their laboratories. The most advanced chemical lexicons and encyclopedias (e.g., Beilstein's *Handbuch*) are already a decade or so in arrears in reporting new compounds and the recently observed properties of substances already known. It is one of the most striking—and seldom remarked—features of modern chemical industry that, of the welling stream of new forms of matter only a very small proportion has even now flowed into chemical technology. Extensive fields of *unapplied* chemistry lie on the outskirts of chemical industry, e.g., that of pressure syntheses from acetylene, associated with the name of its leading exponent, Dr. Walther Reppe. Reppe chemistry offers the possibility of

---

* I.e., man-made, as opposed to natural.

synthesizing on an industrial scale many hundreds, and possibly thousands, of hitherto inaccessible acetylene derivatives; technology has so far, however, found application for only a few of these. What may be called the low " application factor " of modern chemical products, i.e., the comparatively small number of known substances that have become industrial commodities would appear to be a subject calling for urgent techno-economic and methodological study. Possible reasons for this low application factor are both too complex and controversial to be pursued here, and the matter has been mentioned at all only to emphasize that, although the chemical industry is manifestly " science-based ", its relationship to pure chemical discovery has certain very strict limitations. At the present time, taking into account the numerous grades and brands of substantially the same chemical, the British chemical industry in all its branches produces some ten thousand chemical commodities, and of these many are produced on a laboratory rather than on a works scale. In other words, about *one per cent of discovered chemical compounds have so far become industrial chemicals.*

3. CHEMICALS AND TECHNOLOGY

Industrial chemicals may be considered as being applied in technology under four principal categories, each characterized by common general aspects.*

(i) *Modification of Materials* by cleaning (detergency, bleaching, metal-pickling), colouring (pigmenting, dyeing, painting), application of protective layers (rust-proofing, lacquering, case-hardening, fire-proofing). The great increase in the use of iron—the characteristic structural

* These categories concern the *end-point* applications and, therefore, exclude the large class of primary and intermediate chemicals consumed in the manufacture of other chemicals. This use of chemicals *within* the chemical industry itself does, in many sectors, greatly reduce the actual output tonnage reaching the consumer; *thus* between two and three tons of calcium carbide are manufactured for every ton of polyvinyl chloride (PVC) sold.

material of the Industrial Revolution—in the construction of bridges, ships, buildings, and in railway transport produced, even before the mid-nineteenth century, a steeply-increasing demand for protective paints, varnishes and lacquers, etc.

(ii) *Energy Production* by slow or rapid combustion or decomposition as fuels (or fuel components), and as propellant or disruptive explosives.

(iii) *Modification of Processes or Conditions in the Natural Realm* by crop fertilization, growth-promotion, pest- and weed-control, pharmaceutical and antibiotic action.

(iv) *Manufacture of Artificial Materials for the Fabricating Industries* by polymerization to macromolecular substances capable of being formed into rigid objects, rubber-like masses, films, filaments, and fibres for textile production.

It is to be noted that in the first category the " modification " is mainly produced at the surface of the material or articles to which the chemicals are applied, i.e., the applications are " two-dimensional " and the weight of chemical employed is small compared with the mass treated—a little, so to speak, goes a long way.

Definition of a fuel as a source of energy releasable by subjecting it to oxidative decomposition justifies placing together in the second category such rocket fuel chemicals as hydrogen peroxide, liquid oxygen, fluorine, and paraffins with the whole class of industrial and military explosives. Explosives are, in effect, *fuels* that burn with very high velocity, the necessary oxygen for their combustion being an integral part of their molecular structure. A gun is, in essence, a single-cylinder internal combustion engine. When Huygens, in 1680, experimented with what was almost certainly the very first internal combustion engine, he exploded gunpowder in a cylinder closed by a piston. Although this gunpowder engine failed to achieve technical realization, its conception marks Huygens' recognition of the common character of fuels and explosives.

Explosives depend for their action upon their capacity for generating, in a small fraction of a second, very large volumes of gas at high temperature; their mechanical effect is thus great in proportion to their original bulk. Having regard to the devastation wrought throughout the provinces of France and Belgium, not to mention the other theatres of combat, it is of interest to recall that during the entire course of World War I, the British chemical industry produced a total of some 750,000 tons of explosives—only about one-third of the present U.K. annual output tonnages of sulphuric acid or of sodium carbonate!

Chemicals of the third category, if we except the agricultural fertilizers and some of the weed-killing agents, are in most instances substances of considerable molecular complexity of high monetary value and produced in comparatively small quantities. Their action results from their participation, in similar fashion to catalysts, in the biological processes in which they are effective. Indeed, many chemicals applied in the natural domain function as negative catalysts, obstructing or poisoning normal processes of the organisms. Again, effect is great in relation to amount. In passing, it should also be remarked that, in the case of the large-tonnage fertilizers and weed-killers, their application is essentially of a " two-dimensional " nature. It has been estimated that, at the present time, the application of fertilizers in the U.K. (in terms of their N, P_2O_5 and K_2O contents) averages 0·56, 0·43 and 0·44 cwt/acre, respectively.

The fourth category of listed applications differs fundamentally from the preceding three in that the chemicals are used *qua materials* and endure as long as the artefacts which they compose, whereas, in their technological applications, the chemicals in the other categories are destined to be irretrievably dispersed (by dilution, evaporation, etc.), destroyed (by combustion or explosion), or otherwise to become uneconomic to recover (e.g., dyes, pigments, etc.). From the United States comes a picturesque example of chemical dispersion: it is calculated that the iron sulphate flowing to waste from American steel pickling plants contains sufficient iron to build 200,000 motor cars a year, not to mention the vast associated loss of sulphur!

Category four applications have brought chemicals into quantitative competition with natural materials (steel, glass, cotton, wool, rubber, wood, etc.). It has been estimated that, in 1962, the U.K. total output of plastics was of the order of 700,000 tons, and that output was increasing at a rate of 14% (compound interest) a year. On present showing, it would appear to be highly probable that this most recent "three-dimensional" entry of chemicals into technology will produce the most revolutionary quantitative changes in modern chemical industry. These new synthetic materials have extended, actually and potentially, the technological function of chemicals far beyond the limitations of their former modificatory, ancillary, and facilitative roles.

4. PRESENT IMPORTANCE OF CHEMICAL INDUSTRIAL HISTORY

A modern historical study of our national chemical industry is not a mere matter of patriotic piety—or an academic techno-ecological essay, but, in this present time, an urgent necessity to present the evolved and evolving facts of a subject still too frequently superficially treated in the growing body of techno-historical writing, and only too often distorted by the ignorance of the politically-engaged or the partiality of prestige copy-writers. Few industries make such slight impact on the public consciousness as does chemical industry and yet have such far-reaching and ubiquitous effects on the texture and modes of everyday life. Even those intrigued by the wonders of "popular" science, often half-confounded in the public mind with science fiction, are generally unequipped to appreciate the part played by chemical technology in almost every aspect of their immediate environment: the motorist, knowledgeable, perhaps, on the mechanical function of every part of the complex artefact that is his pride and social symbol, is totally ignorant of the score or so chemicals that have been ancillary to its manufacture or compose parts of its final structure; women shoppers gazing at fashionable displays deployed in great city stores do not see the subtle webs of macromolecules tinted by colours never before known even in the prolific natural universe!

One of the principal obvious reasons for the general ignorance of chemical industry is that many chemicals are applied in processes in the privacy of factories by specialist trade consumers. Even those chemicals which reach the public generally do so as the constituents of functional mixtures, packaged and brand-named (only in certain instances is there a statutory requirement to declare their chemical composition). The field of modern pharmaceuticals presents an array of impressive-sounding neologisms; for example, the trade names of a number of modern drugs taken at random and ranging from sedatives to anti-smallpox agents give little or no indication to an organic chemist of their chemical nature. The naming of modern drugs and many other chemicals used by the public has not, in the main, been a matter of commercial obscurantism but has arisen as a necessity of *communication.* Actual systematic chemical names of many chemicals in wide industrial, medical, or domestic use would occupy a long line of print and, in any case, imply a whole series of difficult, highly-specialized technical concepts which render them meaningless to all but those trained in the art.* To the layman, in what the Spanish author Ortega y Gasset has described as this "Aspirin Age", it is of no significance to be told that he is dulling his aches with *acetyl-salicylic acid*, or that, when he repels insects with a dispenser discharging DIMP, he is employing dimethyl phthalate. When the motorist is informed that the properties of his petrol have been improved by addition of *tetra-ethyl lead* (TEL) he is little wiser than before. The end-consumer of chemicals in the public domain tends to think of them as "things" possessing specific desirable effects; he fails

* It is no adverse criticism of medical practitioners or pharmacists to assert that most of them are quite unacquainted with the *chemical nature* of the great number of pharmaceutical agents they daily prescribe and supply!

to discern " chemicals " as forming a pervasive, entirely distinct element in our modern environment.

Only by an historical study of the development of processes, modes of change in relationship to the pure sciences, and the functions of its products in general technology can a meaningful picture of modern chemical industry be put before the educated non-specialist reader.

CHAPTER I

Origins and Rise of the Industry

1. THE CHEMICAL PHASE OF THE INDUSTRIAL REVOLUTION (1750–1850)

The mid-eighteenth century saw the beginnings of a change in human affairs which, at least in its material aspects, was radically to differentiate all subsequent history from what had gone before. This transformation, as important in human affairs as that which changed man from an errant hunter to a settled crop-grower, had its first and, for long, its principal site in the United Kingdom, the first European country to liberate itself from feudal institutions and having, at that time, the most assured access to the mercantile potentialities and raw material resources of the New World. It has been said that the most important contribution ever made by Britain to the world is the steam engine. It was undoubtedly this source of power in association with the well-known constellation of textile inventions that primarily brought into being the Industrial Revolution, with its rapid effects on the concentration and habits of the population. The chemical phase of the Industrial Revolution followed inevitably on the mechanical phase: urbanization of the population produced new foci in the manufacturing economy. Whereas, in 1785 there were in the U.K. 971 hard soap-makers producing an annual average of 16 tons each, by 1830 there were only 309, with an average annual output of 170 tons. Between 1800 and 1835 increase in hard soap production was at a compound interest rate of 2·4%; over the same period the cotton textile industry's output increased at an annual compound interest rate of 6·4%. Both soap and textile manufacture, being consumers of alkali, increased the demand for that chemical commodity. Glass manufacture, the other alkali-consuming technology, although its output was maintained, showed no firm tendency to expand over the first three decades of the nineteenth century. Almost to the end of the second decade of the century, alkalis for soap, glass and textile manufacture were supplied almost wholly from the processed ashes of barilla (a littoral plant of the goose-foot family) and kelp (the ash residue of seaweeds); some sulphuric acid was employed as a sour in textile bleaching and in the small metal trades in the English midlands: the chemical industry, as a new and developing consequence of the Industrial Revolution, was just beginning.

2. SULPHURIC ACID

All processes for the manufacture of sulphuric acid involve oxidation of sulphur to sulphur dioxide, followed by catalytic oxidation to the trioxide and reaction of the latter with water:

$$(1)\quad S \xrightarrow{(O_2)} SO_2 \xrightarrow{(\frac{1}{2}O_2)} SO_3$$

$$(2)\quad SO_3 + H_2O \rightarrow H_2SO_4$$

The actual mechanism of the reaction is still a matter of controversy. As a marketed commodity,* sulphuric acid became available early in the eighteenth century when Joshua Ward of Twickenham manufactured it on a small scale by burning a mixture of brimstone and saltpetre above a shallow layer of water under a glass bell. In 1746, Dr. Roebuck and Garbett, at Birmingham, took the momentous step, on a hint from their

* Boswell records that Dr. Johnson, in 1772, requiring the acid for an experiment, observed to a friend: "You will see there (at Temple Bar) a chemist's shop at which you will be pleased to buy for me an ounce of oil of vitriol; not spirit of vitriol, but oil of vitriol. It will cost a penny-halfpenny." Other witnesses have reported that Johnson was an enthusiastic amateur chemist, but the reason for this interest (except perhaps by intellectual infection from his dissipated friend Topham Beauclerk) and the object of his experiments appear to be equally unknown.

reading of Glauber's works, of employing sheet lead, supported on a framework of wood, in the construction of the reaction chamber. Roebuck and Garbett's process liberated sulphuric acid production from the limitations of glass apparatus and permitted the increase in plant scale necessary for large-tonnage manufacture of the acid. The size of acid chambers was increased from a few hundred cubic feet ultimately to sequences of structures each with the capacity of a large concert hall. For a century and more these great cathedrals of lead became the characteristic feature of the chemical industrial skyline.

At an early stage the practice of passing a draught through the chambers made the sulphuric acid process continuous. With the invention, in 1859, by John Glover (who entered the Newcastle chemical industry as a plumber repairing lead chambers) of his tower, which combined the functions of concentrating the acid and conserving the nitrogen oxide catalysts, the classical sulphuric acid process was fully evolved. The Gay-Lussac tower, invented in France in 1827, to recover catalytic oxides in the exit draught from the chambers, only came into use in Britain after the Glover tower had become an established component of sulphuric acid plant. The first description of Glover's tower appeared in a German periodical (Dingler's *Polytechnic Journal*): the advantages of the invention were so manifest to German acid-makers who until then had been unaware of its existence, that it was immediately adopted by them!

It is of interest to mention at this point that, in 1831, Peregrine Phillips junior, of whom nothing is known except that he was sometime a vinegar merchant of Bristol, patented a process " to save saltpetre (the source of the catalytic nitrogen oxides) and the cost of vitriol chambers " by reacting air with sulphur dioxide in the presence of finely-divided platinum. The techniques for operating such a process industrially did not exist in 1831, and Phillips' process had to wait till 1875 before it became practicable.

Much of the history of sulphuric acid is concerned with the supply of its only solid raw material—sulphur. In the eighteenth and early nineteenth century sulphur for acid-making was imported almost entirely as volcanic brimstone from Sicily. British acid manufacturers, in attempts to achieve at least partial independence of this original sulphur source, sought indigenous sources. As early as 1820, William Hill of Liverpool and, somewhat later, James Muspratt, also of that city, experimented with pyrites from Ireland and Wales. Wharfingers' account-books, preserved in the Library of University College, Bangor, show that Hill, who owned the Cae-coch mine in North Wales, brought by sea to Liverpool, between 1826 and 1835, 3716 tons of pyrites from there. Both Irish and Welsh pyrites have a low sulphur content and were soon superseded by the much richer mineral from Spain, in the mining of which several British acid makers acquired interests. Brimstone, ousted from its original monopoly position, continued to be an important acid raw material. From about 1850 there was a close connection between copper smelting and sulphuric acid manufacture, the sulphur dioxide from the roasting of the ore being utilized in associated acid chamber plants. Tandem technology of this type was to be found in South Wales and in the hinterland of Liverpool. Although Leming had introduced the iron oxide process of removing sulphur from coal gas in 1847, it was only in 1870 that the " spent oxide " from gasworks began to be used by the chemical industry as a supplementary sulphur source.

As the nineteenth century advanced, the key position of sulphuric acid in chemical technology and, by consequence, in the wider field served by that industry, became abundantly apparent, occasioning Liebig's well-known dictum: " It is no exaggeration to say that we may fairly judge of the commercial prosperity of a country from the amount of sulphuric acid it consumes." In the early 1820's the U.K. annual production of sulphuric acid (as 100% H_2SO_4) was of the order of 10,000 tons; by 1900 it touched the million mark—one-quarter of the world output, and greater than that of any other country up to that time.

3. THE SULPHURIC ACID – SALT NEXUS

The demands for soda alkali for the expanding soap and textile industries until the 1820's were mainly met by the toil of many thousands of Scots kelpers and Spanish peasants producing materials of highly variable composition and physical crudity. Frederick Accum recorded in 1802 that U.K. soda resources were being supplemented by double-decomposition of imported American potash with sodium sulphate, the latter then a waste from the soap works and from production of chlorine (for bleaching) by reaction of common salt with sulphuric acid and manganese dioxide. Before the mid-1820's it was apparent that demands for soda alkali would soon permanently outstrip the capacity of the traditional natural sources to supply them. It had long been recognized, both in Britain and in France, the countries principally concerned, that a more efficient method of producing soda was both possible and desirable. The establishment, in 1737, by Duhamel du Monceau of the common base of salt and soda by his synthesis of the latter from the former *via* the intermediate stage of sodium sulphate (saltcake) was, in due course, to link for more than a century sulphuric acid manufacture with that of alkali—the link which, in effect, established the chemical phase of the Industrial Revolution. The evolution of the process that was to found the synthetic alkali industry has a complicated history which, even now, has not been completely elucidated.

Despite the discovery of the common basis of soda alkali and common salt, neither in France nor in Britain were attempts made for many years thereafter to exploit the discovery technologically. In this country over forty years elapsed before patents began to be granted for processes applying this piece of chemical knowledge, although attempts to employ it secretly may have been made at an earlier date. About 1768, members of the Lunar Society of Birmingham were discussing the commercial possibilities of winning " fossil alkali " from sea salt, and at least one of their members, James Keir, conducted experiments, which later (1781) enabled him to produce sufficient soda to supply his small soap-making enterprise at Tipton. In 1769, James Watt, of steam engine fame, in collaboration with Prof. Joseph Black of Edinburgh, was attempting to develop a salt-based soda process. There can be little doubt that long before the end of the eighteenth century, soda alkali supplies in Britain were being significantly augmented by double-decomposition processes of salt or sodium sulphate with potashes. This circumstance would explain the various patent-filings for fully synthetic processes from 1780 onwards. The revolt of the American Colonies had reduced the supply of imported ashes sufficiently at least to turn the minds of the more enterprising alkali-makers in the direction of total synthesis from salt. In 1775 natural soda was quoted on the Amsterdam market at 9·33 guilders/100 lb; before the end of the American war it rose to 22·75 guilders; potashes in the same period rose from 12·55 to 23·40 guilders. Another factor which could have stimulated interest in alkali synthesis from salt at this period was that the price of sulphuric acid—the other essential starting material—had then fallen to about £30 a ton. When Roebuck and Garbett first produced the acid by their chamber process in 1746, the price had been £280 a ton.

It is probable that most of these early alkali processes never progressed beyond small-scale and crude experiments. John Collison, it is known, used the process of his 1782 patent to produce soda for Fry's soapworks at Bristol, and there is evidence that alkali, produced by the divers procedures invented and patented by William Losh and his partner Lord Dundonald, was put to effective use in glassworks in the neighbourhood of their Tyneside factory. However, the fact remains that there was no sustained urgency in a mercantile economy to embark upon a thorough-going industrial operation for which a wholly satisfactory recipe had not yet been evolved. That partial success had been achieved by some British alkali synthesizers is attested by a passage on common salt in the Frenchman Fourcroy's textbook of chemistry, published in 1789: "One feels today the need of a more important use than all these—the extraction of soda, which becomes every day more and more rare, and which is necessary in all the arts. *Many persons possess this secret in England and withdraw on a large scale soda from the salt of the sea.*"

Almost all histories of the early British chemical industry perpetuate the erroneous claim that the slow development of synthetic alkali manufacture in this country is attributable to the very heavy tax on salt: the official records would not support this view: governmental concessions, amounting to almost complete rebate of the salt tax, were granted in cases where the government of the day was satisfied that a feasible process had been invented and would be operated.

4. DEVELOPMENTS IN FRANCE AND BRITAIN

It may be assumed that the factors which determined the timing of early alkali developments in Britain paralleled those which were effective in France. In the latter country, however, private interest in the potentialities of synthesis attracted institutional and even governmental attention. In 1782 the Academy of Sciences offered a prize of 12,000 livres for a practicable process. Malherbe, a Benedictine monk, in 1777, and Athenas, in 1778, put forward processes for consideration by the French Government, but neither process, at that time, recommended itself to authority. Meanwhile, other experimenters were at work, and, in 1789, Nicolas LeBlanc, ex-surgeon to the Duke of Orleans, devised a procedure which was to become the central process operation in the alkali industry of the world for more than a century to come. On 25th September 1791, Louis XVI, now King not by divine ordination but " by the Constitutional Law of the State ", granted to LeBlanc a patent under a law passed by the Assembly on 7th February of that year. By present-day standards the process for which LeBlanc was granted protection would appear to be patentable in one respect only, namely, its claim, as advantageous, for the *use of the reactants in certain stated proportions*.* By contrast, the British patent speci-

* It is, however, to be emphasized that LeBlanc's process was an empirical recipe and *not* an application of the quantitative laws of chemistry—these were still over the scientific horizon.

fications, already referred to, are in general vague in the matter of relative amounts of raw materials to be used: most of them state no quantities at all.

LeBlanc's process consisted in furnacing a mixture of one part each of sodium sulphate and chalk (calcium carbonate) with half a part of coal. The actual reactions which occur have even now not been fully elucidated, and are certainly more complex than the reduction and double-decomposition stages represented by the equations:

$$(1)\quad Na_2SO_4 + C\ \text{(coal)} \rightarrow Na_2S + 4CO$$

$$(2)\quad Na_2S + CaCO_3\ \text{(chalk)} \rightarrow Na_2CO_3 + CaS$$

Interesting features of LeBlanc's description of his process are the proposals he makes for recovery of hydrochloric acid (from the salt–sulphuric acid reaction), either as an aqueous solution or as sal ammoniac by reacting it with ammonia: in short he envisaged alkali-making as the centre of a system of chemical manufacture. " One result of the discovery ", his specification states, " . . . is that France, which consumes a prodigious amount of soda each year for soap and glassworks, laundries, etc., and which exports a considerable currency to buy soda from abroad, will save the foreign exchange without further suffering the arts and industry to go short of this basic necessity through the vicissitudes of war or through the poor harvests of the plant (i.e., *salsola*) which up to now has provided soda . . . One dare *add* even that, because of the abundance of the raw materials and their low price in France, neighbouring nations will become in a short time, dependent on us for these different products . . . "

Considerable obscurity surrounds the origins of the LeBlanc soda process. Its essential features were certainly commonplace to industrial experimenters in Britain, and one view is that it was from this country that LeBlanc indirectly received the germinal notions for his invention. In the year 1789, J. C. de la Métherie, professor of natural history in the Collège de France, in consequence of contacts he had made in England with the members of the Lunar Society, in particular with James Keir, published in the *Journal de Physique* a communication on the synthesis of soda. In another publication, in 1788, again originating in his discussions with the Birmingham " Lunatics ", de la Métherie refers to the use of charcoal and limestone in the separation of iron from its ore. It is suggested that LeBlanc, who was at the time immersed in his soda-making experiments and doubtless a student of relevant contemporary literature, was stimulated by these reports of English technological table-talk to consider that, if carbon and limestone (or chalk) freed iron from its ore, they might in some analogous fashion liberate soda from sodium sulphate. The author of this interesting theory sees LeBlanc's reverberatory soda furnace as an adaptation of Wilkinson's and Dundonald's coke-ovens.

On the first day of February 1793 France and Britain went to war, and on the 9th of the following month hostilities broke out between France and Spain: French supplies of natural alkali were suddenly reduced to kelp and glasswort from her own shores and ashes from forests in territories overrun by her armies. LeBlanc put his process at the disposal of his country and Commissioners were appointed in 1794 by the Committee of Public Safety to review all possible methods of maintaining synthetically the soda supply of the embattled Republic; fifteen synthetic processes in all were reviewed by the Commissioners; these were of three types:

(1) those using natural alkali (potash or kelp) as one of the raw materials;
(2) furnace and aqueous reactions of salt or of sodium sulphate made from salt;
(3) oxidation of salt with lead oxide.

It is of some interest, as illustrating parallelism of developments in Britain and France up to this time, that precisely those three classes of process had been engaging the attention of British alkali synthesizers.

Synthesis of soda in France, which was begun on a small scale by LeBlanc in Paris, shifted mainly to Marseilles, then the centre of the French soap industry. During the years of the Republican and Napoleonic Wars with Britain, France eventually succeeded in achieving independence of natural alkali sources. Four years after the Commissioners of the Committee of Public Safety made their survey of synthetic processes, Chaptal, Napoleon's confidential minister and one of the most important pioneers of the French chemical industry, was urging the necessity of introducing large-scale growing of Spanish *salsola* (the source of barilla ash) into France. In 1820, 17,641 tons of soda were produced in France by the LeBlanc process. The wars which had compelled France to establish a synthetic alkali industry adequate to its national needs, produced no parallel effect in Britain: in this country, in 1820, barilla and kelp were still by far the principal alkali sources.

An indication of the peculiar British lack of interest in alkali synthesis may be seen in the fact that William Nicholson omits all mention of the LeBlanc process from the 1808 edition of his *Dictionary of Theoretical and Practical Chemistry*, then a standard source of industrial recipes. Yet, one moderately-sized LeBlanc factory would have been equivalent in *chemical effect* to the results of the labours of the thousands of kelp-gatherers and burners seasonally toiling along hundreds of miles of Scotland's shores. There must have been many informed persons aware of that potentiality, even before 1808: surely there can be few more striking historical examples of socio-technological inertia!

5. THE LEBLANC PROCESS IN BRITAIN

According to his biographer, Lonsdale, William Losh, who had for some years been experimenting with alkali synthesis on the Tyne, visited Paris in 1802 " to ascertain what was doing in chemistry and . . . was able to make himself master of some of the French chemical discoveries, and their modes of manufacturing soda". From the fact that Losh & Company of Walker-on-Tyne are on record as having paid duty on salt used for alkali manufacture from 1800 onwards, it seems probable that Losh's operation of a LeBlanc-type process began earlier than his 1802

Paris visit. Losh, and indeed anyone in Britain, could have become acquainted with the details of the LeBlanc process in 1797, when the *Annales de Chemie*, a journal with an international circulation, published the report of the Commissioners to the Committee of Public Safety. From its outset, Losh's venture in soda-making was assisted by remission of the salt tax to a mere thirtieth of the standard rate. The fact that Losh & Company, with the combined advantages of tax remission and the then most advanced technique, did not rapidly command the entire soda field occasioned some contemporary surprise. Thomas Carr, Solicitor to the Board of Excise, failed to elicit a satisfactory answer to his questions in this connection in the course of the proceedings of the Select Committee on the Salt Duties in 1818. Charles Tennant, a linen bleacher, who, in 1797, established his St. Rollox works in Glasgow to make sulphuric acid and chlorine for bleaching, may have been producing some synthetic soda as early as 1803; in 1816, when he decided to make it on a more substantial scale, he consulted the French chemists Chaptal and D'Arcet. In 1818, some 100 tons of LeBlanc soda crystals (equivalent to 37 tons 100% soda alkali) were produced at St. Rollox. When Losh, who may have been purchasing his acid from Tennant, decided, in 1821, to produce his own requirement, it was plumbers from St. Rollox who erected his sulphuric acid chamber plant.

James Muspratt, the son of a Dublin artisan cork-cutter, acquired some rudiments of chemical manufacturing know-how in the Irish capital before his migration to Liverpool in 1822. In that year an Act of Parliament reduced the import duty on barilla, causing consternation amongst the soap-boilers on Merseyside, who had been deriving considerable advantage, *vis-à-vis* their London competitors, from the use of kelp, which was tax-free. Muspratt could not have chosen a more propitious moment to start up in Liverpool as a maker of LeBlanc soda. What was new in Muspratt's operations was the *scale* on which he worked; this circumstance has, erroneously, led to his being regarded as the first to introduce the LeBlanc process into this country and as the " father " of British heavy chemical industry. Liverpool and other Mersey towns, in the 1820's, were building up a hard soap trade rivalling in size, and soon to surpass, that of London. With such a market within carting distance of his factory gates Muspratt was able to embark upon alkali synthesis on a scale of many hundreds of tons a year. The 112-ft long and 24-ft wide acid chamber he erected at his Vauxhall Road works far exceeded in dimensions any hitherto built. In less than a decade Muspratt's two factories in Liverpool and St. Helens drove natural alkali from the soaperies of the whole Mersey region.

While Muspratt and others were developing alkali manufacture on the Mersey, the industry was making headway on the Tyne. In the period 1808–33 no fewer than six alkali concerns, in addition to Losh's pioneer factory at Walker, were supplying the requirements of the glass and soap works of the region. In 1830, William Gossage, later to become a leading British soap manufacturer, established with Fardon, at Stoke Prior, the first LeBlanc soda works in the Midlands. Several years later the firm of Chance erected a LeBlanc works at Oldbury, near Birmingham, to supply the saltcake (sodium sulphate) and soda required by their glassworks at Spon Lane, Smethwick. About 1830 several small sulphuric acid works were established in Bolton, which had been a centre of textile bleaching since the eighteenth century; several of these Bolton works subsequently made LeBlanc soda.*

The Chance LeBlanc works, as has been mentioned, was set up primarily to make alkali for the same firm's glass-making operations; Gossage and Fardon's alkali factory used the soda for their own soap manufacture, and at least one soapery at Runcorn made its own alkali requirement, probably before Muspratt became active at Liverpool. This " captive " production of soda, which, in the early years of alkali manufacture, may also have been practised by other firms lost to record, never became

* The Great Lever works founded by John Smith (1771–1851) before 1834 operated the LeBlanc soda process up to 1938, and was then almost certainly the last works in the world operating the classical alkali process.

a stable pattern in the nineteenth century chemical industry. Manufacture of alkali as a commodity for the market was generally a distinct operation: even firms that had originally produced captive soda soon began to produce alkali for sale. Thus, from its outset, alkali manufacture was mainly established as an activity separated from the technologies it supplied. The principal cause of this separation was that the chief outlet for alkali was in the rapidly expanding cotton textile industry. Glass-making and soap-boiling operations allied in nature to those of chemical manufacture, lay, as it were, within the same constellation of processes—textile production was an entirely distinct complex of operations.* Before 1830 the manufacture of cotton textiles had firmly assumed the role of meristem† of the Industrial Revolution—the principal determinant of both the rate and direction of its growth. By the early 1830's the annual import of raw cotton into the U.K. averaged 280 million lb a year; in 1835, 169 steam-powered new cotton factories were being erected in the Lancashire region alone. In the 1850's an average of 884 million lb of cotton were annually passing through Britain's textile factories. That professional wit, Sydney Smith, remarked: " The great object for which the Anglo-Saxon race appears to have been created is the making of calico! "

6. THE LEBLANC SYSTEM: RISE OF BRITISH ALKALI INDUSTRY

During the first half of the nineteenth century, and even later, the alkali industry *was* the chemical industry, the term frequently used in referring to it was the " alkali trade ". The " alkali trade " was, however, not a one-product industry, but a closely-integrated group of processes, conspiring, one might almost say, to serve the demands of the all-important textile manufacture.

---

* Captive production of sulphuric acid and bleaching powder was practised by some bleachers in the early years of last century. Some of these, e.g., Tennant, abandoned their original trade of bleaching and became chemical manufacturers only; the others ultimately became customers of the established chemical industry.

† I.e., the active point of growth as in a plant.

Central to this group of processes was LeBlanc's for the production of soda alkali (sodium carbonate); the other processes were either in preparation for, or a consequence of, the soda-making stage. Nineteenth century alkali manufacture and its accompanying operations may, therefore, be most appropriately referred to as the LeBlanc system. The brave expectations, expressed in LeBlanc's patent specification, that France would supply neighbouring countries with French-made soda had proved ill-founded: Britain, at least, had imported not the product, but the *process*.

The original LeBlanc system in its essential aspects had fully emerged before 1840: it was already the nucleus that gave pattern to the heavy chemical industry; it was to continue in ever-growing measure to determine its course of development for decades to come. Its primary operation was the manufacture of sulphuric acid by the chamber process; the acid from this stage was partly marketed as such to the metal trades and to the textile bleachers (as a sour) ; the greater part of it was, however, reacted in furnaces with common salt to produce saltcake (sodium sulphate). Part of the saltcake was (after 1835) sold to the glass industry, which used it to replace all or part of the soda formerly employed. The main bulk of the saltcake went into the LeBlanc soda furnaces where, by reaction with limestone and coal, it was converted to " black ash ", a foul-smelling mass containing, in addition to soda, unreacted limestone and coal, and calcium sulphide. From the black ash was extracted by means of water passing over it in a series of vats, the soluble soda, which was finally recovered by evaporation of its solution.

Alkali waste—the calcium sulphide, etc.—from this extraction process presented one of the first disposal and (later) recovery problems of the chemical industry. As the alkali trade developed hundreds of thousands of tons of waste were piled up on agricultural land around the chemical manufacturing sites. It is to be noted that the sulphide in the waste represented a total loss of the sulphur originally present in the sulphuric acid. This was not the only loss: in the 1840's, and for many years thereafter, most of the

FIG. 2. SITE OF JAMES MUSPRATT'S NEWTON WORKS TODAY

These pre-sulphur-recovery waste mounds, near St. Helens, in places 50 ft high, contain, now largely in the form of sulphate, all the sulphur used in the course of 21 years of soda manufacture on this classical chemical industrial site. Here, in 1837, Muspratt attempted without success to operate Gossage's sulphur recovery process.

hydrochloric acid (muriatic acid) from the saltcake-making stage, even when recovered in Gossage towers, had to be run to waste. The LeBlanc soda process was at that time a method of making one chemical (soda) and wasting two (sulphur and hydrochloric acid*). Most LeBlanc works used part of their byproduct hydrochloric acid for making bleaching powder. At this stage in its history, then, the LeBlanc system produced four chemicals, namely, sulphuric acid, saltcake, soda, and bleaching powder. The importance of the last-named in the politics, economics and technology of the industry was such that its history demands, at this point, special consideration.

The traditional wind-and-weather bleaching was a process requiring weeks for completion, being dependent, too, on seasonal conditions. As the cotton and other textile industries expanded, more and more agricultural land was being used for bleaching-greens. It became apparent that, unless a more rapid bleaching technique could be found, the time and land demands of the classical process would soon set stricter and stricter limits to the expansion of the all-important textile industry. The bleaching action of chlorine in the presence of moisture had already been exploited to a certain extent on the Continent, when James Watt and Charles Tennant began experimenting with the technique at Glasgow.† In 1799, Charles Macintosh, one of Tennant's partners at St. Rollox, invented the method of fixing chlorine by absorbing it in slaked lime, thus producing a " solid ", transportable chlorine (about 36% by weight of actual chlorine), free from the objectionable characteristics of the gas. Although Macintosh and his partners were only later aware of the fact, a certain Robert Roper, in 1791, had produced an entirely similar bleaching commodity. Despite the manifest benefits the new bleaching agent could confer on the textile industry, it was not immediately or widely adopted. In 1820 St.

---

* In effect the chlorine " half " of the sodium chloride molecule.

† Chlorine was also almost certainly being used, before 1800, by bleachers in Lancashire. The history of the introduction and early application of chlorine bleaching in the U.K. is almost entirely undocumented.

Rollox works produced a mere 330 tons of it, and Customs records show that, in 1821, England exported to Ireland, where linen manufacture was a national industry, less than *half-a-ton* of bleaching powder; indeed, in that same year Ireland *exported* to England and Scotland some 66 tons! Before 1850, U.K. production of bleaching powder had probably reached an annual output of about 10,000 tons. Bleaching powder, by reducing the bleaching time from weeks to hours, facilitated greatly the expansion of textile manufacture and thereby increased the demand for its coproduct, soda alkali.

Up to 1840 the LeBlanc system produced chlorine by heating a mixture of salt, manganese dioxide and sulphuric acid. During the first decades of the LeBlanc manufacture of soda vast quantities of hydrochloric acid (hydrogen chloride) were allowed to pass freely into the atmosphere and large sums had to be paid from the profits of the industry to farmers and landowners in compensation for damage to crops and woodlands. William Gossage, who later became renowned as a soap-maker, invented a method of absorbing hydrochloric acid in a film of water passing over twigs, broken bricks, coke, or other materials packed in a tower. The Gossage tower was one of the first and not the least important inventions contributing to the final development of the LeBlanc phase of the chemical industry. Most of the alkali-makers adopted Gossage's invention, with the result that, having solved one problem they were faced by another—that of finding profitable outlets for the hydrochloric acid recovered. A part of the recovered acid was used in place of salt in the production of chlorine for bleaching powder, but by far the greater part was run to waste—as discreetly as possible—into nearby streams and rivers.

7. THE WELDON AND DEACON-HURTER PROCESSES

With the passing of the first Alkali Act in 1863, which enforced condensation of 95% of the hydrochloric acid from the saltcake furnaces, bleaching powder production became even more firmly established as a department of the LeBlanc system, and more efficient means were sought of producing chlorine from by-product hydrochloric acid: these means were forthcoming in the Weldon (1866) and Deacon–Hurter (1868) chlorine processes, both essentially processes in which the oxygen of the air, in the presence of a catalyst, oxidised the hydrogen chloride to water and chlorine:

$$4HCl + O_2 \rightarrow 2H_2O + 2Cl_2$$

Weldon's process, which involved only minor departures from established liquid phase techniques and had the advantage of producing concentrated chlorine, was rapidly adopted by the industry. Deacon's method, a gas phase process, produced dilute chlorine. It raised the unprecedented problem of catalysing reaction in a large flow of mixed gases, controlled at a temperature sufficiently low to favour production of chlorine in economic quantity.* The finally effective form of Deacon's process was the result of ten years of research by Ferdinand Hurter, Deacon's works chemist and pupil of R. W. Bunsen. It came into wide use only towards 1880. Together the Weldon and Deacon–Hurter processes constituted the bleaching powder sector of the LeBlanc system; they marked the beginning of large-scale chlorine production, which, by other processes, has continued as a key chemical industrial operation into modern times. The LeBlanc factories of the Midlands never produced chlorine commercially; Albright & Wilson, in that region, however, produced chlorine captively on a small scale by the Weldon process from 1877 to 1916.

Recovery of hydrochloric acid continued greatly to exceed the demand for chlorine for bleaching powder manufacture. In 1867 utilization of hydrochloric acid (calcd. as 100% HCl) in the U.K. is estimated to have been approximately as follows:

* Deacon began his experiments in the same year that Guldberg and Waage enunciated the law of mass action, but clear understanding of chemical equilibrium between opposing reactions lay a score of years ahead.

Production of bleaching powder	62,800 tons
Production of carbon dioxide from limestone (for making sodium bicarbonate from the carbonate)	8,000
Metal industries (pickling, etc.)	50,000
Loss in works operations	12,000
	132,800

Since the salt decomposed in that year was 425,630 tons, actual hydrogen chloride output was 244,000 tons; thus, despite the large-tonnage applications, 45% of the acid was still being put to waste. Long after the invention of the Gossage absorption tower, the chemical manufacturers were relying on dissipation of a great part of their muriatic fumes into the atmosphere from lofty chimneys, which also served for the production of draughts in the alkali furnaces, acid chambers, etc. Some of these chimneys were famed features of the landscape, as was, for example, "Tennant's Stalk", which was 435½ ft high (erected 1842). Other chemical chimneys were Townsend's at Port Dundas, 454 ft (1859); Blinkhorn's at Bolton, 367½ ft (1842); and Muspratt's at St. Helens, 276 ft (1835). These costly structures brought certain prestige to the firms owning them; thus, Townsend's chimney was made to overtop Tennant's by being surmounted by a 20 ft iron crown!

In the early 1860's esparto grass was first introduced in this country as a raw material in paper manufacture; imports of the grass increased from 192 tons in 1859 to 22,957 tons in 1863. Demand for bleaching powder to treat this new material raised bleaching powder prices from £11 to £14 and even £18 a ton, affording the manufacturer a profit of £5 to £8 a ton. As the LeBlanc system rose to its zenith in the 1880's, production of bleaching powder rose to a final peak of 150,000 tons a year. The chief centres of its production were, in the earlier period, at Glasgow and on the Tyne; later Lancashire, particularly the Widnes area, became pre-eminent. Absence of a regional textile industry and remoteness from ports prevented establishment of bleaching powder production in the Midlands. Considerable quantities of the commodity were made at Netham, near Bristol, during the latter part of last century.

8. HISTORICAL GEOGRAPHY OF THE LEBLANC SYSTEM

The factories of the LeBlanc system were mainly co-sited with soap, glass and textile production, i.e., the principal alkali-consuming industries. To a lesser extent local availability of salt, limestone and coal were contributory siting factors. Once natural alkalis were displaced by synthetic soda, the soap industry declined in London and became increasingly important on the Mersey, where the vast salt-fields of Cheshire and the limestone quarries of Derbyshire and North Wales supplied two of the raw materials of the LeBlanc system: the third—brimstone (or pyrites)—was imported through Liverpool. Glass manufacture in the Mersey region, at St. Helens, as on the Tyne, became another consumer of LeBlanc soda and saltcake. Until the discovery of local salt deposits in 1874, salt was transported by road and rail from Cheshire to the Tyne chemical works. Salt had to be conveyed by sea from Liverpool to Tennant's St. Rollox works, which by the mid-century was the largest chemical factory in the world. It was undoubtedly the absence from the geology of Scotland of both salt and good, accessible limestone formations that limited development of heavy chemical industry in that country. In the Midlands, from 1830 onwards, and at Bristol, from about 1865, the LeBlanc factories drew on the Staffordshire salt-field for their primary raw material.

In 1861 the LeBlanc system in South Lancashire, its principal site, was decomposing 2600 tons of salt a week to produce 1800 tons of soda (sodium carbonate, Na_2CO_3), 180 tons saltcake, 170 tons soda crystals, 225 tons sodium bicarbonate, and 90 tons caustic soda (NaOH). (The last named product, made by re-acting sodium carbonate with slaked lime, was not marketed by the U.K. alkali trade until 1853; the usual practice was for the soap-boilers to " causticize " purchased soda at their own works.) In 1862 the entire U.K. LeBlanc industry decomposed 250,000 tons of salt to produce a similar total weight of products, valued at £2,500,000 and employing some 10,000 workers in their manufacture, and about double that number indirectly in mining raw materials and producing casks and other containers for the transport of the commodities.

9. THE FIRST ALKALI ACT (1863)

In 1860 the British alkali-makers discovered their first need for collective action in the negotiation of new duties on chemicals under the terms of a commercial treaty with France. The Alkali Manufacturers Association, with a permanent secretariat, was established in London in 1862 to deal with matters concerning the interests of the industry as a whole. A Bill for " the more effectual condensation of muriatic gas in alkali works " had been introduced in Parliament: the alkali-makers saw trouble ahead. The appointment of a Select Committee of the House of Lords to take evidence on the matter of noxious vapours (in fact, muriatic or hydrochloric acid) was to a considerable extent a political move instigated by the landed gentry. In Lancashire the landowner's cause was championed by Lord Derby, the Committee's chairman. " You will remember ", wrote one of the alkali-makers to a fellow-manufacturer, " . . . that all the Alkali-Trade voted with the Liberal Party, and this may not have been unobserved and commented upon by electioneering agents." By co-operative action, inspired chiefly by John Hutchinson and Henry Deacon of Widnes, the alkali-producers were able to prevent the alkali industry having imposed upon it crippling legislation, framed by those who were not financially interested in it, or were actually antagonistic towards it. The Alkali Act of 1863, which came into effect at the beginning of 1864, required absorption of 95% of the byproduct hydrochloric acid—a requirement well within the technical means then available. This enactment was the first recognition by the Legislature of the special character of chemical manufacture. Under the Act, Alkali Inspectors were appointed; the annual reports submitted by these Inspectors provided an important documentation of the LeBlanc system as it rose to its zenith. The Chief Alkali Inspector's Report for 1864 shows that, at that time, LeBlanc works were regionally located as follows:

Lancashire	36 works
N.E. England	20
Scotland	8
Ireland	5
Midlands	4
S.E. England	3
S.W. England	3
N. Wales	2
S. Wales	2
Total	83

By about 1840 the LeBlanc factories, although some barilla continued to be imported up to the mid-century, were supplying the total requirement of the consuming technologies in the U.K. Sulphuric acid, on the other hand, had been manufactured on a small scale in the U.S. since 1793 and chlorine bleaching was established by 1830 in paper mills in that country; several firms, which were later to become important chemical concerns, were in business there before 1870. The LeBlanc system, however, never became established, and before the Civil War America was mainly dependent on imports for chemical requirements. Chemical industry in the U.S. began its real development in the modern phase.

James Muspratt, in 1838, visited the U.S. to explore the potential chemical market. Two years later Muspratt and Tennant shipped the first cargoes of soda alkali across the Atlantic. An American Tariff Act of 1842, imposing a duty of 20% on chemicals, may be interpreted as a reaction to the arrival in that year of British bleaching powder at the quays of New York and Philadelphia and an attempt to give some protection to what domestic chemical industry there was. That fiscal measure notwithstanding, the export of chemicals from the U.K. to the American soap, glass, paper and textile industries rapidly expanded. By 1860 between 30 and 40% of the bleaching powder annually produced in this country crossed the Atlantic, and up to 60% of the soda output accompanied it. Before 1880 as much as 70% of British bleaching powder found its way to the American market. Liverpool and Glasgow were the chief chemical exporting ports to the U.S., and " Liverpool prices " became almost universally quoted. The LeBlanc factories on the Tyne found their markets in Europe, particularly in the Baltic countries and Germany, before the political (and technological) unification of the latter in the 1860's. In the Midlands the LeBlanc system produced mainly for the industries of its region.

10. INVENTION WITHIN THE LEBLANC SYSTEM

The still comparatively elementary technology of the industry, the facilities afforded by the Companies Act, and above all, its manifest profitability were factors co-operating to make heavy chemical manufacture an attractive enterprise in the 1860's and early 1870's. Some firms formed during this period carried out only one stage or department of LeBlanc manufacture, purchasing their starting material from larger concerns in the trade. Processes in some of the smaller factories were carried on at low and variable levels of efficiency, which could be sustained only so long as profit margins remained wide. Family firms commonly showed a decline in efficiency in the second and subsequent generations. The mechanical engineer, if not the plumber, dominated operations in many of the works, where often the crudest empiricism reigned supreme. Kingzett, in his book on the alkali trade (1877), quotes a contemporary view that " a manager can be too much of a chemist! " One powerful factor in maintaining technological inertia in the old heavy chemical industry was the general secrecy, which militated against generalization of experience and the evolution of practice into inventions.

FIG. 3 A BARREL-TYPE BLACK ASH REVOLVER

Batch reaction furnaces of the barrel and cylindrical forms were the heart of the classical alkali-manufacturing system. The cylindrical form was that commonly used in the Lancashire region; the barrel form shown here was used in the Tyneside chemical area.

FIG. 4. A CYLINDRICAL-TYPE BLACK ASH REVOLVER

This revolver was operated for a number of years in a factory in Widnes. The product of purely empirical design, it was found in practice to be too long in relation to its diameter to give entirely satisfactory results. So far as is known, this was the only revolver ever to be made with three discharge-ports. There were strict limits to scaling-up the classical batch processes.

Some chemical proprietors discouraged their chemists from attending scientific societies and clubs, lest they should inadvertently betray some technical secret of their employer's business. Only in the larger companies were the processes under the surveillance of chemists. What process inventions there were, not surprisingly, generally had more of engineering in them than they had chemistry.

Elliot and Russell's revolving black ash furnace (1853) was a fairly simple expedient for eliminating most of the arduous manhandling from the saltcake-to-soda conversion stage. Some twenty years passed before " revolvers " became universal in the alkali factories. The Weldon and Deacon–Hurter chlorine processes, already mentioned, were, perhaps, the most important inventions to be made within the LeBlanc system. It is significant that the latter, which bears more resemblance to the continuous gaseous reaction processes of the modern era, required the labours of a graduate chemist of Heidelberg for nearly ten years to bring it to industrial utility. The various mechanical saltcake furnaces, devices for stirring at high temperature mixtures of salt and sulphuric acid, invented in the late 1870's and early 1880's, were eminently engineering conceptions: no notions of physics or chemistry dictated their design. For many years it was attempted to make saltcake directly from salt without intermediate production of sulphuric acid. The obvious advantage of such a short cut was that it would obviate the need for acid chambers, which had a high capital cost and required an extensive site for their accommodation. In 1870, Hargreaves and Robinson at Widnes, the latter a foundry engineer long associated with the local chemical industry, devised a method for production of saltcake by passing a mixture of sulphur dioxide and steam downwards through a furnace packed with briquettes of salt. This process produced a saltcake of high purity and raised hopes amongst British (and foreign) LeBlanc manufacturers that an economically significant advance had been made in the technology of their system. Up to the beginning of the 1870's the attention of continental LeBlanc manufacturers was focussed on developments in this country. Mond wrote in 1871 to his friend Merle at Salindres: "You will find here (Widnes) all the important inventions, the Deacon, Weldon and Hargreaves processes, the revolving Kilns, etc. . . . The journey from Paris here takes only 18 hours." British LeBlanc manufacturers led in the field of chemical industrial inventions at that time. The last invention within the LeBlanc heavy chemical industry was Chance's solution in 1887 of the problem of substantially recovering the sulphur content of the alkali waste, so that it could be re-converted to sulphuric acid and recycled through the system. More than twenty years earlier Ludwig Mond had developed a process for recovering part of the sulphur, but this, because of its great imperfection, had not been generally adopted throughout the industry. Chance's recovery process owed its success, where others had failed so often in the past, to the more advanced state of engineering in connection with the pumping of carbonic acid gas. With the emergence of Chance's recovery process, which was adopted throughout the industry with the greatest rapidity, the LeBlanc system realized its full capacity to be improved and extended by its contemporary engineering techniques. Chemical science had hardly been applied to it in any systematic way. Black ash revolvers were designed by rule-of-thumb, acid absorption towers likewise; often they were too big or too small to function efficiently in the particular case; the heat economy of none of the operations had received organized attention. Only during the last quarter of the nineteenth century did a few continentally-trained chemists, notably George Lunge and Ferdinand Hurter, initiate the application of physical chemical methods to the study of the more important processes of the LeBlanc system, e.g., the reaction of chlorine with lime in bleaching powder chambers and the mechanism of absorption of hydrogen chloride gas in water.

It would be unhistorical to claim that inventions within the LeBlanc system were forced out of it by pressure of economic competition alone, or even mainly: most of them were solutions of problems which it had been attempted to solve from the earliest years of alkali manufacture. In 1881, Prof. Sir Henry E. Roscoe, in his Presidential Address to the Society of Chemical Industry, commented: " Whilst it is true that the most general and most important discoveries in the alkali trade have been made by Englishmen, it still remains the fact that many improvements and suggestions due to foreign skill and enterprise are not taken up in England as they deserve to be, and this is to a great extent due to the want of communication between the countries on these important points." Within the LeBlanc system there was remarkably little competition between the principal firms—price agreement was the rule. Until about 1880, when production of alkali by the Solvay process* had become important, there was no competition from outside the system. With the accelerating rise of heavy chemical manufacture in Europe and America and, in particular, the expansion of the Solvay process, the LeBlanc inventions served to maintain the viability of the old heavy chemical processes in the U.K. sufficiently to prolong their declining existence into the second decade of the present century. By 1883, 45% of French and German soda was being produced by the Solvay process; Belgium had ceased LeBlanc production entirely, and even in Russia the new process was already in operation. Six years later, in Britain, 219,000 tons of soda were made by the Solvay route, and 584,000 tons by the old process. In face of these developments the LeBlanc manufacturers could still be hopeful of the survival of their soda industry, because it was also the purveyor, through its Weldon and Deacon processes, of the chlorine " half " of the salt molecule, which the Solvay process had perforce to waste.

11. MOVES TOWARDS UNIFICATION: THE UNITED ALKALI COMPANY

In 1883, the LeBlanc manufacturers of Lancashire formed the Bleaching Powder Association to regulate both the domestic and export markets for what had now become the key commodity of

* See p. 83

the LeBlanc system. The Association, more importantly than its achievement of its explicit object, served to foster amongst its members a sentiment of co-operation, which gathered strength with the competitive crisis. Although there are, as far as is known, no memoirs of the later deliberations of the members of the Bleaching Powder Association, it can be inferred from other statements of the period that at least three circumstances led them ineluctably to consider some type of far-reaching merger of their enterprises; these were: the widely differing levels of technical and commercial efficiency of many of the smaller LeBlanc firms, which was weakening the competitive strength of the trade as a whole; actual and prospective tariffs against their chemicals in their traditional markets; final realization that Solvay soda was not the limited threat many had originally believed it to be.* Late in 1890 came news from the U.S. of the impending McKinley Tariff. Despite the background of national financial crisis, occasioned by the failure of the banking house of Baring, the British LeBlanc manufacturers determined to proceed with the promotion of what was then to be known as the United Chemical Company. The United Alkali Company, as it was finally decided to designate the merged firms, was incorporated on 1st November 1890, and capitalized at £6,000,000. In all, the new company acquired forty-eight factories, located on the Tyne, in Scotland, Ireland, and Lancashire. Included were three salt-works, guaranteeing some measure of control over that primary raw material. The merger, at its formation, had an annual production capacity for some 1·5 million tons of LeBlanc chemicals, which then included sulphur (recovered by the Chance process) and potassium chlorate (an outlet developed for Weldon and Deacon chlorine). At the time of its incorporation the Company was negotiating the purchase of the LeBlanc departments

* Mond himself, in 1877, saw the Solvay soda process, because of its inability to co-produce chlorine, only as a limited auxiliary operation within the orbit of the LeBlanc system. In May, 1885, Walter Weldon wrote to a Widnes alkali-maker " my own profound conviction is that the further manufacture of LeBlanc alkali will very soon become impossible."

of five more firms. When the amalgamation was complete, the United Alkali Company, the first important merger in the history of the world's chemical industry, was in effect the unification of the greater part of Britain's old chemical industry. Only one of the old Le Blanc firms, the Oldbury Alkali Company,* remained outside; archives of that concern, however, show that from the outset it co-operated with the United Alkali Company in matters of markets and prices.

The over-capitalized United Alkali Company was never a rewarding investment for its shareholders; its role, however, in preventing disintegration of the British heavy chemical industry in the last years of the nineteenth century and early years of the twentieth, has never been sufficiently recognized. Although both the modern soda and chlorine processes were in operation in this country before 1900, their scale of operation was still inadequate to meet the requirements of the domestic and competitive foreign markets. Brought into existence to secure the economics of the old technology, the Alkali Company's historical role was, in the event, largely technical: it was the first example in the industry of what came to be known as rationalization; it enabled the LeBlanc system, without bankruptcies and disastrous closures, gradually to be replaced by the modern technology. Its great dependence on exporting, rather than the increasing obsolescence of its principal processes, was the chief source of economic weakness of the United Alkali Company. By 1898 the effect of the McKinley and Dingley Tariffs reduced British alkali exports to the U.S. to about 20% of the 1890 tonnage: the trans-Atlantic market for British alkalis, which had been a chief factor in the nineteenth century expansion of the industry was vanishing forever. The British Solvay soda process users, despite some manufacturing cost advantage, were also economically afflicted by the rapidly changing conditions of world trade in chemicals which marked the opening of the modern era.

---

* Later (1898) Chance & Hunt Ltd.

12. SULPHURIC ACID-DEPENDENT HEAVY CHEMICAL DEVELOPMENTS

In the course of the nineteenth century, manufacture of a small but important group of chemicals became linked with sulphuric acid production; these were alum, nitric acid, ammonium sulphate and super-phosphate fertilizers, the nitro-explosives, and copper sulphate. Generally, manufacture of these sulphuric acid-dependent chemicals lay outside the LeBlanc system factories, or was a peripheral operation therein. Originally, however, the chamber plants of the LeBlanc alkali works provided the acid.

Alum

Alum-making has, with great measure of truth, been called the oldest chemical industry; certainly, apart from common salt, this complex of aluminium and potassium sulphates was the first substance to be produced as a chemical commodity approaching a modern standard of purity. It owed its early and continued importance to its application in textile technology, in which it was used as a mordant to fix and diversify natural dyes on fabrics; it was also important in the processing of leather.

The traditional production of alum, in its first stage, made use of a natural process. Aluminous shale from waste mounds at coal-workings, or alum slate was exposed to weathering over a long period, during which their iron sulphide content was oxidized by air and moisture to sulphuric acid, which formed aluminium sulphate with the alumina present. The aluminium sulphate, containing quantities of iron sulphate, was extracted with water and potashes or impure potassium chloride (the latter a waste from glass manufacture), added to react with the iron sulphate. On concentrating, a crystalline mass of potassium alum ($Al_2(SO_4)_3 \cdot K_2SO_4 \times 24H_2O$) separated, the iron remaining in solution. In the eighteenth century the weathering process was replaced by a roasting stage in which the shale or slate, mixed with brushwood and heaped in long mounds resembling potato

FIG. 5. TYPICAL CHAMBER SULPHURIC ACID WORKS

This factory, established in 1855, operated for over a century. Large chambers of this type were a familiar feature of the old chemical industrial landscape.

clamps, was ignited and allowed to smoulder. This advance shortened the alum-making process by many months. The new technique was used in Scotland by Charles Macintosh and his partners from 1795 at Hurlet, near Paisley, and from 1805 at Campsie, north of Glasgow. At Whitby, on the Yorkshire coast, alum was produced by roasting alum slate. Some 100 tons of shale was required to yield 3 tons of commercial alum; thus the process was highly wasteful of handling energy, as of time and heat for concentrating. In 1841, 3237 tons of alum were produced at Whitby and 1200 tons at Hurlet. Macintosh's inventive contribution to the alum process was a type of furnace in which a flame was drawn by chimney draught across the surface of the alum solution to concentrate it. This minor invention has the particular interest that it was the first application of a method which came to be universally employed in the LeBlanc factories to concentrate soda solution from the extraction vats.

Robert Hervey, in 1839, patented a process for treating calcined shale with sulphuric acid; Joseph Cliff, in 1845, patented a similar procedure, anticipating by some months Peter Spence's alum-making process. Spence finally established sulphuric acid treatment of clays as a viable industrial process. A similar alum process was operated at Felling works, Newcastle-on-Tyne, from the early 1830's to 1886; small quantities of alum were produced at other LeBlanc factories. Spence, in 1847, left Cumberland, where he had been producing alum on an experimental scale, and set up works in Manchester, where he was close to the principal market for his product—the Lancashire cotton trade. By 1882 Spence had factories in operation at Birmingham and Goole and had become the largest alum producer in the world and enjoyed a near monopoly of its manufacture in the U.K.

Nitric Acid

Until modern times nitric acid was made industrially by distilling a mixture of concentrated sulphuric acid and potassium or sodium nitrate. The sodium sulphate co-produced in this process was known as " nitre cake " and was used to augment the supply of saltcake for soda manufacture; thus nitric acid production linked-up with the LeBlanc system. It is not known when nitric acid entered technology, but its application in the metal trades is assumed to have followed upon sulphuric acid becoming commercially available at low cost. The acid was made in Birmingham as early as 1759. Before discovery of the Chile saltpetre beds in 1825, nitrates for nitric acid-making were recovered from the soil. Between 1804 and 1825 some nitric acid was produced from Peruvian guano (also imported as a soil fertilizer). From 1825 to 1925 nitric acid in Britain was made almost entirely from Chile saltpetre. In the mid-1870's U.K. nitric acid output was about 10,000 tons annually.

Synthetic Fertilizers

The Peruvians had been aware of the agricultural value of the guano (bird dropping) deposits in their country long before its conquest by Spain. Humbolt brought a sample of this natural nitrogenous and phosphatic fertilizer to Europe in 1804, when its composition was investigated by a number of famous continental chemists. Guano began to be imported into Britain in 1840, thereafter its use rapidly expanded, imports reaching their zenith of 280,000 tons in 1870. From that year onward the Peruvian deposits declined in quality and availability; by the late 1880's annual imported tonnages had fallen to less than 20,000.

In 1827, 40,000 tons of bones for use in agriculture were imported into England; Huskisson valued this import at between £100,000 and £200,000. Concern with the possibility of improving the fertility of the land by modifying it mechanically and by treating it with growth-promoting substances was a direct consequence of the population increase resulting from the Industrial Revolution. The restrictions on food imports and consequent high prices of corn, encouraged the cultivation of wheat, barley and clover.* A scientific approach to soil fertility made its appearance about the middle of the eighteenth century with the experiments of men like Francis Home, the Earl of Dundonald, and Charles Macintosh. When Liebig visited England he became deeply interested in the crude expedients of its agriculturists. Referring to the practice of liming he wrote: "In October the fields of Yorkshire and Oxfordshire look as if they were covered with snow." He witnessed the drastic procedure of furnacing the surface soil in the expectation of rendering it more fruitful. The English geologist, William Buckland, showed Liebig the deposits of fossilized saurian dung (coprolites), lying in a brown seam for miles along the course of the Severn—a deposit which the great German chemist forecast would give the country "a means of supplying the place of recent bones" as a source of phosphatic fertilizer.

At the 1837 meeting of the British Association in Liverpool Liebig was asked to prepare a report on the state of organic chemistry, a commission which led, in 1840, to the publication of his *Die Chemie in ihrer Anwendung auf Agrikultur und Physiologie*, in which Liebig propounded his theory that plants obtained the minerals they require from the soil and their supply of nitrogen from ammonia in the atmosphere. He believed that the minerals, since they are removed with every crop taken, have to be replaced, while, on the other hand, the atmosphere could serve as an ammonia (nitrogen) source indefinitely, artificial nitrogenous fertilizers only accelerating a process that, in time, takes place naturally. From then dated Liebig's experiments to discover the nature and amounts of substances removed from the soil by growing plants, a practical consequence of which work was his proposals for their artificial replacement. In 1845, James Muspratt, who had struck up a friendship with Liebig at Liverpool in 1837, took out a patent on the latter's behalf for the manufacture of soil additives. At Muspratt's LeBlanc soda works

* This increase in crop production resulted in an agricultural depression when the Corn Laws were repealed in 1846.

near St. Helens his sons Richard and James (" Sheridan "), who had been Liebig's pupils at Giessen, made considerable quantities of their professor's patent fertilizers; these consisted of various formulations of plant ashes, gypsum, calcined bones, silicate of potash, phosphates of magnesium and ammonia, as well as common salt. Since the Liebig fertilizers had been subjected to furnacing, they proved to be too insoluble to be effective in the soil. The object of fusing the mixtures had been precisely to render them less soluble, so that rain would not wash away " the soluble parts of the manure ".

The support given by the landed gentry to the founding in London of the Royal College of Chemistry in 1845 was in part stimulated by the hope that it would bring the science of chemistry to the rehabilitation of agriculture, then menaced by impending importation of cheap grain and by the potato blight. This was too much to expect from a metropolitan institution controlled by Liebig's pupil, A. W. Hofmann, who doubtless respected the almost Aristotelean authority of his master's views. An article in the *Farmer's Magazine*, in 1856, claimed that " agricultural chemistry has suffered much from having a Pope (Liebig) believed to be not only infallible, but beyond the reach of criticism ". It was important, nonetheless, that Liebig's theories were in part well-founded.

When John Bennett Lawes inherited his father's estate in 1834, he had some knowledge of chemistry from a course he had taken at University College, London. If he already did not know of Liebig's suggestion that bones might be made more effective as a fertilizer by treatment with sulphuric acid, he doubtless learned of it from Joseph Henry Gilbert, who joined him, in 1835, in his experiments with phosphatic fertilizers: Gilbert had worked in the world-famous laboratory at Giessen. At about this time Edward Packard, James Murray, Henry Bell and others also began solubilizing bones with acid on a small scale. Later, Lawes bought Murray's patent for producing superphosphate from coprolites. In 1841 Lawes patented his process for producing superphosphate from bones and established works at Deptford.

The erection of this factory marked the effective beginning of the world's superphosphate fertilizer industry, which was to become the biggest technological consumer of sulphuric acid. Packard & Company's full-scale superphosphate works began at Bramford in 1854; earlier Packard had been producing fertilizer on a small scale from local coprolites at Snape in Suffolk from 1843. In Scotland, in 1846, J. Pointer & Son, and Tennant & Company (at Carnoustie), and Miller & Company (at Aberdeen), in 1852 also set up superphosphate works. Acid from chamber plants (about 70% H_2SO_4), which was strong enough for saltcake manufacture, had to be concentrated to 80% strength (" brown oil of vitriol ") before it could be used for solubilizing phosphatic raw materials: fertilizer production was, thus, the origin of the first large-scale demand for more concentrated sulphuric acid. The concentration was carried out in lead pans supported by cast-iron plates protected from the fire by refractory bricks.

Lawes soon found that the use of bones as a raw material set a limit to his enterprise; he therefore, in 1842, extended the scope of his process to mineral phosphates. The first mineral phosphate he used was the coprolitic deposit to which Liebig's attention had been brought by Buckland and of which the latter wrote: " In the remains of an extinct animal world England is to find the means of increasing her wealth in agricultural produce, as she has already found the great support of her manufacturing industry in fossil fuel . . . the remains of a vegetable world." Later (1857) Lawes used Norwegian apatite; this was the first mineral phosphate to be imported by this country. In the 1870's Belgian phosphate was used as a starting material and, in the 1880's, importation of phosphate rock from the U.S. began; North African phosphate was introduced in the late 1890's.

By 1860 exports of superphosphate from the U.K. were beginning and the home market was rapidly expanding. In 1861, 40,000 tons of phosphatic materials were solubilized. A decade later there were some eighty superphosphate works in the country; these were often situated at ports, so that raw material could be obtained without the cost of an overland haul, and the fertilizer could be more readily shipped abroad. Where there was no LeBlanc works in the vicinity, from which acid could be purchased, the fertilizer manufacturers set up their own chamber plants—a development which ultimately led to permanent national over-capacity for sulphuric acid production. Lawes' connection with the industry in which he had played the chief part in establishing was severed in 1872, when he sold out his interest to the Lawes Chemical Manure Company. Thereafter he devoted his life to experimental work on his estate at Rothamsted, where he founded the research establishment from which much of modern knowledge of soil fertility has originated.

Production of the synthetic nitrogenous fertilizer, ammonium sulphate, set up an important nexus between the coal-gas and chemical industries. This fertilizer was first produced by the Gas, Light & Coke Company in 1815. Its adoption was slow; more than a decade elapsed before it was in established demand. By 1879 the national annual output of ammonium sulphate was 40,000 tons; in 1889 it was 117,000 tons.

The beginning of potassium fertilizers in the U.K. is not on record, but production of the sulphate from potassium chloride imported from Germany probably began about 1850. John Miller of Aberdeen was marketing the sulphate in 1852; twenty years later production was on a considerable scale.

Up till 1890, manufacture of fertilizers, particularly superphosphate, was largely a British chemical industrial monopoly; from that date onwards, however, production in other countries, in parallel with the general development of their heavy chemical industries and the rapid decline of guano sources, became very rapidly competitive. Superphosphate and ammonium sulphate manufacture continued into modern times to be the most important sulphuric acid-dependent chemicals. Lawes and others in the 1840's were marketing compound fertilizers in the form of guano and superphosphate mixtures; modification of the physical form of the fertilizers, to facilitate their application to the soil, was limited to conventional crushing and milling operations.

Nitro-Explosives

The Industrial Revolution was far advanced and its canal and railway eras fully emerged without development of any alternative to muscle-power or gunpowder as a means of disrupting and displacing large quantities of geological matter; there was no advance in military explosives from Waterloo to the Crimea. In the construction of the South Eastern Railway, in 1843, 18,500 lb of gunpowder was used to blast the cliffs at Dover; this was considered a notable operation at that time. Manufacture of gunpowder is not a chemical process, but the simple mechanical mixing of the constituent combustibles (charcoal and sulphur) with the oxidizing agent (potassium nitrate); the operation is one of formulation, not of *synthesis*. Gunpowder production is properly not a department of chemical industry. That industry became concerned with explosives only when certain chemicals were found to behave explosively, often many times more powerfully than the black powder mixture which had served the needs of war and mining technology for an uncertain number of centuries.

The discoveries which led to the establishment of the explosives branch of chemical manufacture were made by continental chemists. In 1846, C. F. Schönbein, professor of chemistry at Basle, reacted a concentrated mixture of nitric and sulphuric acids with cotton, and, in the same year, Ascanio Sobrero, an Italian chemist, treated glycerine with that acid mixture. Both cotton and glycerine have the molecular structure of alcohols and the substances obtained from them by treatment with nitric acid are *nitrates*, although, by long-established usage, they are known respectively as nitro-cellulose and nitro-glycerine. The explosive nature of both nitro-cellulose and nitro-glycerine derives from the fact that they contain, intimately related within their molecular structure, carbon, hydrogen and the powerfully oxidizing nitrate group. Guncotton (as nitro-cellulose early became known) and nitro-glycerine were the first synthetic explosives and their production involved eminently chemical processes.

Nitro-cellulose attracted attention as a potential artillery explosive almost immediately after its discovery, and trials were made in the arsenals of a number of continental countries. Schönbein wrote in a letter to Faraday: "I think it might advantageously be used as a powerful means of defence or attack. Shall I offer it to your government?" Whatever Faraday replied, Schönbein came to Britain in 1846 and granted the sole rights to manufacture guncotton under his British Patent to John Hall & Sons of Faversham in Kent. In July 1847 an explosion destroyed the Faversham factory, killing twenty-one workers. Explosions also took place in other countries and the manufacture of guncotton was discontinued in France and Britain. In 1855 Alexander Parkes of Birmingham produced on an experimental scale a modified form of guncotton, said to resemble cordite; a report was made on the effect of a half-ton shell propelled by its means. Frederick Abel,* in 1863, discovered that guncotton could be rendered stable by thoroughly pulping the cotton, so that residual acid could be readily washed away after the nitration process. From that time onwards guncotton was manufactured in this country.

Probably because it is a liquid, the possibility of developing nitro-glycerine as an explosive was investigated much later than in the case of nitro-cellulose; it continued until 1863 to be a laboratory curiosity. In 1863 the Nobel family began laboratory experiments in Stockholm and produced some nitro-glycerine as a commercial explosive. When an explosion killed one of the Nobel brothers and four assistants, the Swedish Government prohibited manufacture in the capital. Alfred Nobel thereafter continued to work on nitro-glycerine in a barge anchored in a lake near Stockholm. By 1864 he had discovered that nitro-glycerine could be stabilized by absorbing it in the diatomaceous earth, kieselguhr, and that this stabilized form could be made to explode with great power by means of a mercury fulminate detonator. Nobel called his putty-like explosive "dynamite".

---

* Abel was one of the first twenty-six students to enrol at the opening session of the Royal College of Chemistry in 1845. He was for five years one of Hofmann's assistants.

Before 1870 numerous dynamite factories were established on the Continent and in America. In 1871 the works of the British Dynamite Company were erected in the sandhills of the Ayrshire coast at Ardeer; this was the first important nitro-explosives factory in Britain. Nobel invented two other nitro-glycerine compositions—blasting gelatine (1875) and ballistite (1887), the latter a smokeless propellant. In 1889 Abel and Dewar, to overcome certain disadvantages of ballistite, developed a propellant consisting of a mixture of guncotton and nitro-glycerine gelatinized with acetone and stabilized with petroleum jelly; this mixture, extruded into cords of various thicknesses, was named " cordite " and was adopted as the standard British propellant explosive. Nobel claimed that cordite was an infringement of his ballistite patent; he lost, however, his action at law against the British Government.

Picric acid (trinitrophenol), produced by the action of a mixture of nitric and sulphuric acids on phenol, was first used as a yellow dye about 1850. Turpin, in 1886, showed that the compound could be detonated, and its potentialities as a high explosive were rapidly explored. The firm of Read Holliday Ltd, a long-established dye-making concern in Huddersfield, received a contract to produce picric acid for use in the Boer War. Trials of the explosive were made at Lydd in Kent, hence the name " lyddite " by which it was subsequently known in this country. Several other dyestuff intermediates were, about this time, converted to higher nitro-derivatives and their explosive capabilities investigated, e.g., TNT (trinitro-toluene) and tetryl (N-nitro-N-methyl-2,4,6-trinitroaniline).

Immediately prior to World War I, lyddite was the established British warlike high explosive; it was being produced at a rate of less than 35 tons a month. A very small tonnage of TNT was being made, this explosive being mainly purchased by the War Office from abroad. Cordite was manufactured at a rate of possibly 75 tons a week. Small quantities of guncotton, tetryl, and several special-purpose gunpowders completed the range of governmental requirements. The output of civil explosives did not then exceed some 300 tons a week, and of this a large part was exported. With the exception of the Government's establishments at Waltham Abbey and Woolwich (the first participation by the State in a department of chemical manufacture), production of explosives was wholly in the hands of commercial firms. The rise of explosive manufacture augmented demand for sulphuric acid of high strength and led, before 1914, to the introduction of the first contact process plant in the U.K.

Copper Sulphate

William Cooper, a Berkshire veterinary surgeon, was one of the first in the U.K. to propose chemical treatment of crops for their protection against disease. In 1868 Cooper began marketing his so-called " vitriol dressing ", consisting of a mixture of iron and copper sulphates. By 1880 he was annually selling dressing sufficient for the treatment of 150,000 acres, and pharmacists in rural areas were also selling copper sulphate to farmers. Following Millardet's discovery that copper salts were effective against the mildew which, in the early 1880's, raged in the vineyards of France, there was a great increase in demand for copper sulphate. The sulphate was used in the production of the " Bordeaux " mixture with which the vines were sprayed; the other chemicals used in this mixture were lime and soda. Before modern control agents were developed for crop treatment, over 50,000 tons of copper sulphate for that purpose were annually produced in Britain. The manufacture in later years became concentrated at Widnes in Lancashire. In modern times demand for copper sulphate has declined owing to development of more sophisticated pest-control agents.

13. OTHER HEAVY CHEMICAL BEGINNINGS

The manufacture of a small number of inorganic chemicals, which have become key products in the modern chemical industry, began in the nineteenth century by processes which

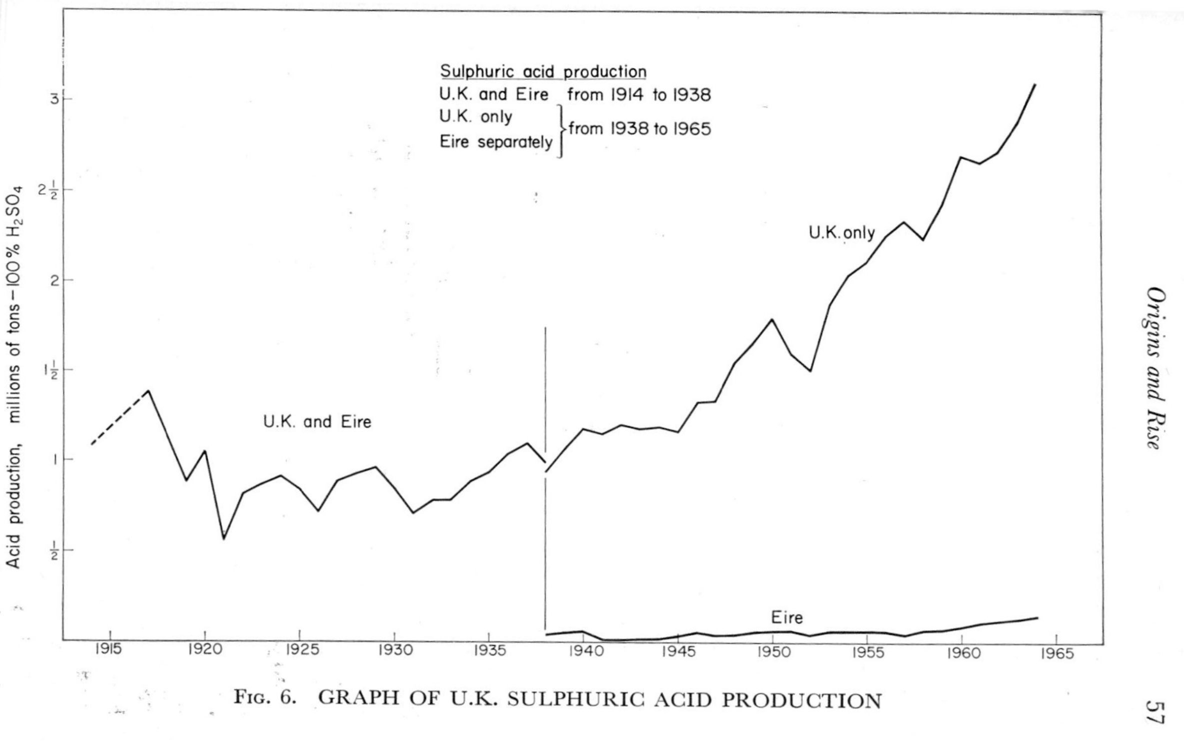

FIG. 6. GRAPH OF U.K. SULPHURIC ACID PRODUCTION

have been wholly superseded by modern developments. Most important of these products were phosphorus, oxygen, hydrogen peroxide, and cyanides.

Phosphorus

The element phosphorus was discovered in 1670 and from that date to 1777 the method of its preparation—reduction of the phosphate in human urine—kept it an expensive curiosity. In the latter year Scheele produced phosphorus by charcoal-reduction of bone-ash, thus opening the way to its industrial production. Phosphorus manufacture in Britain, by Scheele's method, effectively began in 1844, when Arthur Albright persuaded his business partner, Edmund Sturge, to employ it in order to supply the growing demand from the match industry, which was then importing the element from continental producers. The partners had the advantage, over their French competitors, of the cheap coal of the English Midlands, which at that period sold at as little as one-and-sixpence a ton, whereas industrial coal in France cost ten times as much. Sturge and Albright established their phosphorus works at Selly Oak, near Birmingham. In 1851 the partners exhibited phosphorus at the Great Exhibition, and in 1852 a Birmingham newspaper advertised their phosphorus for " improved Lucifer matches ". The firm had also begun to export small quantities of phosphorus. Albright, in 1854, broke away from his association with Sturge and took over the phosphorus works (by then moved to Oldbury) on his own account. Two years later he formed a partnership with a fellow Quaker, John Edward Wilson. From that time up to 1914, the firm, as a monopoly producer of phosphorus, grew steadily but not spectacularly in importance. During that long period the only outlets for phosphorus were in match manufacture and for making marine flares. Albright & Wilson's output of white phosphorus in 1855 was 54 tons; by 1881 it had increased to 446 tons.

The phosphorus process employed at Birmingham consisted

in treating bone-ash (later phosphate rock) with sulphuric acid and concentrating the liberated phosphoric acid, which was then mixed with charcoal, dried, ground, and finally heated in bottle-shaped fire-clay retorts. To prevent its spontaneous combustion the phosphorus was condensed under water. The melted phosphorus was then treated with chromic acid to improve its colour, filtered through chamois leather, and cast into sticks.

Anton Schrötter, of Vienna, who discovered the stable, non-poisonous red form of phosphorus in 1845, was at the British Association Meeting in Birmingham in 1849. There he met Albright who purchased his patent and set about devising an industrial process for red phosphorus manufacture. The change from the ordinary to the red form is brought about by heating in an inert atmosphere. Explosions which may occur, because of the exothermic nature of the reaction, had persuaded Schrötter himself that his discovery was of no industrial interest. Albright, using a sufficiently thick-walled cast-iron pot, fitted with a pressure-relieving device, was able to produce red phosphorus safely. It was only after 1855, when the Swedish brothers Lundström invented their safety match, that red phosphorus found a large-tonnage market.

Oxygen

Oxygen was the first of what are now known as the " industrial gases ". In 1880 the French brothers Leon and Arthur Brin, developing a discovery made by Boussingault in 1851, invented a process for fixing atmospheric oxygen by making use of the temperature–reversible reaction between that gas and barium oxide ($2BaO + O_2 \rightleftarrows 2BaO_2$). The Brins Oxygen Company was incorporated in 1886 to produce oxygen, by the Brins' process, for metallurgical applications. This company became the first successful industrial producer of oxygen in the world. Use of the oxy-hydrogen or oxy-coal gas flame for welding and of oxygen for cutting steel was first made in this country by Thomas Fletcher of Warrington. It is recorded that one of the first " industrial " applications of the last-named technique was made by safe-breakers! Although cylinder-packed oxygen had been made available by the Brins Company, application of oxygen in welding and cutting developed only slowly in the early 1900's. In 1887 the Company produced only 142,116 cu. ft of oxygen; in 1889, output reached a million cu. ft. When the Brins Company started oxygen production the main outlet was for limelight. Opticians selling or hiring optical lanterns acted as retailers of the gas and the Brins Company adopted the magic lantern as its trade mark, a mark still used on cylinders by its modern successor, the British Oxygen Company. The barium peroxide process for oxygen continued in use in the U.K. until 1906.

Hydrogen Peroxide

The first process for hydrogen peroxide consisted in reacting barium peroxide with dilute acid. Interest in the industrial potentialities of this reaction was stimulated by the fact that the Brins' oxygen process had made barium peroxide more readily available. There had been some hydrogen peroxide production on the Continent before Bernard Laporte, an importer of textile chemicals, set up a works at Shipley in Yorkshire, in 1888, to make hydrogen peroxide from imported barium peroxide. The dilute hydrogen peroxide produced was used as a bleaching agent, particularly for straw in the hat trade; in 1898 Laporte opened a branch works at Luton, the chief site of the straw-plait industry.

Cyanides

Following the accidental discovery of Prussian Blue by Diesbach early in the eighteenth century, quantities of prussiates (complex cyanides) for textile printing were made by furnacing organic matter with iron salts and alkali. Charles Macintosh, at Campsie, and James Muspratt, at Liverpool, were amongst those who manufactured prussiates in the first years of last century. Early in the history of coal gas production it was found that hydrocyanic acid (HCN) was present in the gas, and some was recovered for cyanide manufacture. The 50 tons or so a year of crude potassium cyanide required by the U.K. electroplating trade prior to 1880 were produced by thermal decomposition of potassium ferrocyanide (yellow prussiate), obtained by the crude procedures already referred to. Hydrocyanic acid itself began to be used in the 1880's as an insecticidal fumigant.

The financial consequences of the Industrial Revolution were, in the 1880's, putting pressure on the currency system: greatly increased supplies of gold were urgently required. That additional gold requirement could be met only by working lower grade ores for which an economic extraction process was lacking. In 1884, an American metallurgist, Henry Renner Cassel, persuaded a group of Glasgow businessmen, including Charles Tennant (grandson of the founder of St. Rollox), that he had such a process. The Cassel Gold Extracting Company was formed to develop Cassel's invention. It was ultimately found that Cassel, who discreetly withdrew to New York after disposing of his shares in the new company, was unable to substantiate his claims. The chemist who investigated and reported adversely on Cassel's process, John Stewart MacArthur, continued work on the extraction problem and made his extremely important discovery that a weak solution of potassium or sodium cyanide removes gold, as a complex cyanide, from low-grade lodes.

MacArthur's extraction process was tried out in New Zealand in 1889 and in the Transvaal in 1890; its complete success brought about the necessary revolution in gold-mining, and demand for cyanide rose steeply. Until 1900 the process employed for cyanide production consisted of heating at 800°C a mixture of coal and sodium carbonate in an atmosphere of nitrogen or ammonia. The cyanide was extracted from the cool reaction mass with water and concentrated to dryness. This process in the U.K. and a similar one operated at a works in Germany produced for a number of years most of the world cyanide requirement which, by the beginning of this century.

reached some 4000 tons annually. From 1899 to 1908 the United Alkali Company produced at Runcorn an important part of British cyanide output by a process starting from carbon disulphide and ammonia (Raschen process).* In 1894 the Oldbury Alkali Company (later Chance & Hunt Ltd) formed a subsidiary, British Cyanides Company, to produce sodium cyanide by a process similar to the Raschen, using ammonia sulphocyanide—a circumstance that later led to its entry into plastics manufacture as British Industrial Plastics Ltd (p. 199).

14. COAL TAR DISTILLATION

Coal tar was the subject of one of those occasional technological reversals, by which the byproduct of a process becomes, because of developments in the applicational field, more important than the product for which it was originally carried on. Coal tar was first produced for use as a protective coating for the hulls of wooden ships and for ropes; the gas was a byproduct for which there was no established use. Although in 1807, gas flames were already flickering nightly in Pall Mall, the founding documents of the coal gas industry were William Murdoch's 1808 paper to the Royal Society entitled " An Account of the Application of the Gas from Coal to Economical Purposes ", and Samuel Clegg's, in the same year, to the Royal Society of Arts on " Schemes of Gas Lighting in Factories and Public Buildings ". These pronouncements brought about a wide public realization of the potentialities of the new means of illumination and, by the early 1820's, gasworks were established in many cities. Rivers were being fouled by the increasing quantities of useless tar poured into them: a solution of this disposal problem was imperative if the gas industry was to continue its expansion. Charles Macintosh is reputed to have originated the idea of using the tar as a fuel to fire the gas retorts—a procedure first adopted at Manchester gasworks. Before 1820 it was known that

* Developed by Dr. J. Raschen, Chief Chemist to the United Alkali Company.

coal tar, on distillation, yielded a light oil (naphtha); this found some application as a lamp-fuel and solvent. When Macintosh required naphtha, in 1823, to dissolve the rubber used in making his waterproof garments (later known as " mackintoshes ") he obtained it from a small tar-distilling firm in Leith, which obtained its tar all the way from Birmingham. Naphtha production, however, provided a very limited outlet for tar, which continued to be used either as a fuel or as a timber preservative; the latter application became of increasing importance with the growth of the railway network, requiring vast numbers of wooden sleepers. Bethell of Birmingham in 1838 introduced the use of heavy tar oils for pickling railway sleepers. In the 1840's Read Holliday of Huddersfield allowed a nearby gasworks to dump its tar on a site he had prudently acquired. Holliday, who in 1848 invented a series of naphtha burning lamps, embarked upon distillation of his cost-free raw material to produce naphtha and creosote oil; the latter he marketed as a substitute for tar in the protection of timber. Before the mid-century a number of tar distilleries were in operation.

Early in its history chemists in several countries turned their attention to coal tar; naphthalene was isolated from it in 1819, anthracene in 1832, aniline, quinol and phenol in 1834. John Dale, in 1838, prepared crude nitro-benzenes by acting on coal tar naphtha with a mixture of sulphuric acid and potassium nitrate. He was unaware of the chemical nature of the product he obtained. At the 1842 meeting of the British Association in Manchester, Leigh exhibited samples of benzene and nitrobenzenes he had prepared from coal tar. None of these discoveries at the time had any technological consequence. In 1846, A. W. Hofmann encouraged one of his students at the Royal College of Chemistry, Charles B. Mansfield, to investigate the liquid hydrocarbons present in tar. Mansfield's 50-page patent specification laid the foundation of the benzene or, as it is commonly known, benzole industry. By this time the essential pre-conditions for organic chemical industry based on aromatic hydrocarbons had been attained.

15. THE RISE OF DYE AND PIGMENT MANUFACTURE

Dyes

Man's demand for colour in his artefactual world is a psychological rather than a technological need. Modern technology would have been considerably simplified if Ford's " choice " of colour, offered to customers for his Model T (any colour so long as it was black), had been generally accepted for all artefacts where colour is ultimately an aesthetic requirement only and serves no differentiating, warning, or other symbolic function. In the technology of colour fashion has been a chief diversifying factor, particularly in the realm of textile dyes. The great expansion of the textile industry, which had been one of the principal factors in calling into being the inorganic heavy chemical industry in the first half of last century, did not stimulate a contemporary rise of a synthetic dyestuffs industry.

The natural dyes were complex organic compounds, or mixtures of compounds; before they could be synthesized, or *imitated*, their chemical nature had to be more fully understood; organic chemistry at that period had not reached the necessary level of development. The natural dyes in common use were ten in number, no fewer than seven of them being in the red range; some colour diversification was obtained by variation of the mordants with which the dyes were applied; green, as such, was lacking in the dyer's palette and had to be produced by double-dyeing with blue (indigo) and yellow (fustic or annatto). From about 1830 onwards, chemists, mainly on the Continent, made tentative investigations on some of these natural products, particularly indigo and madder, e.g., aniline was obtained by destructive distillation of the former, and madder was shown to contain two distinct chemical compounds. In 1856 over 75,000 tons of natural dye materials, valued at £2·15 million, were imported duty-free into the U.K. By the early 1850's it appeared that, since natural dyes were obtained from crop plants, a situation might soon be reached in which shortage of dyes could apply a brake to the rapid expansion of the textile industry. In the Manchester region, in the early 1850's, an artificial purple dye, murexide, a French discovery made some years previously, was being produced on a commercial scale by treating uric acid, from Peruvian guano, with nitric acid.* Picric acid was used to a small extent in this country as a yellow dye from 1849.

In 1849, A. W. Hofmann commented in a report on the possibility of synthesizing quinine from naphthalene. At the stage organic chemical theory had then reached, it appeared that naphthylamine differed in constitution from quinine only by the elements of two "equivalents" of water. This suggestion, in 1856, prompted William Henry Perkin, one of Hofmann's students, at the age of nineteen, to attempt the synthesis of quinine in a laboratory at his home. Perkin prepared the sulphate of allyl toluidine and oxidized it with potassium dichromate; the product he obtained was not the sought-for alkaloid but a reddish-brown precipitate. Surmising that this could be a general reaction of aromatic bases, he investigated the action of dichromate on aniline sulphate. In this instance the produce was a black precipitate, which on drying and extraction with alcohol yielded a brilliant purple solution. It is to be emphasized that, although Perkin was aware of the nature of his starting material, the organic chemistry of the time could not enable him to ascertain the chemical nature of his product; indeed, the dye had fallen out of commerce before the advance of science, in 1888, fully elucidated its structure. Perkin's discovery was a scientifically observed accident, a fortunate consequence of applying simple organic chemical techniques to test an erroneous guess: the process he worked out to exploit it was of necessity empirical.

Within little more than eighteen months, with the help of his father and brother, the former a boat-builder and carpenter, the latter a builder, Perkins established a factory for commercial

---

* This, of course, was *not* a synthesis, and the chemical *structure* of murexide lay well beyond the horizon of the organic chemistry of the time.

manufacture of his aniline colour, subsequently named mauve. The pre-conditions for chemical industry based on organic raw materials having emerged, Perkin had taken the initiative, not with a synthetic replacement of one of the natural dyes, but with a coloured substance quite unknown in the realm of nature.

If the chemical structure of the products had perforce to remain obscure, the techniques for producing what became known as aniline colours were, although different, little more difficult in practice than those of the contemporary inorganic heavy chemical industry. Those sufficiently experienced technically profited by Perkin's example: by 1860 four firms, in addition to Perkin & Sons, had entered the field. Two years later, Hofmann was forecasting that Britain would become the chief dye manufacturing and exporting country in the world, because of its overwhelming output of the starting material—coal tar. In this country and in France new aniline colours came into production in rapid succession: magenta (1856), violet imperial (1860), Bleu de Lyons (1861), sulphonated aniline dyes (1862), Hofmann's Violet and aniline black (1863), and methyl violet (1866). Perkin added to the coal-tar dye palette with " dahlia ", a red-purple obtained by methylating mauve, aniline pink by oxidizing mauve, Britannia Violet by brominating magenta, and Perkin's Green by acetylating Britannia Violet. Ten years after its discovery mauve had fallen out of fashion.

The chemical structure of alizarin, one of the red-coloured substances occurring in the root of the madder plant, was shown by Gräbe and Liebermann, in 1865, to be based on anthracene. Two years later, Wurtz and Kekulé discovered a method of replacing an aromatic sulphonic group by a hydroxyl. Perkin, acquainted with these developments, sulphonated anthraquinone and fused the sulphonic acid with caustic alkali; when the melt was dissolved in water the colour of alizarin appeared in its dramatic beauty. When Perkin applied for a patent for this synthesis of a natural dye, he found that he had been anticipated by a single day by Gräbe and Liebermann. Five months later he had worked out an alternative patent-free route and, in 1869, before his second patent was filed, marketed synthetic alizarin. In 1873, Perkin & Sons produced 435 tons of the dye; on the Continent no significant competitive alizarin manufacture took place until after that year.

In the first twenty years of the dye industry the inventive initiative lay almost wholly in Britain and France; developments in Germany were relatively insignificant. German chemists trained in the organic chemical schools of their country, sought experience in the application of their science in Britain. Caro, Martins, Leonhardt, Schad, and Witt put new dyes on the range of their British employers, and Griess,* while at the Royal College of Chemistry, discovered his diazotizing reaction—still the key step in the synthesis of a large part of the world's dye output. Ivan Levinstein, who received his chemical education in Berlin University and Technical High School, set up his small works, where he was to produce magenta, aniline violet, Manchester Brown and other dyes, in Manchester in 1864.

The advent of aniline dyes did not bring a sudden halt to the importation of natural colouring materials; in 1859, 19,800 tons of madder were landed at Liverpool alone; in 1870 the U.K. import of that dye was 10,750 tons (equivalent to about 1075 tons of actual alizarin). Large tonnages of the traditional dyewoods continued to be consumed in British dyeworks. Synthesis of alizarin, however, brought about profound changes in the situation. Madder constituted almost 60% of the tonnage of natural dyes used in the calico-printing works in the U.K. in 1870. To replace this natural dye by its synthetic equivalent a rapid expansion of the British dye industry became pressingly necessary. Perkin and his brother, whose factory had capacity to supply less than half of the potential domestic alizarin requirement, decided not to risk their capital in a battle with continental,

* Griess did not enter the dye industry, but passed his working life as a chemist in Alsopp's Brewery. He was, however, retained as a consultant by the dye-making firm of Dale & Company, Manchester.

particularly German, competition. They sold their business to Brooke, Simpson & Spiller.* Already, in the year of Perkin's retirement, German alizarin production was about double that of the U.K. Until the appearance of synthetic alizarin the German dye industry had followed closely on the heels of the British, taking advantage of the absence of a unified German Patent Law before 1871 to exploit foreign dye discoveries. Alizarin profits became an important source of funds for German dye research; this economic factor, coupled with its resources in the matter of highly trained organic chemists, made Germany, after the mid-1870's, the site of the principal dyestuff discoveries and development. Most of the German chemists who had sought the British factories as the centre of advance returned, taking their experience with them, to work in the swiftly expanding dyestuffs industry of their own country.

Only when the German dye-makers, in 1883, forced up the price of alizarin to British consumers and threatened to stop supplies unless sales contracts were continued, did the dyers and printers in this country react by forming the British Alizarine Company, with a production capacity equivalent to about one-fifth of the then total U.K. requirement. Otherwise, the British dye consumers cared little who made the dyes, so long as they got the quality and quantities they required at the best prices. The passing of the technical initiative in the new branch of chemical industry to Germany was not a matter of commercial, or indeed, governmental concern: the philosophy of *laisser-faire* and Smilesean self-help governed trade and enterprise.

Britain's primacy in the Industrial Revolution had established a vast commitment and tradition in the textile, iron and coal industries. Dye manufacture was an ancillary to one of the main British industrial themes: it must have seemed eminently satisfactory and reasonable to the dye-users in this country that, by a providential specialization of an international division of labour, this service should be so efficiently supplied by foreigners. After

* Formerly Simpson, Maule & Nicholson, established in 1860.

all, the value of the dyes was about 1% of the value of the textiles to which they were applied. As an industry dependent for its advance on the chances of research, the dye industry was not an attractively safe field for investment compared with the huge, long-established and still highly-profitable technologies. There was no failure of nerve or mysterious malady afflicting British enterprise. Nor was lack of trained British organic chemists a factor in the situations: in its early decades the U.K. dye industry, as we have seen, availed itself of a " brain drain " from Germany; had the industry become a focus of investment, there can be little doubt that the German universities and technical high schools would have been adequate sources of manpower for technical management and research. Levinstein himself employed German chemists at his Manchester works. It was largely Ivan Levinstein's attribution of the failure of the British dyestuffs industry to keep pace with that of Germany to the attitude of the government, the lack of technical training in the universities and general absence of vision that gave rise to the unhistorical view, still widely accepted, that there was stupid and culpable neglect of the nation's future and security—only in the light of the disaster of 1914 could that even seem so. The *economic* picture was, in fact, very much more favourable than the technical one; thus in 1913, the U.K. imported German dyes to the value of £1·893 million, but *exported* heavy inorganic chemicals (including fertilizers) to the value of £14·349 million.

After what may be called the " alizarin crisis ", the British dyestuffs industry continued its slow and erratic expansion. Occasional inventions and discoveries made their appearance, e.g., the discovery of chrysoidine (London Orange), the first azo wool-dye, by Witt in 1875, in the laboratory of Williams, Thomas & Dower at Brentford, and the development by Read Holliday & Sons, in 1880, of the ingrain dyeing process, whereby azo colours were formed *in situ* in the fibre or fabric. In 1890 the U.K. dye industry employed some 1000 workers, one-twentieth of the number then employed by the five largest German dye firms.* In 1898 an association of dye-consumers, the Bradford Dyers' Association, was formed; this forced a merger of the still important suppliers of natural colours, but the producers of synthetic dyes took no similar action—a step which could have significantly modified the pre-World War I history of their industry.

British tar production rose from 175,000 tons in 1870 to 640,000 tons in 1880. A large part of this output was distilled and the benzole, etc., exported to Germany as raw materials of dye manufacture. In effect, this trade was a kind of British participation in the prosperity of the German dye industry. In the early 1880's gas users began to demand gas of higher calorific value for use in cookers, industrial processes and illumination with the new incandescent mantles. Accordingly, the gas industry raised the gasification temperature, with consequent reduction in tar output. This sudden tar shortage occasioned the setting up of a species of symbiotic trade with Europe and North Africa, whereby British ships taking coal to the gas industry in these countries brought back tar which was distilled for export to Germany. Ultimately, Germany achieved independence of British tar supplies by means of benzole recovered from the improved coke ovens of her steelworks.

Probably because he was aware of absence of native enterprising *élan* in the industry, Charles Dreyfus, with Swiss financial backing, established the Clayton Aniline Company, in Manchester in 1876, primarily to produce intermediates for export. In 1890 the German chemical firms of Agfa and Bayer made an incursion into the British dyestuffs industry by each acquiring a one-third share in the issued capital of Levinsteins, which at that date became a limited liability company with Ivan Levinstein as managing director. Five years later, this German enclave was wound up and Levinstein Ltd continued in existence with British capital. After the passing of the Patent Law Amendment Act in

---

* It is not known how many native chemists were employed in the U.K. dye industry, but it is on record that, in the 1880's, they were paid at similar rates to the manual workers.

1907, by which foreign patentees were required to work their inventions in Britain or grant licenses to others to do so,* several continental firms established works to produce dyes in this country. In 1907, Badische Anilin und Soda Fabrik (BASF) purchased a 24-acre site for their operations at Bromborough on the Mersey. Four years later CIBA, the principal Swiss dye firm, acquired the Clayton Aniline Company. Meister, Lucius & Bruning built a factory at Ellesmere Port, also on Merseyside; here, in 1913, they produced 146 tons of indigo. It is doubtful whether the 1907 Patent Law Amendment, for which Levinstein in particular had campaigned with Cato-like insistence for many years, alone would have brought about a far-reaching development of the British dyestuffs industry; indeed, by inducing the more developed German industry to operate subsidiaries in this country, its effect could have been the reverse of its intention.

Immediately prior to the outbreak of World War I the U.K. dyestuffs industry was being carried on by twelve independent firms, including the two German concerns on the Mersey. The outputs of seven of these firms were on a small scale. In general, because of their small size, the British dye companies were not vertically integrated technically; they began manufacture with the organic intermediates and purchased their inorganic reagents (sulphuric and nitric acids, caustic soda, etc.) from the heavy chemical industry, between 3 and 5 tons of these latter being required for production of one ton of dye. From 1900 onwards the U.K. had become increasingly dependent on Germany for dyestuff intermediates and chemical plant, such as enamelled pans, filter presses and pressure vessels.

When World War I broke out, Britain was importing from Germany every month 250 tons of alizarin, 100 tons of indigo, 1200 tons of aniline dyes, not to mention smaller quantities of a large number of other dyestuffs. The abrupt cessation of

---

* It seems probable that this Act would have had a greater effect on the *scale* of foreign dye manufacture in the U.K. than it did, had its intention not been partly frustrated by Mr. Justice Parker's interpretation of it as not requiring manufacture sufficient to meet the national demand.

these imports in August 1914 found the British dyers unable to carry on a large part of their business; they were not even in a position to dye the uniforms of the soldiers who were to fight Germany! So acute was the situation that Royal Warrants were issued to permit trading with the enemy, and for some time an exchange of dye intermediates was carried on in neutral Holland. Recourse was also had to natural dyes, a large proportion of the Allies' requirement of which was supplied by the British Dyewood Company, then a subsidiary of the United Dyewood Corporation of New Jersey.

Pigments

Manufacture of pigments, i.e., insoluble coloured substances for application to surfaces* with a medium (oils, waxes, etc.), did not develop as a branch of dye-making, but as a peripheral activity in heavy chemical works or as the concern of independent specialist firms; thus, for example, the firm of Johnson, Mathey & Company, founded in 1743, specialized, in addition to its business in rare metals, in the manufacture of vitreous colours and pigments, in particular for the pottery trade. Many of the most important pigments are inorganic oxides or insoluble salts, others are lakes formed by precipitation of vegetable or synthetic dyes with aluminium hydroxide.

One of the oldest and most important pigments is white lead, which was in use before the Christian era. In the nineteenth century white lead continued to be produced by what was known as the Dutch process, whereby metallic lead in earthenware pots containing vinegar was exposed for a number of weeks to the combined action of the warmth and carbonic acid gas from the horse-dung in which the pots were buried. Chemically, the product was a basic lead carbonate. In 1853, Britain exported 3000 tons of white lead, about half being destined for the U.S. and the

* Pigmentation in *volume* is almost entirely a modern requirement, arising out of the development of synthetic materials. Coloration of foodstuffs with dyes is an older form of bulk pigmentation.

colonies. In 1890 a firm was established in London to produce white lead by a rapid process (MacIvor's), which consisted in reacting lead oxide (litharge) with ammonium acetate, then precipitating lead carbonate by passing in carbon dioxide gas. H. L. Pattinson's white, oxy-chloride of lead, which he made for many years at his alkali works at Felling in Gateshead, proved inferior in use to white lead. Because of its toxic nature, and its tendency to be discoloured by sulphurous atmospheres, a number of attempts were made to replace lead white by zinc or barium salts; thus J. B. Orr, in Glasgow in 1872, began manufacture of "lithopone", a complex of zinc sulphide and barium sulphate, originally made by Thomas and William Griffiths in 1869. Eventually, Orr's lithopone was manufactured in large tonnages on Merseyside.

The manufacture of Prussian Blue (ferric ferrocyanide) was introduced into this country in 1756 by Louis Steigenberger from Frankfurt. His London firm was later known as Lewis Berger & Sons. Potassium ferrocyanide (yellow prussiate of potash) and Prussian Blue were made by Macintosh at his Campsie alum works; James Muspratt was producing the pigment at Liverpool in 1822–3, before he embarked upon alkali manufacture; it was also in the 1830's a product of the Heworth alkali works on the Tyne. Lead chromate came into use as a yellow pigment at the beginning of last century. Andrew Kurtz, who later became an important alkali manufacturer in St. Helens, produced the chromate at a small works in Liverpool about 1835. Kurtz's chrome yellow enjoyed a considerable vogue after its use to paint the carriage of Princess Charlotte;* it was an exhibit in the chemical section at the Great Exhibition of 1851. By 1845 there were 54 pigments in commercial use for printing wallpaper. As the century advanced, iron came more and more to be used on bridges, ships, and other structures exposed to the weather; this greatly increased the demand for paint. In 1850 there were some 250 paint-making firms in the

* Daughter of George IV.

U.K., 130 of them in London; by 1900, there were 400, employing in all 13,800 workers. Most of these firms were users of pigments made by other concerns; their number and their numerous employees, however, indicate the importance pigments had assumed by the beginning of the twentieth century.

16. PRE-SYNTHETIC PHARMACEUTICALS

In Britain, as in other European countries, pharmacy was for centuries in the hands of the apothecaries, whose principal business was preparation by traditional recipes and simple physical procedures of medicaments from vegetable sources; they also dealt in a small number of inorganic substances, such as sulphur, mercury compounds, and Glauber's salt. A *materia medica* cabinet, sold in many hundreds by the firm of Evans early last century, used to train pharmacists to identify crude drugs, contained 270 samples of roots, barks, leaves and chemicals. Developments in the traditional field of pharmacy were slow and discoveries occasional, there being no rational theory of medicine until the fifth decade of the nineteenth century, in which Liebig published his *Organic Chemistry in its Application to Physiology and Pathology* (1842) and Dorpat founded his institute of experimental pharmacology in Estonia (1849). Prior to that period, experiments on the effects of substances on the living organism were blindly empirical. The Pneumatic Institution, founded at Bristol, in 1798, by Dr. Thomas Beddoes, to investigate the physiological action of gases and their possible effects on disease, was, in intention at least, an anticipation of the continental developments; it made, however, only one contribution to the subject—the discovery, by Humphry Davy, its superintendent, of the anaesthetic action of nitrous oxide. From 1850 onwards advances in medicine were made in the sure light of the conviction that the processes of life were chemical in nature and might be expected to be modified by chemical reagents, although many of these reagents were complex organic substances the chemical nature of which was elucidated many decades later.

In 1785, Dr. William Withering, one of the Birmingham "Lunatics" published his classical account of the effects of extract of foxglove (*digitalis*) on cardiac endema—one of the most important contributions to medicine in the eighteenth century. Thomas Morson, who had studied the chemistry of cinchona bark in Paris, introduced the preparation of quinine into this country in 1821: this was the beginning of one of the most important British contributions to pharmacy—the isolation and purification of alkaloids.

Although modern anaesthesia was largely an American development, it is of interest that the effects of the common anaesthetics appear to have been first observed by British chemists. Davy's discovery of the action of nitrous oxide has already been mentioned; in 1818 his pupil Faraday discovered the action of ether, and, in 1847, on the suggestion of a Liverpool chemist called Waldie, J. Y. Simpson used chloroform as a total anaesthetic in obstetrics. Duncan Flockhart & Company of Edinburgh exhibited chloroform at the 1851 exhibition. Before the end of the nineteenth century ethyl chloride had some application both as a general and as a total anaesthetic in the U.K.

In 1841 pharmacists embarked on the organization of their profession by forming the Pharmaceutical Society, confirmed by Royal Charter two years later. In addition to protecting the interests of pharmacists, the Society existed to advance scientific pharmacy and to establish uniformity of training as the foundation of their professional practice. The Pharmacy Acts (1852, 1868, 1908) gave to holders of the qualifications required by the Society certain legal privileges and placed on the Society the duty of administering these Acts. A further organizational step was taken in 1864 when the London, Edinburgh and Dublin pharmacopeias were replaced by a single national publication. Some indication of the slow development of British pharmacy may be had from the fact that the 1898 Pharmacopeia was still in use in 1914; the earlier issue contained descriptions of only four synthetic drugs; the 1914 edition, however, listed eighty synthetic pharmaceuticals—all but a few were imports from Germany. Not having a highly diversified dyestuffs industry and, therefore, only a limited development of applied organic chemistry, the British chemical manufacturers did not parallel the rise, which began in the early 1880's, in Germany of synthetic drug manufacture.* Up to the outbreak of World War I there was no synthetic pharmaceutical industry in the U.K., although some progress had been made in the production of sera and vaccines. The great majority of medicaments were preparations from naturally-occurring materials; chemical synthesis was limited to a few simple operations, such as the conversion of morphine to codeine. Wholesale druggists, the fore-runners of the modern pharmaceutical industry, grew crops of some of the medicinal plants, but most of their drugs, natural and synthetic, were imported.

17. SEMI-SYNTHETIC MATERIALS

In the whole history of technology from the beginning of civilization to the nineteenth century, no new type of material came into use for the making of artefacts. The first new materials to be produced were not fully synthetic but were chemically-modified forms of the natural materials such as the sap of the rubber tree, cellulose, and protein. These semi-synthetics continue to be useful in their own right; their historical importance is that they were a bridge between traditional natural materials (plant and animal fibres, horn, stone, wood, metals), and fully synthetic modern plastics; they served as pathfinders for the synthetics that were to follow. The semi-synthetic prelude lasted from about 1800 until the first decade of the twentieth century, when the phenol-formaldehyde resins entered the field.

The sap (latex) of the rubber tree (*Hevea brasiliensis*) was the first natural organic substance in historical time to be chemically modified to make it structurally useful. Before the end of the

* E.g., antipyrin (1883), antifebrin (1885), phenacetin (1888), pyramidon (1893), and aspirin (1898).

eighteenth century various solutions of rubber had been used occasionally to render balloons gas-tight, and surgeons had used small rubber tubes for draining wounds. Macintosh, in 1823, patented his invention of waterproofed fabrics; this was the first industrial application of rubber. Thomas Hancock's mastication process, made public in 1835, and his discovery in 1839 of the method of making rubber non-tacky and highly flexible, or hard and horn-like, by treating it with sulphur, a procedure later known as " vulcanization ", completed the emergence of the fundamental technology of rubber. Manufacture of mackintoshes did not represent a very large consumption of rubber, being a " two-dimensional " application. In the year of the Great Exhibition of 1851, the importation of rubber for all purposes into the U.K. amounted to about 700 tons.

Alexander Parkes of Birmingham, who began his technical career as an apprentice in the metal industry of that city, became interested in rubber and invented his sulphur chloride process for cold vulcanization, which Hancock enthusiastically hailed as " one of the most valuable and extraordinary discoveries of the age ". Parkes in 1853, extended his vulcanization process to vegetable oils, producing " factice ", a substance still used in rubber technology. It is not known when Parkes became interested in Schönbein's nitro-cellulose, but it appears probable that he first became acquainted with it as collodion about 1855. In a lecture, delivered in 1865, Parkes said: " For more than twenty years the author entertained the idea that a new material might be introduced into the arts and manufactures and, in fact, was much required." He had by that time been developing for some years his " new material "—Parkesine, a formulation of nitro-cellulose, naphtha, nitrobenzene, camphor, vegetable oil, and colouring matter. Parkesine was exhibited in London at the International Exhibition of 1862, and described by its inventor as useful for making " medallions, salvers, hollow-ware, tubes, buttons, combs, knife handles, pierced and fret work, inlaid work, bookbinding, card cases, boxes, pens, penholders, etc." By 1865, Parkesine could be produced at a shilling a pound, and in 1866 the Parkesine Company was formed to develop the material commercially. Daniel Spill, who for some years had been engaged in the rubberizing of fabrics, became works manager to the Parkesine Company. The new venture did not prosper and was liquidated in 1868; Spill, however, obtained financial backing for a new concern, the Xylonite Company, to continue the development of Parkes' invention. After the Xylonite Company also went into liquidation, Spill for three years continued the enterprise as Daniel Spill & Company until, in 1877, with L. P. Merriam, an American, he co-founded The British Xylonite Company. By this time the rising British lower middle class was making manifest its respectability in celluloid* cuffs and collars, thus providing a profitable market for British Xylonite. The celluloid market was greatly extended in 1884 when it was adopted by the Sheffield cutlery trade as a material for knife handles. During the first few decades of its use celluloid found a great many of the applications to which modern synthetics are familiarly put. In 1877 the Xylonite Company had 29 employees and was producing celluloid at a rate of eight hundredweights a week; by 1888, 500 workers were employed, and in 1902, 1160. Celluloid production in the U.K. and U.S. reached its zenith in the later 1920's.

When, in the 1840's, the electrical industry was beginning to emerge, particularly in America, Joseph W. Swann, a druggist, conceived the notion of an incandescent electric lamp having a filament of carbon as its incandescent element. For many years, with the collaboration of C. H. Stearn, a Liverpool bank clerk interested in high vacua, and F. Topham, an expert in glass-blowing, he experimented with modified cellulose filaments of various kinds. In 1883, Swann produced a reasonably efficient lamp, using nitro-cellulose to form the pre-carbonized filament. Swann was aware of the textile potentialities of his filaments, but did not pursue that aspect of his invention. Ultimately both

* The name "celluloid", for compositions of the Parkesine or Xylonite type, was registered in 1872 by the brothers Hyatt, who developed a nitro-cellulose in the U.S.A.

Stearn and Topham embarked upon manufacture of lamps. Meanwhile, de Chardonnet, in France, was experimenting with nitro-cellulose filament for textile-making. In England, C. F. Cross and E. J. Bevan, in 1882, had begun their joint research on the chemical modification of cellulose. Later, in association with Clayton Beadle, they discovered a syrup-like product (cellulose xanthate), resulting from the action of carbon disulphide on soda cellulose; to this they gave the name "viscose" and patented it in 1892. Cross was a friend of Stearn, who suggested the use of viscose filaments in his lamp manufacture. Surprisingly, Cross and Bevan had not considered the textile potentialities of their discovery; they were, however, aware of its possibilities in other directions. In a paper to the Society of Chemical Industry in 1893, they clearly recognized in " this raw material of the natural world (i.e., cellulose) . . . 'plastic' capabilities of a high order, and requiring only the right treatment, i.e., chemical modification, to enable us to follow nature in working it up into endless varieties of form and substance." Amongst the applications Cross and Bevan suggested were adhesives, sizing and filling textiles, and the production of casts and moulds. It may have been a consequence of a suggestion by James Swinburn, later an important figure in the British plastics industry, who saw the filaments in use at Stearn's works, that Cross, Bevan, Stearn and Topham set about developing a process for commercial production of textile fibre from viscose. To that development Topham contributed the invention of the spinning box, the essential mechanical step to successful viscose filament manufacture. In 1905, the long-established silk-spinning firm of Courtauld began production of viscose on a small scale at Coventry. The semi-synthetic fibre industry had begun.

Although Schutzenberger had acetylated cellulose about 1865, cellulose acetate was first fully investigated in 1894 by Cross and Bevan. Cellulose triacetate—Cross and Bevan's product—was found to be soluble only in such costly and toxic solvents as chloroform. An American, G. W. Miles, in 1905, discovered that the diacetate was readily soluble in acetone, a non-toxic solvent in established industrial use (e.g., in cordite manufacture). It appears that industrial development of cellulose acetate in the U.K. prior to World War I was confined to the operation of a single small concern, the Safety Celluloid Company, which had a make-shift works at Park Royal, near Ealing, originally to produce cellulose acetate to replace celluloid in the manufacture of sheet and film. This company came to specialize in making acetate " dope " for aeroplane wings and textile-supported film for replacing shattered windows in the battle zones of France; they also made clear panels of the acetate for insertion in the upper wings of fighter biplanes, so that pilots could see an attacker approaching from above. The brothers Henri and Camille Dreyfus, who since 1910, at Basle, had been making cellulose acetate film for Pathé Frères, to meet the cellulose acetate requirements of the British Government during the last years of World War I formed the British Cellulose and Chemical Manufacturing Company, with works at Spondon. On the cessation of hostilities, the demand for acetate dope came suddenly to an end and, with great rapidity, the Dreyfus brothers developed Celanese fibre.

The production of a new material by the modification of milk protein (casein) with formaldehyde was carried out on an industrial scale before the end of last century in Germany and France. In 1909, Victor Schutze of Riga patented a variant of this process and the Syrolit Company was formed to exploit it in Britain. The Syrolit works were a former cloth-mill at Stroud in Gloucestershire. When, in 1913, after ruinous trials, Schutze's process had proved impracticable, Erinoid Ltd was formed to acquire the Syrolit Company and, after modifying the process, successful manufacture was finally established. Erinoid is still widely used in the production of buttons and similar small artefacts.

CHAPTER II

Transition to the Modern Industry

1. GENERAL NATURE OF THE TRANSITION

At a late date in its history as a defined field of technology, the chemical industry underwent a transformation as profound and far-reaching as that which took place in the textile industry in the eighteenth century. In both instances the metamorphosis was due in principal part to a change in the basic manufacturing processes. The transition from the earlier chemical industry, in which many of the modern chemical industrial branches existed in their proto-forms, took place over a period, dating from the importation of the effective ammonia–soda (Solvay) process in 1872 to the introduction of the Haber ammonia process and the formation of the second large chemical industrial merger in the mid-1920's. Within this period the transition accelerated very steeply.

For the greater part of last century the chemical industry had entirely the role of an ancillary to several dominant traditional technologies. Some idea of the status of the heavy chemical industry at the middle of last century may be had from the fact that of the 132 chemical exhibitors at the Great Exhibition of 1851 only three were alkali firms, two on Tyneside, one in Manchester (Tennant, Clow & Company): none of the large LeBlanc concerns of Lancashire apparently judged their products likely to stimulate the interest of the public. Following the transitional epoch, the industry has penetrated into fields with which it was hitherto without connection and also from its own substance, so to speak, it has created wholly new technologies, e.g., modern pharmacology is a comparatively recent offspring of dyestuffs manufacture, and man-made fibres have dramatically changed the former ancillary role of the industry *vis-à-vis* textile manufacture. The mid-nineteenth century was historically characterized as the age of power, the mid-twentieth has become the *age of chemistry*: through the modern chemical industry has come the main impact of the applied forms of the great positive sciences (chemistry and physics) on human affairs. It is wrong, however, as is sometimes done, to regard the chemical industry as a *product* of the application of science to industry, as science in economic action.* Holders of that view—too often educationalists and more often ill-informed politicians—regard scientists as scientific manpower, shock-troops of a particularly effective technique, not recognizing that technology is more than science, and science very much more than technics.

The principal factors in the transition were both qualitative and quantitative. Of the former, the more important was the change-over to *flow* processes, i.e., procedures in which there is continuous movement of substance through all stages from raw material to finished product, the movement being sustained either by the force of gravity, draught, or by mechanical impulsion such as pumping. Only two important flow processes existed in the earlier industry—the chamber sulphuric acid and Deacon–Hurter chlorine processes. Advantages of processes of this type—now the commonest in chemical industry—are that they eliminate a great part of the former need for manual labour in handling materials between stages and they enable more effective control to be maintained over the chemical reactions and, consequently, over the uniformity of products. Because they facilitate return of heat from one stage to another, flow processes are very much more efficient thermally than batch (discontinuous) processes. Thus, in purely economic terms the LeBlanc process became gradually uncompetitive with the ammonia-soda process because of the chronic increases in costs of fuel and labour. The

---

* E.g., James Taylor, in his *Cantor Lectures*, 1961, defines the modern chemical industry by its *scientific content*. This in our view is misleading.

large-capacity plants of modern chemical industry would be economically impossible, except on the basis of flow processes and of closely integrated systems of such processes. Other factors promoting the transformation of the industry were introduction of electrical energy for effecting chemical change, organization of the new discipline of chemical engineering and application of physical chemistry to the fundamental operations. At this period organic chemistry was still finding its main application in the dyestuffs branch of chemical industry, in which its practice was producing increasing diversification of products *by established processes.*

World War I was the chief non-technical factor in the establishment of the modern U.K. chemical industry. In 1902, Ivan Levinstein lamented: "It is difficult to get the House of Commons to consider any commercial question of national importance." Long before 1918 the Government was brought to fullest recognition of the significance of chemical industry for the technology of war, and, in the immediately post-war years, similar recognition was given to its importance in the national economy by a change in Government policy with regard to protective tariffs. Two not unrelated consequences of World War I were the establishment in the U.K. of nitrogen fixation on a massive scale and the formation of the second and largest merger in the industry.

2. THE AMMONIA–SODA PROCESS

The historical importance of the ammonia–soda process, generally known as the Solvay process, is that its introduction began the transition from the classical to the modern heavy chemical industry. Chemically, the process is simple, involving reaction between carbonic acid (carbon dioxide) and a concentrated solution of common salt saturated with ammonia. The primary product is sodium bicarbonate. Following its discovery in 1811 by the French physicist and engineer, A. J. Fresnel, a number of years elapsed before numerous attempts were made on the Continent and in this country to base a production process on the reaction. The first attempt in the U.K. to develop an industrial process was made in Scotland in 1836 by John Thom, who for some time, by a somewhat crude procedure, produced a small weekly tonnage of soda crystals at the Camlachie works of Turnbull & Ramsay by whom he was employed as a chemist. For two years (1840–42) James Muspratt, at his Newton (St. Helens) works, where James Young (later of paraffin fame) was manager, attempted to operate a variant of the ammonia–soda process, which he had to abandon after a fruitless expenditure of some £8000. For some years, in the 1840's, a small plant was operated commercially at Leeds, but it, too, did not prove competitive with the LeBlanc Establishment. In 1854, Henry Deacon, in an attempt to produce the high quality sodium carbonate required by the glass industry in which he had been till then employed, erected a pilot plant to operate a form of the process at Widnes. Deacon's weekly output of sodium carbonate never exceeded three tons. He, too, had perforce to abandon his experiment because of difficulties in conserving ammonia and in producing sufficiently pure carbon dioxide. These abortive essays at establishing an economic ammonia–soda process occasioned Solvay's comment that "never before was the industrial realization of any process attempted so frequently and for such a long period of time . . ."

The problem of converting salt to soda (sodium carbonate) efficiently by means of ammonia on a large scale was effectively solved in 1863 by Ernest Solvay, son of a salt-refiner and nephew of a gasworks manager. "My infancy was spent amidst sodium chloride," he once remarked, and later he assisted his gas-making uncle, in whose gasworks he began his ammonia–soda experiments. Solvay approached the problem from an engineering stand-point, as a process demanding the absorption and recovery of gases with minimum losses: therein lay the key to his ultimate success. In an address to the Society of Chemical Industry in 1885, Ludwig Mond said: "We have to give to Mr. Solvay the honour of being the inventor of the apparatus, which alone has made this process of value to the public, and of having thus become the founder of the ammonia–soda industry." In the 1860's, Solvay and his brother developed the new process commercially at Couillet in Belgium. At this time Ludwig Mond was living in Widnes, maintaining himself modestly on the fringe of the LeBlanc industry by free-lance trading in chemicals and by acting as a chemical consultant. In 1872, Mond was contemplating starting on his own account a small works for the recovery of sulphur, by his own process, from LeBlanc alkali waste. At this *Sternstunde* in his career, Mond learned, quite casually from Henry Deacon, of the success of the Solvay operations in Belgium—Deacon had never lost his interest in the ammonia route to soda. "Mr. Deacon," Mond wrote to a continental contact, "has told me that you spoke to him about the rapid development in Belgium of the manufacture of soda salts by the ammonia process. You will understand that this process, which produces no waste,* is very disturbing to me . . ." Hurriedly abandoning his LeBlanc sulphur project, Mond visited the Continent to ascertain the facts for himself, and, in April 1872, acquired a two-year option on the exclusive rights to work Solvay's process in the U.K. With his Widnes colleague, John Tomlinson Brunner, former book-keeper in Hutchinson's alkali works, he formed a partnership and purchased with borrowed capital a mansion and a 130-acre site in Cheshire. By 1874, Brunner, Mond & Company were in production; producing in that year 800 tons of soda; six years later the annual output was 18,800 tons, and by the end of the century exceeded 200,000 tons. In the meanwhile, several other concerns had established ammonia–soda production in the Cheshire salt region; these were eventually taken over and shut down by Brunner, Mond & Company, as was the Schloesing-Rolland type ammonia–soda process operated at Clarence Salt Works, Middlesborough, by Bell Bros. for about twenty years. Incidentally, it should be noted that this long period of operation of the Schloesing-Rolland process amply

* I.e., waste of the kind produced by the LeBlanc process (largely calcium sulphide). The waste from the Solvay process is calcium chloride.

refutes Mond's assertion that this was unsuccessful. The United Alkali Company developed the Mathieson variant of the ammonia–soda process, first at Widnes and subsequently at Fleetwood, where brine was locally available from the Presall salt-field. Early in this century the ammonia–soda process had all but completely displaced the LeBlanc soda process in U.K. alkali works: it survives to the present time as the only process used for producing sodium carbonate in this country.

3. INTRODUCTION OF ELECTRICAL ENERGY

Electrolytic Processes

Until the opening year of the nineteenth century, no means, other than the application of heat, were known of bringing about changes in the chemical composition of matter. The first known application of electrical energy for this purpose was made in 1800 by Nicholson and Carlisle, who used current from a voltaic pile to decompose water into hydrogen and oxygen. In the same year, Cruikshank decomposed common salt by passing an electric current through brine, and observed the evolution of chlorine gas. This type of decomposition is known as electrolysis. Metallic sodium was originally isolated in 1807 by Davy, who electrolysed molten caustic soda; in 1833, Faraday liberated that metal by electrolysis of common salt, in the course of a research which led him to elucidate the nature of the electrolytic process and to formulate its quantitative laws. More than half a century elapsed before electrolysis became an industrial process. The principal reason for this long hiatus was absence of generators for the very heavy electric current required. Emergence of electrochemical processes had to await developments in electrical engineering, which began in the late 1860's with the evolution of the modern dynamo. Although dynamos were available in the U.K. about 1867, it was only towards 1880 that large, reliable machines came on to the market. During the years 1883–93 all the basic inventions required for the introduction of electrical energy into

FIG. 7. THE TRANSITION BEGINS

The first industrial-scale mercury cathode caustic-chlorine installation in the U.K.—H. Y. Castner's plant at the works of the Aluminium Company at Oldbury in 1894. The cells were of the " rocking " type, which were superseded by the so-called " long " (Castner-Solvay) form early in the twentieth century. Production of chlorine and caustic soda by electrolytic processes marked the end of the LeBlanc system of manufacture.

chemical industry were made, constituting one of the most important aspects of the transition from the old to modern chemical manufacture. All the processes based on the use of electricity, either electrolytically or electrothermally (see below), are of the continuous flow type.

Salt, when electrolytically decomposed in solution (brine), yields caustic soda and chlorine; when fused salt is electrolysed, the products are metallic sodium and, again, chlorine. Industrial processes were evolved for both types of decomposition and these form the electrolytic sector of the modern heavy chemical industry. The problem which had to be solved in the electrolysis of brine was that of maintaining separation of the chlorine and caustic soda, which would otherwise react with one another. This was achieved in two ways: by insertion of a porous, chemically-resistant diaphragm between the electrodes at which the products are formed, or by use of mercury as one of the electrodes. The vessel in which the decomposition is effected is known as a "cell", those in which a diaphragm is used being known as "diaphragm cells", while those in which the negative electrode is mercury are styled "mercury cells". Diaphragm and mercury cells arose as two distinct process departures that have developed alongside one another throughout the history of modern heavy chemical manufacture.

The first steps in the development of diaphragm cells took place in Germany (1886–90) and the U.S. (1888–93). In the U.K., in 1890, James Hargreaves and Thomas Bird at Widnes, in Lancashire, invented a diaphragm cell in which the caustic soda liquor was drained away from the back of a perforated cathode, thus preventing the alkali reacting with chlorine in the electrolytic region. This British cell was the prototype of all cells operating on the so-called "drained-cathode" principle. No purely British diaphragm cell has been developed since the Hargreaves–Bird, which was used in this country in only one works, closed down in 1927. The U.S. has been the site of continued development of the diaphragm type of cell, and today two American cells of that type, the Hooker and Dow, produce some 70% of the total U.S. chlorine output. Two types of diaphragm cell are in use in the U.K., but both of these were developed in the U.S.

Although earlier attempts were made in Canada and Austria to develop a mercury cell, the first industrially important mercury cathode process was invented in the U.K. in 1892 by Hamilton Young Castner, an American engaged in the manufacture of aluminium at Oldbury, near Birmingham. The aluminium process employed used metallic sodium made by electrolysis of fused caustic soda. Believing—erroneously—that a caustic soda of exceptional purity was necessary Castner considered that a process using a mercury cathode gave the best prospect of producing soda suitable for his purpose.* Castner solved the problem of circulating the mercury cathode from the electrolytic region to that in which it was brought into contact with water, so that caustic soda could be formed with the sodium it contained, by imparting a rocking motion to the cell. The Castner rocking cell was first used on an industrial scale in America—after the United Alkali Company and Brunner, Mond & Company had both rejected opportunities offered to them of undertaking its exploitation. On the Continent the rocking cell was adopted by Solvay & Company, which concern had a large interest in the Castner-Kellner Alkali Company, formed in 1895 to develop Castner's invention in Britain. The Castner-Kellner Company began production at the Runcorn (Cheshire) factory early in 1897.

In 1893, Castner, whose constellation of electrolytic inventions laid the foundation of the world's electrochemical industry, patented an electrothermal process for graphitizing carbon, rendering it very much more durable as the material for the positive electrodes of diaphragm and mercury cells. This invention was of key importance to the technology of industrial electrolysis.

Until 1900 the rocking cell was the only one of the mercury type in large-scale operation. About 1899, Solvay & Company developed a long trough-like cell which remained stationary

* Caustic soda from diaphragm cells contains a substantial quantity of undecomposed salt.

while the mercury, in a shallow stream, flowed continuously along it. The Castner-Kellner Company introduced this " long " cell into the U.K. in 1902; from that date the history of the mercury cell process has been one of slowly accumulating minor improvements of the Castner–Solvay cell.

The process devised by Castner for production of metallic sodium by electrolysis of fused caustic soda produced oxygen as its co-product, and the current efficiency was less than 50%. Because of the difficulties in developing a process for electrolysis of fused salt, Castner's sodium cell held the field until 1920. The last Castner cell installation in the U.K. ceased operation as late as 1952. Electrolysis of *fused salt* was mainly an American development, although a practicable cell, using a molten lead cathode, was operated in the U.K. on a pilot plant scale in 1905 by E. A. Ashcroft. The United Alkali Company made a thorough study of Ashcroft's invention, but decided not to adopt it—mainly, it appears, because of their conviction that it could not produce sodium competitively with that made by the established Castner process. Between 1915 and 1920 the various functional elements leading to a successful cell were invented in the U.S. and Europe; in 1924 James Cloyd Downs, of the Niagara Electrochemical Company, organized these into a single invention. During the last four decades the Downs sodium cell, with some modifications, has come into general use as the modern process for manufacturing metallic sodium. It was introduced into the U.K. by ICI in 1937.

While the technical processes which form its modern matrix were evolving, over the past seventy years, electrolysis of brine underwent a gradual shift of importance from the alkali to the chlorine side. The original enterprisers regarded their electrochemical undertaking as part of the alkali industry, chlorine being a generally profitable by-product: by the mid-1930's it had become the *chlorine industry.* The alkali industry increasingly became the separate province of the Solvay process.*

---

* Recent developments, however, indicate the possibility of the Solvay process being ultimately ousted by brine electrolysis.

Electrothermal Processes

Electrical energy is used in two important electrothermal processes—manufacture of phosphorus and calcium carbide. In an electrothermal process, resistance to the passage of a current through materials contained in a furnace transforms the electrical energy into heat, bringing about chemical reaction in the same fashion as heat from combustion of a fuel: the electricity as such does not, as in electrolysis, bring about the chemical reaction. Despretz, in 1849, invented the electrical resistance furnace, but its further development had to await the evolution of the dynamo; it was only in 1887 that Héroult devised the first electrical furnace of the modern type, marking the effective beginning of the new field of electrothermal technology.

In 1888, Readman, Parker and Robinson, of Wolverhampton, were granted patents for a process of phosphorus manufacture in which the heat was generated by passing a current through phosphatic materials in a furnace. The introduction of the electrical furnace process was—and is—the most revolutionary change in phosphorus manufacture in the course of its century or so of history. What George Albright called "the almost brutal simplicity of the electricide method" replaced the old system of multiplicity of thin fireclay retorts, requiring individual charging with a specially prepared mixture, by a *continuous* process, using phosphate rock, unmodified except by crushing to convenient size, mixed with silica and carbon.* For some forty years the economics of phosphorus production had depended on cheap British industrial coal: dependence was now on cheap electricity, a circumstance which was to make its manufacture more profitable in countries where there is abundance of cheap hydroelectric power.

Readman's patent was acquired by the Electric Construction Company of Wolverhampton, which, in 1890, formed The Phosphorus Company and established a factory at Wednesfield, Staffordshire. This latter concern was purchased by Albright &

* The reaction in the furnace is: $Ca_3P_2O_8 + 3SiO_2 + 5C \rightarrow 3CaSiO_3 + 5CO + \frac{1}{2}P_4$.

Wilson, which subsequently maintained an unchallenged monopoly of phosphorus production in Britain. After operating the Wednesfield plant for two years, Albright & Wilson transferred manufacture to their Oldbury site. Until 1902 the electrothermal plant at Oldbury produced about 200 tons of phosphorus a year. In 1907 a larger electrothermal plant was installed; this, because of scarcity and high cost of electrical energy in the Midlands, was provided with its own power station. Seventy-five per cent of Albright & Wilson's production in 1908 consisted of phosphorus for the match industry, for which output there were about ten customers. (In the 1950's phosphorus production was at the same level, but constituted less than 3% of the firm's chemical output.) In 1899, phosphorus sesquisulphide was introduced as the actual incendiary constituent of matches, and was adopted in 1900 in the U.K. by Bryant & May. Use of the sulphide ended the industrial hazard of "phossy jaw" (phosphorus necrosis) in the match industry. During World War I phosphorus production rose to a peak in 1918 with an output of 2860 tons, because of the demands for smoke production on the battlefronts; the 1920 production was only 715 tons, reflecting the effect of the post-war depression and the cessation of military use. When the war was over, considerable quantities of phosphorus were contained in shells and other munitions lying useless in France and in other battle areas. Setting an upper limit of 1000 tons, Albright & Wilson contracted to buy up this idle phosphorus at 9½d. a pound, and eventually shipped it to England. In 1920, phosphorus manufacture ceased at Oldbury, and for the following fourteen years the U.K. requirements of the element were imported from Albright & Wilson's associated concern, the Electric Reduction Company, Quebec, where cheap electric power was available. Aided by the Key Industry Duty (see p. 110) on various phosphorus derivatives, phosphorus manufacture was begun again in the U.K., first at Widnes, in 1934, on the site of one of James Muspratt's former works, and later at Oldbury.

In 1892, Thomas Leopold Willson, a Canadian, using a 70 kW

Héroult electric furnace, produced 300 lb of calcium carbide in 24 hours. It had not been his intention to produce that compound, but to obtain metallic calcium by reduction of lime. Willson was instantly aware of the significance of his accidental discovery and proceeded to exploit it commercially. Moissan, in France, had discovered the same reaction in the same year; Willson, however, was first to put electrothermally-produced calcium carbide on the market, and is therefore rightly regarded as the founder of the modern carbide industry. In 1895, the Acetylene Illuminating Company, which had acquired rights to operate Willson's patents in the U.K., built an experimental plant at Leeds, using coal-generated steam to produce its electricity. The British Aluminium Company was then erecting its works at Foyers in Scotland, and the Acetylene Illuminating Company arranged to use part of the hydroelectric power for carbide manufacture. Carbide output at Foyers probably never exceeded 500 tons a year. Finding, in 1903, that they could not prevent the use of Willson's process by others, and disappointed in their expectation that acetylene illumination would prove competitive with coal gas, the Acetylene Illuminating Company ceased carbide manufacture. At that time the United Alkali Company made a brief excursion into carbide manufacture, producing in the years 1902–4 a total output of 300 tons.

Carbide was not produced in the U.K. between 1904 and 1908, when a factory using hydroelectric power was established at Askeaton, near Limerick, and the Imperial Automatic Light Company set up a works at Thornhill in Yorkshire. The Irish plant was an early failure, but the Thornhill works continued for a decade to be the sole U.K. producer. In the last year of World War I the Government encouraged the British Cellulose & Chemical Manufacturing Company in its project of setting up the first large carbide works in this country. That plant, however, although it had a claimed capacity of 20,000 tons a year, never exceeded about one-third of that output. Cost of production was £33 a ton, while imported carbide was selling at £26 a ton. Deprived of the protection of the 33⅓% import Key Industry

Duty on synthetic organic chemicals, because of a legal ruling that calcium carbide is *inorganic*, the Cellulose & Chemical Company abandoned its unprofitable enterprise. For twenty years thereafter, until World War II had already began, all U.K. carbide requirements were imported, mainly from Scandinavian countries. Britain was for many years the world's principal carbide importer. During the inter-war period the electrothermal carbide process underwent rapid development, both in technique and scale of production. The Söderberg continuous electrode, the essential element in the modern high-output carbide furnace, came into use in the early 1920's. Although the British chemical industry did not contribute to the development of the modern carbide process, it obtained the immediate benefit of the progress made abroad when the most advanced types of Norwegian furnace were erected here in 1941 and 1943.

The attempts to introduce carbide manufacture in Britain in the early years of this century may have been, in part at least, prompted by the general search, evident about that time, for new fields for financial investment, and, in chemical industry, for diversification. Failure to establish indigenous manufacture of a commodity for which there was a growing domestic demand (about 5000 tons a year in 1908–10, 7000 tons in 1918) bears a certain resemblance to the case of dyestuff production, particularly in view of the existence, by 1900 in Germany, of five carbide factories contributing to a total national output of 9000 tons a year. As subsequent events have demonstrated, difference between the costs of hydroelectric and coal-produced power has not proved a decisive inhibiting factor. Power generation developments have, over the past few decades, progressively decreased coal consumption per kilowatt-hour. It seems probable that total reliance on imported carbide supplies retarded development in this country of such acetylene derivatives as polyvinyl chloride (PVC), polyvinyl acetate (PVA), butadiene (for rubber synthesis), acetaldehyde and other heavy organics.

In 1896 the Scottish Cyanide Company established a factory at Leven in Fife to produce cyanide electrothermally by a process patented in 1894 by Readman. From the outset the company was in difficulty, in part because their production cost was more than twice what they had estimated, and in part because at that period the price of cyanide was falling. By 1900 the concern had accumulated debts to the extent of £50,000 and had to make a provisional settlement with its creditors. In 1907, matters having reached a crisis, the company made an approach to the United Alkali Company; this approach was without result and cyanide production ceased shortly afterwards. The United Alkali Company had been finding the chronic financial embarrassment of the Scottish company " a menace to the legitimate and profitable cyanide trade ". At Leven the electricity was generated by the water-power of the small River Leven. The primary product was barium cyanide made by heating in electric furnaces a mixture of barium carbonate (imported from Germany) and pitch in a nitrogen atmosphere; other cyanides were produced by double-decomposition processes. It appears that the high manufacturing cost derived from the labour-consuming working-up stages of the process. This Scottish episode is of interest as an example of the economic, as opposed to technical, difficulties besetting early electrothermal enterprises.

4. THE RISE OF CHEMICAL ENGINEERING

It would not be too much to say that, until almost the end of last century, chemical industry, particularly in its heavy branch, was largely engineering: chemical plants were viewed as comparatively simple machines or structures, furnaces and pots in which chemical changes were brought about. Practically all the advances made in heavy chemical manufacture were improvements of *means*, mechanical and structural: the engineers, in the old chemical industry, had something like the impact of Ransome, Croskill and Tull on the advance of agriculture.

The engineering in chemical factories was cognate with that employed by the builders of engines and civil works; the first recognition of the possibility that a special engineering training might be required could fully emerge only with a transition from the empirical attitude to the chemistry involved and to the application of the principles of physics. In the 1880's a few men began to discern that engineering in relation to chemical technology was becoming a specialized activity which could, with probable advantage, be organized into an instructional discipline. Henry Edward Armstrong, in 1884, at the Central Institution in London, planned a four-year course that included chemistry, engineering, mechanics, mathematics, physics, drawing, manufacturing technology, workshop practice, modern languages, and applied art; he was, however, not permitted to put this scheme into operation. What Armstrong was proposing was a balanced course of instruction in a *range* of subjects, rather than a closer integration of conventional engineering with chemistry and physics.

George Edward Davis, who, after beginning his career at Windsor and Eton gasworks, had considerable experience of the LeBlanc system as employee, Alkali Inspector and consultant, took the first important steps towards defining the subject-matter of chemical engineering and establishing its basic concepts. In 1887, Davis gave a course of lectures at Manchester Technical School; because of the novelty of their treatment of chemical technology, these attracted considerable attention. Instead of describing *in extenso* the processes of the contemporary chemical industry, Davis analysed the conduct of them into a series of basic operations. He had discerned that chemical manufacturing processes may be regarded as combinations or sequences of a comparatively small number of procedures.* Davis did not set out his novel approach synoptically, but allowed it to emerge in the course of extended exposition; for that reason some technical historians have either overlooked or denied Davis' priority: American writers, in particular, give the credit for originating the notion of unit operations to Arthur D. Little, who first explicitly employed the term in a 1915 Report to the Massachusetts Institute of Technology (MIT). Instruction in chemical

* Davis nowhere used the modern term " unit operations ".

engineering at the MIT, as embodied in W. K. Lewis' textbook, was explicitly organized on the basis of Davis' system. The American journal, *Chemical Engineering Progress*, allows that " George E. Davis presented the essential concept of unit operations, and particularly furnished an understanding of its value for education ". Invention of the unit operation approach gave chemical technology its proper idiom: for the first time it could be systematically presented in instructional courses and thought about in a more effective manner.

After publishing several of his 1887 lectures in the *Chemical Trade Journal*, which he founded, Davis abandoned their further publication until 1901, when he produced the whole course in volume form as his *Handbook of Chemical Engineering*. A second and enlarged edition of the Handbook appeared in 1904; this latter ran to more than a thousand pages. Davis' book, which laid the foundation of chemical engineering education, is a classic of the empirical tradition. The unit operations are not referred to the underlying physics, but are related to procedural categories and practical utility: they are wholly *artificial*, and physics cross the frontiers of most of them, thus the laws of heat flow are operative in electrothermal furnacing as they are in heat-transfer, and those of viscosity are involved in heat-transfer as much as in settling and filtration. The general transition to continuous and catalytic processes has greatly increased both the possibility of and the need for applying physical theory to the treatment of chemical industrial operations. While the modern sophisticated treatment of chemical technology penetrates to more fundamental levels, the general disposition of the subject matter recalls Davis' pioneering *Handbook*.

Although Davis' ideas found fertile soil in America, for many years the only scene of far-reaching chemical engineering development, they had little impact in this country during the first quarter of the century. Before 1914, courses in chemical engineering were offered at the Imperial College of Science and Technology and at Battersea Polytechnic, but these were limited in scope and did not lead to professional qualification. Chemical shortages and plant difficulties during World War I called some attention to the need for training chemical engineers. In 1918 a number of industrial chemists, meeting in London, formed the Chemical Engineering Group of the Society of Chemical Industry; four years later the professional body, the Institution of Chemical Engineers, was established. At that time no university in the U.K. was conducting courses in the subject. The first university degrees in chemical engineering were awarded in 1929. While Britain, in common with some other European countries, including Germany, was slow to build up a body of trained chemical engineers, the surveillance of chemical processes by application of physical chemical principles was well advanced by the end of the first quarter of this century, and, as Lord Fleck recalled in 1961, " Once physical chemistry took control . . . then the whole energy concept cleared things up and we moved toward a very much better and more regulated industry."

5. PRIMARY EFFECTS OF THE FIRST WORLD WAR

War, which removes the normal cost-restraints and profit considerations from the economics of the relevant technologies, urgently escalates demands and prodigally consumes natural and manufactured supplies; it has been in almost every generation a prime accelerator of technical development. For the first time in history, war became a potent factor in the rise of the chemical industry in 1914. The incidence of war had three principal effects: it greatly increased the scale of production of a number of common chemicals, modified some existing processes and speeded up the introduction of several modern processes, and caused the State to enter chemical manufacture. Surprisingly, apart from temporary production of poison gases (other than chlorine), the war was substantially without effect in diversifying the products of the industry.

On the outbreak of hostilities the instant and most urgent demand was for explosives. The full weight of this demand fell on the heavy chemical and explosives branches of the industry. Demand for sulphuric acid was suddenly and enormously increased. Much of this requirement was met by the United Alkali Company and other concerns by diverting acid from fertilizer production. Most of the acid produced prior to 1914 was of too low strength for use in manufacture of nitro-explosives. Such plants as existed for concentration of sulphuric acid were of the evaporator type, constructed of volvic stone, iron-silicon alloy, or silica. Within eighteen months concentrated acid output, using these methods, had been increased to 7000 tons a week. Steps were also taken to introduce contact sulphuric acid production by several established continental processes (BASF, Grille, Tentelew, and Mannheim) at Government factories at Queen's Ferry, Gretna and Avonmouth, a United Alkali Company works in Widnes, and at the works of Staveley Coal & Iron Company (now Staveley Iron & Chemical Company). By these processes an output of 7000 tons a week of oleum* was ultimately achieved. Before World War I, U.K. production of oleum amounted to only a few hundred tons a week. This last development marked a permanent addition to the technology of British sulphuric acid manufacture. Because of demand for Chili nitre for explosives production, this import ceased to be used for producing the nitrogen oxide in the " nitre pots " of the sulphuric acid chamber process; instead, use was made of catalytic oxidation of ammonia—a method which has continued to the present.

At the outbreak of war the Government had only recently adopted trinitrotoluene (TNT) as a high explosive; picric acid was then the standard shell-filling. The only source of toluene then available was coal tar. It was early realized that this would be quite inadequate to meet wartime needs. In Holland the Asiatic Petroleum Company had large petroleum distilleries, one of the products from which was toluene. Before the war the Petroleum Company had supplied mononitrotoluene to the

---

* Oleum may be regarded as concentrated sulphuric acid (H_2SO_4) containing sulphur trioxide (SO_3) in solution. It is produced in several standard strengths.

German dyestuffs industry. Having regard to their neutrality, the Dutch Government would not permit export of mononitrotoluene to Britain for production of TNT; arrangements were therefore made for the removal of the petroleum plant to Portishead, near Bristol, and plants were set up at Oldbury and Queen's Ferry for nitrating the crude toluene. These developments have the historical interest that they represent the first nexus between British chemical industry and petroleum refining. The total U.K. output of TNT during the war was 238,000 tons (in modern terminology, about a quarter of a megaton!). In the U.K. the Nobel explosives industry, in 1914, was producing some 9000 tons of explosives a year; in 1918 the output was 50,000 tons. By far the major part of explosives production was carried out in Government factories, erected during the war; these works were in many instances staffed by employees seconded from Nobels. A total of almost half-a-million tons of " organic " explosives was produced—a manifold transformation in the scale of manufacture of organic chemicals.

At an early stage it was realized that toluene resources could be conserved by making use of high-explosive mixtures of TNT with ammonium nitrate, the so-called amatols, provided an adequate supply of ammonium nitrate could be established. Two sources of nitrate ion (NO_3) were available—calcium nitrate imported from Norway and sodium nitrate from Chile. Brunner, Mond & Company, because of their long experience in handling ammonia in the Solvay process were invited to investigate processes for production of the nitrate. That company explored three routes, namely, modification of the Solvay process to use sodium nitrate in place of salt as its raw material, double-decomposition of sodium nitrate with calcium chloride (the by-product of the Solvay process), and double-decomposition of sodium nitrate with ammonium sulphate. All three routes were put into large-scale operation, but the last named became the principal process employed. It is of more than passing interest that the great success of the ammonium sulphate–sodium nitrate process was achieved only after its intensive investigation on the

principles of the phase rule* by Freeth. In the course of the war ammonium nitrate became the principal explosive chemical produced; the total output was 378,000 tons.

Picric acid requirements were supplied by the United Alkali Company, and several dyestuff concerns such as L. B. Holliday & Company. This manufacture soon threatened to exhaust supplies of phenol from coal tar distillation and processes for its synthetic production from benzene were established by the South Metropolitan Gas Company and Brunner, Mond & Company. Picric acid manufacture in the U.K., owing to the gradual change-over to amatol as the charge for shells, did not reach the great scale it attained in France. Total U.K. output of picric acid during the entire war period was 68,500 tons.

World War I had opposing effects on U.K. chlorine manufacture: it maintained Deacon and Weldon plants in barely economic production for the duration, and it accelerated development of the modern chlorine processes. At the outset chlorine, in the form of bleaching powder, was required at the war fronts for sanitary purposes only. One chemical manufacturer commented, probably with some justice, that " the lives saved by chlorine in its disinfectant capacity have been infinitely greater than the lives destroyed by chlorine as a weapon of war ". At the outbreak of the war, the Castner-Kellner Company, the Electrochemical Company (at St. Helens), and Brunner, Mond & Company were producing chlorine electrolytically. The last-named company had taken over the Hargreaves–Bird diaphragm cell plant at Middlewich in 1913. In August 1914, the United Alkali Company was preparing to import the Gibbs diaphragm cell process.† Erection of the Gibbs cell plant at Widnes began in October 1914; it was augmented in 1916 and 1918 and other Gibbs installations were established at the Alkali Company's St.

---

* The physical law governing the number of phases and components in chemical mixtures under equilibrium conditions.

† This cell was the invention of an Englishman, A. E. Gibbs, who emigrated to the U.S. His cell, patented in 1907, was developed in the works of the Pennsylvania Salt Manufacturing Company.

Helens and Newcastle (Allhusen) factories. At the end of the war these plants, collectively, formed the largest diaphragm cell installation in Europe.

In 1893, Rudolph Knietsch, of the Badische Anilin und Soda Fabrik, was granted the first patent relating to liquid chlorine. He claimed: " A new article, a merchantable package of liquid chlorine enclosed in an iron or steel vessel and sufficiently anhydrous not to attack the iron or steel of the vessel . . . " That Germany in 1915 was in a position to mount a chlorine gas offensive is sufficient evidence that chlorine liquefaction and transport were in a well-advanced state in that country. Following the first gas attack in April 1915, the British Government resolved to retaliate in kind. Arrangements were made for the Castner-Kellner Company to expand its liquefying capacity to 150 tons a week; since 1908 that company had been producing liquid chlorine intermittently in a plant having a capacity of about 5 tons a week. By late September 1915, 6000 cylinders, containing 180 tons of chlorine, had reached the front in France. In April 1918, the U.K. chlorine liquefaction capacity, brought into being mainly to supply chlorine as an offensive weapon, was 250 tons a week. This war-engendered development established liquid chlorine as *the* chlorine commodity, instead of bleaching powder. Before the end of the war, mustard gas, phosgene, etc., had taken the place of elementary chlorine in the chemical offensive. During the period of the war decline in consumption of chlorine for civilian applications more than outweighed the demand for it for military purposes.

As has already been shown, private chemical industry, although it made a very large contribution to explosives manufacture and to production of chemicals for that purpose, was not in a position to take the whole weight of steeply rising warlike demand. The Government had to mount a very large-scale chemical industry; this it did by operating its own factories, by having private explosives firms operate State factories as agents, and by advancing capital to private firms. In the dyestuffs field a state-assisted joint-stock company was formed; this project, in face of the attitude of private industrialists, was not proceeded with; instead, in 1915, the Government gave its financial assistance to the promotion of an ostensibly private concern, formed round the nucleus of Read, Holliday & Sons—British Dyes Ltd. Meldola, overlooking developments in the explosives branch, commented: " For the first time, we have a distinct proposal in this country for the establishment of a State-aided industry." In 1919, under the chairmanship of Lord Ashfield, wartime President of the Board of Trade, British Dyes Ltd was merged with Levinstein Ltd in the British Dyestuffs Corporation, in which the Government took 850,000 preference and 850,000 preferred ordinary shares to replace the £1·7 million loan originally made to British Dyes Ltd. State aid was also given to some other dye firms, e.g., Scottish Dyes Ltd. During the war, in 1917, Levinsteins had taken over Claus & Company, the surviving lineal descendant of Perkin & Sons and several other early U.K. dyestuff enterprises.

The U.K. Alkali Inspector, in his 1891 report, commented: " By no means at present known . . . can ammonia be formed on a commercial scale although the composition is simple, but Nature is coy and resists our attempts to promote the union of three molecules of hydrogen with one of nitrogen." Almost a score of years were to pass before Nature's coyness was overcome. In 1911, the Badische Anilin und Soda Fabrik withdrew from their interest in the Norwegian electrical nitrogen fixation industry: Fritz Haber had by then succeeded in developing his process for the production of ammonia by reaction of nitrogen and hydrogen under pressure in the presence of a catalyst. This was the first of a revolutionary group of modern processes of synthesis from gases and the most notable achievement in the heavy chemical industry of the pre-World War I era. Operated in tandem with the Kuhlmann–Ostwald process for catalytic oxidation of ammonia to nitric acid, the Haber process made an important, but not decisive, as is often asserted, contribution to the Central Powers wartime nitrate requirements. With the incidence of submarine warfare, the British Government, anticipating a serious reduction of Chile nitre imports, sought chemical advice and formed the Nitrogen Products Committee to receive it. To that Committee, the Chemical Society, in a report reviewing the various nitrogen fixation processes, commended consideration of the Haber ammonia process: " The Haber process has for this country the great advantage that almost no electric power is required and that the raw materials nitrogen and hydrogen gas can be produced here about as cheaply as anywhere else." The report added: " With some foresight a plant erected primarily for a military purpose might be easily adapted in peacetime to agricultural objects." A research team, led by one of Haber's former assistants, carried on research on the ammonia process in the Ramsay Laboratory of University College, London. When the Armistice came the problems of purifying great volumes of hydrogen and of the general chemical engineering of the process had not been fully solved. No pilot plant was in existence. The Government, however, was already preparing to put the process into operation on a large scale in a State factory; work had begun at a 266-acre site at Billingham. The Nitrogen Products Committee advised that " nitrogen fixation is essentially a new industry requiring for its initiation and development the actual support of the Government ". The end of this governmental excursion into chemical industry was signalized almost a year later, by an advertisement in *The Times*,* offering for sale " H.M. Nitrate Factory Billingham ".

6. NITROGEN FIXATION

In his Presidential Address to the British Association at Bristol in 1898 Sir William Crookes forecast that by 1931, the wheatlands would, because of declining fertility, become inadequate to meet the needs of the bread-eaters of the world. At the same time Crookes proposed that the limited supplies of fertilizing nitrogen might be augmented by fixing atmospheric nitrogen artificially, and the onset of famine thus avoided. He had himself, in 1892,

* 19th November 1919.

demonstrated the combination of atmospheric nitrogen and oxygen at high temperature to form nitrogen oxides. " This unconsiderable experiment ", he suggested, " may not unlikely lead to the development of a mighty industry destined to solve the great food problem." In the event " a mighty industry " has arisen, based, not on the direct interaction of the atmospheric gases to give nitrogen oxides, but on the reduction of nitrogen to ammonia by the Haber process. During the first decade of this century, nitrogen fixation by the electric arc and cyanamide processes underwent considerable development on the Continent. Both the electric arc and cyanamide processes were large consumers of electrical energy and, at that period, were unlikely to be profitable where cheap hydroelectric power was not available. At that time, with adequate sea-borne supplies of nitre from Chile and calcium nitrate from Norway, Crookes' native land had no immediate cause to be disturbed by his dramatic prophecy. Undoubtedly this reliance on imported nitrates would have continued for a number of years had the circumstances of war not accelerated introduction of ammonia synthesis.

Within a day of the signing of the 1918 Armistice a proposal was conveyed from Brunner, Mond & Company to Lord Moulton, Director General of Explosives Supply, that a commission of chemists should be sent as soon as possible to the BASF works in Germany to obtain information on the Haber process. Sir Keith Price, the member of the Munitions Council to whom Moulton referred this suggestion, hesitated to send a commission consisting solely of employees of Brunner, Mond & Company. In the end, some chemists from other concerns were included, and the commission set out, under the leadership of a Brunner, Mond director, to visit the ammonia works at Oppau and the ammonia oxidation plant at Höchst. A second mission, again largely composed of Brunner, Mond personnel, visited Oppau for some weeks in April 1919. The works had been shut down and by a variety of ingenious acts of sabotage its staff did its utmost to prevent the visitors from attaining their objective to the full. Because the plant was not in production, much of the essential quantitative information had to be inferred or guessed. " It is doubtless of great advantage," the commissioners concluded, " to have seen the thousand and one things embodied in a working plant and we feel disposed to place a high value on the experience . . . " The activities of the commission, which was regarded as sponsored by the Government, were not made public.

In the political climate of late 1918, after the Lloyd George " Coupon Election ", there was little prospect of a State nitrogen industry being established. Meanwhile, Brunner, Mond & Company formed The New Nitrogen Syndicate Ltd to investigate the Haber process; this concern was wound up after a few months of apparently fruitless existence. Roscoe Brunner, Chairman of both Brunner, Mond and of the disbanded Syndicate, proposed to Lord Moulton that a more broadly-based concern should be formed to include as participants such companies as Armstrong, Whitworth; Johnson Matthey; Vickers; Brotherton; and, of course, Brunner, Mond & Company: the Government, too, might participate financially. This proposal is of historical interest as the first attempt to establish *corporate* manufacture in the British chemical industry. By the time the Government advertised their Billingham factory for sale, only Brunner's own company was ready to negotiate its purchase, the contract for which was completed on 24th April 1920. The purchase price was £715,000, and the sale included that of " all relevant Government information ". Two months later Brunner, Mond & Company formed its subsidiary, Synthetic Ammonia & Nitrates Ltd.

The Billingham " factory " which Synthetic Ammonia & Nitrates took over, consisted of a number of largely undrained fields on which only contractors buildings had so far been erected; no actual nitrogen fixation plant had been delivered; entirely new arrangements had to be made with supplying firms. An £80,000 research laboratory was set up and an experimental Haber plant erected at the Castner-Kellner works,* where high-

* The Castner-Kellner Alkali Company was merged with Brunner, Mond & Company in 1920.

purity hydrogen was available as a by-product from brine-electrolysis. After many set-backs, production of ammonia began at Billingham early in 1924, although some sulphate of ammonia had been made in the previous year from ammonia purchased from Brunner, Mond & Company.

In 1925, Synthetic Ammonia & Nitrates Ltd became a public company and received a Government loan of £2 million for twenty years. This loan, made under the Trade Facilities Act, had the instant effect of making the chemical industry news and of bringing its conduct into politics. What became known as " The Nitrogen Fixation Affair " became a matter of press and parliamentary controversy. The fact that Brunner, Mond had purchased, along with the beginnings of the Billingham factory, the information on nitrogen fixation obtained at Oppau was brought into the discussion. The *Daily Herald* saw the sale of the " secret information " as a lost opportunity by the Government of freeing the farmer from the fertilizer monopoly, adding that " it would have forced Brunner, Monds to recognize that they are no longer the only chemical pebble on the beach ". In the same newspaper appeared a cartoon depicting the *Herald*, personified as a " worker ", wielding bellows to blow away a smoke-screen which top-hatted Sir Alfred Mond (later Lord Melchett) is attempting to maintain round Billingham.* The reply was given to a demand in the House that the names of the Oppau commissioners should be made known, that they were entitled to the same protection as civil servants. The " secret German information ", of which so much was made by the opposition, was described by the Financial Secretary to the Treasury as " practically useless and no use is being made of it ". The Billingham factory erected by Synthetic Ammonia & Nitrates was much larger than that originally planned by the Government in 1918–19, and the design of its plants was almost entirely based on the work of the company's own research laboratory and engineers. Its output greatly increased the U.K. supply of

---

* Issue of 9th April 1925.

fertilizer nitrogen at pre-war price, and finally made this country independent of Chile nitre imports.

7. INTRODUCTION OF PROTECTIVE MEASURES

World War I produced a revolutionary change in scale of operation in the chemical industries of victors and defeated alike. Although this expansion had been occasioned by demand for a limited range of warlike chemicals, it had, because of the nature of the industry, in fact brought into being enormously increased potential for making, particularly, dyes and organic chemicals. If the great chemical combines in Germany and America were not soon to assert their competitive power in the U.K. home market, and if the new British capacity was now to be profitably employed, immediate protective measures were urgently required. The Government, acting under the Customs Consolidation Act of 1876, prohibited, by simple proclamation, importation of dyes into the U.K., except under licence. In 1919, because of a ruling by Mr. Justice Sankey, this measure was held to be invalid. Foreign chemical manufacturers availed themselves to the full of this remarkable conflict between legislature and judicature: in the year following the decision some £7 million worth of dyes was imported into this country! This situation, so favourable to dyestuff consumers but so disastrous to dye-makers, was aggravated by arrival cost-free of German dyes, taken as reparations under the Versailles Treaty. To prevent the threatened destruction of the U.K. dyestuffs industry, the Government introduced the Dyestuffs (Import Regulation) Bill, which received the Royal Assent towards the end of 1920. The new Act prohibited for ten years* importation of " all synthetic organic dyestuffs, colours and colouring matters, and organic intermediate products used in the manufacture of any such dyestuffs, colours, or colouring matters ". Provision was made for admission under licence of products of this kind not made in the U.K. or in insufficient supply. An advisory committee, with

* Extended in 1931.

FIG. 8. AMMONIA CONVERTERS (BILLINGHAM)

The titanic engineering of these converters, with bolts as thick as a man, dramatically conveys the constructional demands of a modern heavy chemical process.

members from the dye-consuming and dye-producing industries was appointed to assist the Board of Trade in administering the Act.

After absorption of the imports made under the Sankey judgment, the U.K. dye industry began, in 1922, to derive benefit from the Dyestuffs Act. During the decade that followed the British dyestuffs industry steadily expanded, improving the quality of its products, and developing its technical service to the consumer. Before World War I the technical assistance afforded by the German dye industry to its domestic and overseas customers had contributed greatly to its commercial success at a time when the British industry offered no similar service.

The second post-war protective measure, although not exclusively concerned with chemical products, had an important sector of application to them. The Safeguarding of Industries Act of 1921 protected so-called key industries by imposition of import duties, which came to be known as key industry duties, or K.I.D. An *ad valorem* import duty of 33⅓% was imposed under the Act on all synthetic organic chemicals,* other than dyes and intermediates dealt with by the Dyestuffs Act; a number of inorganic products, e.g., compounds of tungsten, thorium, and cerium, were also made dutiable. There was considerable resistance, both inside and outside the House, to the passing of these protective measures. In Parliament resistance, in part due to the traditional British fear of the bureaucrat, may to some extent also have been the result of ignorance of the technological implications of preserving free trade: in 1919–20 not a single member of Parliament had had a scientific education. Outside the House there were what the Editor of *Nature* (4th Dec. 1919) called the " Bourbons of the Manchester school " who found " nothing in the changed conditions of the world, in the circumstances of the Empire, or in the influence of the war on our home industries to induce them to modify their conviction in the smallest degree. To them the basic principle of Free Trade . . .

* The Board of Trade listed some *thousands* of organic chemicals to which the duty was applicable.

has all the force of a natural law as fixed and immutable as seemed to them the law of gravitation." Fortunately for the future of the British chemical industry a common sense, if not a scientific, appraisal of recent experience prevailed and behind the protection of the Dyestuffs (Import Regulation) and the Safeguarding of Industries Acts the U.K. home market was sufficiently secured to the modern chemical industry.

Later a measure of protection for British industry in general was provided by the Import Duties Act 1932. This imposed a 10% ad valorem *revenue* tariff on imports, subject to certain exceptions as regards raw materials and foodstuffs. Additional *protective* duties were imposed on the recommendation of the Import Duties Advisory Committee. As a result of the representations of the Association of British Chemical Manufacturers (ABCM), formed in 1916 as a trade association to promote co-operation in the industry, many additional duties were imposed, but in no case did the total duty approach the KID level. Twenty to 25% was the usual maximum. The protection given to its customers was of great benefit to the chemical industry, since the bulk of its products goes to industry including agriculture.

8. MERGERS

Although World War I accelerated formation of the mergers now to be described, these originated primarily in the necessity for transition from the old to the modern type of chemical industry; just as the United Alkali merger of 1890 was the economic and organizational development necessary for transition from the LeBlanc system to modern heavy chemical manufacture. As has been shown, the dyestuffs industry in consequence of the necessities of war, was by 1919 largely merged in the British Dyestuffs Corporation. In 1918, the various explosives firms were amalgamated as Nobel Industries round the nucleus of the descendant of Alfred Nobel's original enterprise. Brunner, Mond & Company had absorbed practically all the alkali-making concerns outside the orbit of the United Alkali Company, and, through its associated firm, Synthetic Ammonia & Nitrates, had established the nitrogen fixation industry. Thus, by 1926, a substantial part of the U.K. chemical industry was focussed in four concerns, each the product of amalgamation. In terms of capital the largest of these was Nobel Industries (£15·975 million); Brunner, Mond had an issued capital of £13·749 million, British Dyestuffs Corporation £4·775 million, and the United Alkali Company only £3·725 million.

Early in 1926, the Midland Bank, which had sponsored an issue of preference shares by British Dyestuffs Corporation, proposed to Nobel Industries that they should absorb the Corporation. Sir Harry McGowan (later Lord McGowan), Chairman of Nobel Industries, rejected this proposal, at the same time expressing his view that a far-reaching amalgamation was urgently necessary to save the British chemical industry from the effects of increasing competition with the great chemical combines of Europe and America. McGowan put his proposal for an amalgamation of the four principal chemical industrial groups to Sir Alfred Mond (later Lord Melchett), the recently appointed Chairman of Brunner, Mond & Company, in July 1926. Mond undertook to consider it, so that a decision could be taken on McGowan's return from South Africa, which he was about to visit. When McGowan returned, Mond, with several directors of his company, was in America; now more than ever convinced of the urgency for the merger, he set off immediately for New York, where on 25th September discussion began and, three days later, a decision was taken.

Many of the initial plans for the new combine were made aboard the Cunard liner *Aquitania*, which took McGowan, Mond and their colleagues back to Britain. In London the planners, having obtained the agreements of the boards of the United Alkali Company and the British Dyestuffs Corporation, set about the financial organization of the new company. Imperial Chemical Industries Ltd shares in the combine were offered to the shareholders in the four constituent concerns; for the privilege of this financial transaction the new company had to pay £1 million Government stamp duty. At its formation Imperial Chemical Industries was capitalized at £65 million. Sir Alfred Mond was elected Chairman and Sir Harry McGowan President and Deputy Chairman. In a statement issued to the shareholders of the constituent companies, the advantages of the merger were set forth. This manifesto began by referring to " large and powerful combinations commanding huge resources in equipment, technique and finance, with which we shall have to deal ", then went on to outline the benefits of unity, both *vis-à-vis* these foreign combines and in negotiations, on behalf of the industry, with the Government. Unnecessary duplication of capital expenditure and of manufactures could be avoided, research concentrated, and interchange of technical knowledge would reduce process costs. The advantages were indicated of internal trade between the constituent concerns using one another's products as raw materials or intermediates. Most important of all, the merger would have " command of adequate financial resources ". The document was, in effect, an extended definition of what has come to be known as rationalization. Sir Alfred Mond, who is credited with having introduced that term, in its industrial sense, on being consulted by a dictionary editor as to what he intended by it, answered: " The application of scientific organization to industry, by unification of the processes of production and distribution with the object of approximating supply to demand."

The structure of Imperial Chemical Industries Ltd was to a very great extent dictated by the nature of its four original constituent companies. In a general sense each of these concerns was the focus of a specific type of chemical manufacture, thus Brunner, Mond was mainly engaged in alkali production, the United Alkali Company principally in manufacture of a range of heavy inorganic chemicals, including the mineral acids, alkali (by the ammonia–soda process) and chlorine (by the diaphragm cell process), the British Dyestuffs Corporation and Nobel Industries Ltd were respectively identified with dye and explosives making. Each of the merging companies, however, was itself the end result of a series of mergers, which circumstance, particularly in the case of Brunner, Mond and Nobel Industries, had led to their carrying on operations related to but not strictly cognate with their central chemical activity. Thus, Brunner, Mond had by acquiring Chance & Hunt, in 1917, entered the manufacture of many of the heavy inorganics also made by the United Alkali Company; through its subsidiary, the Castner-Kellner Company, merged in 1920, it was a large manufacturer of chlorine and caustic soda (by the mercury cell process); it was also, through its association with Synthetic Ammonia & Nitrates, the monopolist in nitrogen fixation. During World War I Brunner, Mond had secured their limestone supplies by acquiring Buxton Lime Firms, itself the result of the merger of thirteen lime-quarrying firms in the Derbyshire Peak District. Nobel Industries, in consequence of its history of mergers, had already, by 1926, organized its operations into five manufacturing groups: explosives, metal and ammunition, artificial leather, collodions and varnish,* miscellaneous. Organization within the United Alkali Company was mainly based on the regions in which its factories were situated. The Dyestuffs Corporation had at that time the most homogeneous field of activities.

It was mid-1929 before executive boards were set up to integrate the new company's activities into groups (later to be known as divisions). In the operational organization of the merger cognate activities were, as far as possible, allocated to the same manufacturing group. Brunner, Mond, which became the Alkali Group, gave up its Castner-Kellner and Chance & Hunt factories to the General Chemicals Group (largely the former United Alkali Company). Synthetic Ammonia & Nitrates became Billingham Group, while the Brunner, Mond lime interests became Lime Group. In accordance with the general principle, the former United Alkali Company's ammonia–soda factory at Fleetwood was taken over by the Alkali Group. From the Dyestuffs Corporation was derived the Dyestuffs Group and

* These were, in effect non-explosive applications of nitro-cellulose.

from Nobel Industries the Explosives, Metals, Paints, and Leathercloth Groups. The manufacturing groups were allowed a large measure of autonomy in their technical planning and capital expenditure, as well as in research and appointment of staff. Most of the groups were large concerns each requiring extensive and specialized management; such large and diverse groupings could not be centrally administered except as parts of a federal structure operating within the limits of general policies while entirely in control of their day-to-day affairs. A consequence of this necessary type of administration was that, for many years, groups (and even the later divisions) maintained a considerable measure of individuality and " atmosphere ". ICI Headquarters in London, at this time, was establishing its central advisory and other services to deal with raw material purchases, engineering, contracts, general policies, legal questions, patents, publicity, and many other matters.

At the time of its formation ICI enjoyed a monopoly of alkali, chlorine, hydrochloric acid, nitric acid, and metallic sodium manufacture, as well as in the production of nitro-explosives and many of the principal types of dyestuffs. It is probable, although actual statistics are lacking, that the combine brought within its ambit between a quarter and a half of the total chemical industry of the U.K. as it existed in 1926. Outside the merger remained such firms as Fisons, Albright & Wilson, Distillers, Laportes, Brothertons, British Oxygen and National Smelting, which in the years shortly ahead were to become increasingly important and, in all cases, themselves nuclei of amalgamations.

CHAPTER III

The Modern Industry

HEAVY INORGANIC CHEMICALS

Heavy inorganic chemical industry is concerned with the tonnage conversion of widely-available, relatively simple naturally-occurring substances into products which may themselves have final consumer uses, but are more often intermediate, facilitative, or ancillary materials. Although this branch of the chemical industry has been characterized by considerable diversification in the modern period, it has, because of its nature as just defined, not shown it to the degree displayed by the organic sectors of chemical manufacture. At the end of the transition period a great many heavy inorganics were in long-established production and their subsequent history has been mainly one of increasing scale and of occasional process innovations: they have, it could be claimed, only slight historical content. At the present time the U.K. heavy inorganic chemical industry produces between 200 and 300 chemicals on scales ranging from the order of 100 to 2 million tons annually. Manufacture of inorganic chemicals is carried on by some 40 firms at over 100 sites. Most of these concerns specialize in production of groups of chemically-related compounds, or in chemicals used by specific industries, such as textile and paint manufacture. The very large tonnage chemicals are produced by a small number of the larger firms. In what follows the history of U.K. manufacture of the principal groups of heavy inorganic chemicals will be traced from the transition period to the present.

1. CHLORINE, INORGANIC CHLORINE DERIVATIVES, CAUSTIC SODA AND SODIUM CARBONATE

The sector of heavy inorganic manufacture concerned with this group of products is the modern alkali industry. Throughout World War I the various Weldon and Deacon chlorine plants of the United Alkali Company had been maintained in operation. The post-war trade depression steeply decreased the demand for bleaching powder and liquid chlorine and these old processes were almost all shut down, although several continued in intermittent operation until 1928. In 1921, liquid chlorine output was less than 2500 tons. From 1920 onwards, however, the U.K. chlorine industry has been based mainly on the mercury cell process. In the immediately post-war years chlorine manufacture was carried on only by the Castner-Kellner Alkali and United Alkali Companies; several smaller firms began production in the early 1920's; their scale of operation, however, did not affect the dominant position held by the two large companies; the ICI merger of 1926 brought these concerns under the unified control of its General Chemicals Group, which then had 95% of the chlorine capacity of the U.K.

In the ten years preceding World War II the chlorine industry moved into a new realm of scale. In that decade ICI greatly expanded its chlorine installations in the Mersey region, and, in 1930, shut down its diaphragm cell plant at Gateshead while at the same time establishing a new mercury cell plant at Billingham, where the co-produced caustic soda was used in the manufacture of sodium by the Castner cell process. It has been estimated that, in 1929, U.K. chlorine output was about 15% of the world total; by 1936 the proportion had declined to 13%. In the home market liquid chlorine became increasingly important, taking the place of bleaching powder which would otherwise have supplied the demand. Liquid chlorine's increase in importance derived mainly from its use in the production of chlorinated organic derivatives, e.g., chloroethylene solvents and chlorinated benzenes. Hydrochloric acid (hydrogen chloride) was also being increasingly produced from chlorine and the hydrogen co-produced with it in the electrolysis of brine (see Industrial Acids below). Technical changes in the manufacturing processes in this period include the replacement of the original concrete troughs of the mercury cells by troughs of rubber-covered steel and cumulative improvements in the structure of both mercury and diaphragm cells.

During the years immediately prior to World War II the Government prepared for the wartime production of chemical supplies, particularly those requiring chlorine in their manufacture. Various chlorine factories were erected, ICI acting as the operating agency for the Government. When war came its demands did not produce the effects on the chlorine industry made by World War I. The State chlorine plants, in the event, were not required to produce to their designed capacity, and outside the Ministry of Supply factories chlorine production gradually increased to 1943 and thereafter declined. World War II did not significantly modify the trends in the U.K. chlorine production and utilization prevailing in 1939; technologically, too, the industry underwent few changes. During the war the price of chlorine was officially controlled, remaining until 1943 at its pre-war level; from that date to the end of the war permitted increases totalled only 20% of the 1939 liquid chlorine price. In 1947–48, ICI purchased from the Government the wartime chlorine factories. By 1950, chlorine demand in the U.K. had returned to the maximum level of wartime output by both the ICI and the State chlorine installations. In 1952–53 the steeply upward trend of chlorine demand was checked and declined. This decline reflected a general trade depression; it is of interest to note, however, that demand remained substantially above the maximum level attained during World War II. It has been estimated that, in this century, U.K. chlorine output has increased at a compound interest rate of about 6% *per annum*. Statistics of total chlorine outputs are not available, but it is probable that at the present time Britain is annually producing some 500,000 tons.* At the beginning of the century British

---

* This estimate includes chlorine co-produced with sodium metal in the electrolysis of fused salt in Downs cells (*see* Metals *below*).

chlorine production was about 47,000 tons a year (approximately half the then world total), of which 10,000 tons was made by brine electrolysis. If the estimate of present chlorine output is correct, there has been an approximately 10-fold increase in total chlorine production in this country since 1900.

In 1950, Murgatroyd's Salt & Chemical Company, a firm jointly owned since 1954 by Fisons Ltd and the Distillers Company Ltd, installed American Hooker-type diaphragm cells at their factory at Sandbach in Cheshire. Six years later this company brought into operation an Udhe (German) mercury cell plant at the same site. Murgatroyd's, while supplying the chlorine requirements of their controlling companies, also market large tonnages of liquid chlorine and the co-produced caustic soda liquor. In 1953, the Associated Ethyl Company (now the Associated Octel Co Ltd) began manufacture of chlorine by the mercury cell process; again the cells were of German design. The chlorine made by this company is wholly captive and is used in the production of bromine and ethyl chloride for tetraethyl lead (TEL) anti-knock mixture.* In the post-World War II years TEL manufacture has become one of the largest single chlorine consuming processes.

During the period of the hegemony of bleaching powder in the chlorine industry many attempts were made to achieve its production by a continuous process; various types of mechanical plants were devised for reacting a continuously moving body of slaked lime with a current of chlorine gas, but none of these found wide adoption. In 1916–18 the Castner-Kellner and United Alkali Companies developed modern rotating kiln processes which operated for many years. British bleaching powder exports, which rose temporarily before World War II to 35,000–40,000 tons a year, sank in the post-war years to less than 10,000 tons in 1953. Apart from its replacement by liquid chlorine, bleaching powder has been further displaced by sodium hypochlorite as a bleaching and disinfecting agent. Sodium hypochlorite is now generally produced from the low-strength gas

* In this petrol additive, tetraethyl lead ($PbEt_4$) is mixed with various proportions of ethylene dichloride and ethylene dibromide.

from chlorine liquefaction plants and the caustic soda liquor from the electrolytic cells; its manufacture, therefore, fits well into the economic pattern of the modern chlorine factory. Sodium chlorate, large-scale manufacture of which ceased in the U.K. towards the mid-1920's, has been manufactured by electrolysis of brine on a moderate scale since 1938 by Staveley Iron & Chemical Company, but a large part of the U.K. requirement of this and of potassium chlorate continues to be imported. Since about 1935, increasing production of peroxygen bleaching agents (hydrogen peroxide and its derivatives) has also displaced the classical bleach from its principal application and from the export market.

The remaining inorganic chlorine products, apart from hydrochloric acid (see Industrial Acids), namely, phosgene, the chlorides of sulphur, aluminium chloride, and the chlorides of phosphorus are all manufactured in quantities of the order of thousands of tons a year. Sulphur chlorides and aluminium chloride have been in established production from about the beginning of the century; phosphorus chlorides (used in the manufacture of plasticizing agents for plastics) were first produced industrially in this country by Albright & Wilson in 1923, although they had been produced in small amounts by that firm considerably earlier. Phosgene was first produced in the U.K. in large tonnages as a poison gas during World War I: its production in the 1950's up to the present for manufacture of organics has been on a steadily increasing scale. Production of inorganic derivatives has caused a steady increase in chlorine requirement; the principal increase of modern demand, however, derives from the great expansion of chlorinated organics manufacture in the post-World War II period.

Caustic soda, produced as a dilute solution (*c.* 10%) in diaphragm cells and as a more concentrated solution (*c.* 50%) in mercury cells, was, up to the period of World War II, marketed as the solid, obtained by evaporating the cell liquor. With increasing cost of fuel, the transport advantage of this procedure became less than the cost of concentrating, and, from the early

1950's onwards, caustic soda increasingly reached the consumer as the liquor (*c.* 46·9% NaOH). This development was favoured by the increase in facilities for transport of bulk liquid chemicals by rail and road. Marketing of solid soda had required the production—often at the chemical factory—of great numbers of non-returnable steel drum containers. At the present time more than 85% of caustic soda is sold as liquor.

Since the 1926 ICI merger, production of soda ash (anhydrous sodium carbonate) by the Solvay process has been carried on solely by the Alkali Division of that company, meeting the total home requirement and constituting a very large tonnage export; exports of soda ash in 1964 were valued at £2·1 million. In 1958 the U.K. output of soda ash was one million tons. During the ninety years of its use in this country the Solvay process has remained fundamentally that imported by Ludwig Mond in 1872. From an early stage in its activities Brunner, Mond & Company, in addition to soda ash and sodium bicarbonate, produced caustic soda by what is known as the " lime–caustic " process, i.e., by reaction of sodium carbonate with slaked lime; this route to caustic soda provides an alternative to that of brine electrolysis. There is an important difference between the two process routes: in the case of the electrolytic process the amount of caustic soda produced is determined by the laws of chemistry (1·127 tons caustic soda/ton of chlorine), whereas the quantity of " lime-soda " produced is a matter of choice to meet demand. In recent years, particularly after World War II, the rising demand for chlorine has greatly increased output of co-produced electrolytic caustic soda with the consequence that, to meet the total caustic soda requirement, a progressively decreasing output of " lime-soda " has been necessary. Maintenance of this symbiotic relationship has been greatly facilitated by the circumstance that both soda processes are operated by the same enterprise.*

* In 1964, ICI Alkali and General Chemicals Divisions were merged as Mond Division, thus unifying alkali and chlorine production.

2. THE INDUSTRIAL ACIDS

Sulphuric Acid

Sulphuric acid is manufactured by some fifty different firms in the U.K. at many more than that number of sites. The wide geographical distribution of acid plants, as has already been shown, originated mainly in the nature of the superphosphate and ammonium sulphate fertilizer industries. In many instances the acid is still used captively at the producing site, e.g., The Associated Octel Co Ltd produce sulphuric acid at Hayle (Cornwall) and at Amlwch (Anglesey) for their process for extracting bromine from the sea. Sulphuric acid in the modern period has not been an important export chemical, and, consequently, demand for it in the home market has sharply reflected the general state of industry as a whole. In 1900 the U.K. output of sulphuric acid was one million tons. By the outbreak of World War I production was almost 1·2 million tons, and during that war it reached a peak of 1·38 million tons; thereafter demand declined sharply, reaching 0·58 million tons at the depth of the post-war depression in 1921. From that time onwards, with some substantial fluctuations, especially during the depression of the early 1930's, demand has trended upward. It is notable that World War II had little effect on the pre-war trend of demand, and that it was only in 1946 that U.K. sulphuric acid output was level with the peak acid production during World War I. The principal difference between the situation prevailing during World War I and World War II was that, in the case of the former, large quantities of sulphuric acid were required by the explosives industry to produce nitric acid from Chile nitre. In 1964 U.K. sulphuric acid output was 3·1 million tons. During the present century the history of sulphuric acid in Britain, as elsewhere, has had two important aspects: changes in the sources of the sulphur raw material and progressive shift from the chamber to the contact process. Apart from the limited introduction of the tower process (essentially a modification of the chamber process) and of the Mills–Packard conical, water-cooled chamber in 1913, there have been no revolutionary developments in the technology of manufacture.

At the present time there are no fewer than six sources of sulphur for sulphuric acid manufacture, namely, natural sulphur (brimstone), iron pyrites, spent oxide, zinc sulphide concentrates (imported primarily as a zinc source), anhydrite (mineral calcium sulphate), and sulphuretted hydrogen (from petroleum refineries). The early history of the first three sources has already been discussed. Prior to 1930 only the first four sources were available. In 1928, for example, the 928,000 tons of acid were principally produced from pyritic sulphur (45%); spent oxide was next in importance as a raw material (25·6%); 20·1% of the acid was produced from brimstone, and only 9·3% from zinc sulphide.

Following the take-over of the Billingham factory by Synthetic Ammonia & Nitrates, extensive anhydrite formations were discovered below the site. After the ICI merger, this mineral was used for the manufacture of ammonium sulphate by a process involving direct reaction with ammonia, thus obviating the need for intermediate production of sulphuric acid. Technical and cost studies made at that time showed that it should be economic to use anhydrite, mixed with coke and sand, to yield sulphur dioxide (for acid manufacture), and, as a co-product, a clinker suitable for making builder's cement. Between 1939 and 1951 about 1·5 million tons of acid were produced by this process from anhydrite.*

At the outbreak of World War II, 37% of sulphuric acid produced in the U.K. was made by the contact process; by 1944 the proportion had risen to 44·8%. If the Government-owned plants are taken into consideration, the shift towards the contact process is more striking. By 1945–46, approximately half of the U.K. acid production was by the contact process. Another

---

* During World War I, Germany, cut off from sources of brimstone and pyrites, produced sulphuric acid from anhydrite. Under the extra-economic conditions of war no attempt was then made to make the process economic by production of cement.

important feature of wartime acid production was the shift away from pyrites as a raw material. Political circumstances, as well as technological considerations, were contributory to this development: there was uncertainty as to the side on which Spain, the principal pyrites source, might enter the war. Brimstone consumption, practically all imported from America, doubled, and the use of domestic pyrites increased from a mere 3000 tons or so a year to some 30,000 tons. These domestic pyrites were mainly colliery residues which had been put out to spoil mounds; their use ceased after the war.

Owing to the great demand by the fertilizer industry, U.K. production of sulphuric acid during the latter half of 1946 reached the highest rate ever recorded. Government plants were maintained in operation solely to meet the high requirements, and not to compete with trade production. In 1947 the Government began to lease its acid factories to industrial manufacturers. New plant capacity being installed at this time was almost wholly brimstone-using contact plant. Italy was still out of the brimstone market and the U.K. was thus increasingly dependent on Frasch-mined* American sulphur. Stocks of spent oxide continued to accumulate at gas undertakings throughout the country. Despite voiced doubts as to the wisdom of making the U.K. acid industry dependent on brimstone from the U.S. Sulphur Export Corporation, sulphur-burning plants, because of their lower capital cost and simplicity of operation, continued to be erected. By 1948 four times as much brimstone was being consumed as in the immediately pre-war period. In 1949, the National Sulphuric Acid Association imported on behalf of U.K. acid-makers 600,000 tons of brimstone and pyrites in 140 ships; freight-money exceeded £1·4 million.

For many years it had been believed that the 700-ft-thick " domes " of sulphur-bearing limestone in Louisiana were virtually inexhaustible. In 1950, surveys appeared to indicate that these formations were geologically unique and could become

* Mined by the Frasch process, whereby super-heated steam is used to melt the sulphur underground so that it can be brought to the surface.

exhausted in from ten to fifteen years. The U.S. Government, which controlled all the exports of brimstone, and which had the Korean war on its hands, made a substantial cut in the supplies to the U.K. in spite of strong representations by the British Government. As an interim measure, the Board of Trade imposed an immediate cut of one-third in supplies of acid made from brimstone, until a more detailed scheme could be worked out to deal with this serious situation. British acid output in 1951 was 200,000 tons less than in 1950. Although it was subsequently discovered that the U.S. brimstone resources were not, as had been feared, in danger of imminent exhaustion, the sulphur crisis of 1951 served to direct attention in this country to the use of anhydrite as a sulphuric acid raw material.

When the 1951 crisis precipitated a situation in which extensive new acid capacity based on anhydrite became essential, ICI, the only acid-producer with a developed process for the use of that raw material, was unwilling to undertake alone the burden of additional capital investment in acid plant. That company proposed that it, along with a number of its largest customers, should establish a non-profit-making consortium for corporate manufacture of sulphuric acid and cement from anhydrite, the process information to be supplied by ICI. The new company, the United Sulphuric Acid Corporation, came into legal existence in September 1951; its authorized capital was £5·4 million, and its planned capacity was 150,000 tons of acid a year. The Corporation's works, which began production in 1954, are at Widnes. Anhydrite for this plant is brought by rail from Cumberland. Also about this time Marchon Products Ltd (now merged with Albright & Wilson Ltd) embarked upon a £2 million project for production of 75,000 tons of acid a year from anhydrite in the Whitehaven district. The Marchon plant, which began production in 1954, was designed by Dr. Kühne, a former managing director of IG Farbenindustrie and a co-inventor of the original anhydrite acid process at Leverkusen during World War I. In 1964, 541,000 tons of a U.K. total of 3,132,700 tons of acid, were produced from anhydrite. Imported brimstone was, however, still the principal acid raw material (590,000 tons imported in 1964).

Since 1951 the sulphur supply situation has been changed by the prospect of recovering great quantities from natural gas, discovered in Canada (Alberta and British Columbia) and France (Lacq). Production at both those sites dates from about 1957. Other natural gas sources have since been discovered in Europe and elsewhere.

After 1946, new plants, with the exception of a few Mills-Packard chambers and a Petersen tower installation, have been of the contact type. In 1964 only about one-tenth of U.K. sulphuric acid was produced by the chamber process. Associated with surviving chamber installations in 1963 there were nine plants for concentrating the weak acid (70% H_2SO_4) to 95–98% by distillation: at one time there were in the U.K. more than a hundred such plants. Contact plant units are generally of much greater capacity than the average chamber or tower plant; in the 1950's many of the surviving chamber plants had capacities of less than 10,000 tons of acid a year.

The use-pattern of sulphuric acid shows between thirty and forty principal industrial applications, of which superphosphate, sulphate of ammonia, and rayon manufacture are by far the most important. Only a few major applications have emerged since 1900; these are oil refining and production of petroleum derivatives, extraction of bromine from sea-water, the manufacture of various plastics, and the large-scale production of hydrofluoric acid (for making fluoro-organic derivatives). From 1900 onwards there was surplus capacity in U.K. sulphuric acid works, and production of superphosphate was increased in order to utilize this capacity profitably. In consequence, competition in the home market became severe and was aggravated by superphosphate imports, which rose from 30,000 tons in 1910 to 205,000 tons in 1927. In the latter year the superphosphate trade attempted to obtain protection by imposition of a duty; this was refused on the grounds that British manufacture did not have sufficient capacity to supply the home market, and that this inability was a result of inefficient working as compared with continental producers. The Import Duties Advisory Committee, in 1934, agreed to the imposition of a tariff of ten shillings a ton on imported superphosphate under the Import Duties Act of 1932. Thereafter, the superphosphate industry slowly recovered, aided by rationalization, changes of governmental policy with regard to agriculture, and supplies of cheap acid.

*Hydrochloric Acid**

As has been shown, hydrochloric acid, in the old chemical industry, was often an embarrassing co-product of the manufacture of saltcake. Before the end of the first decade of this century it was evident that this century-old state of affairs was at an end. In 1912 the Castner-Kellner Alkali Company, by then operating the largest electrolytic caustic-chlorine plant in the world, began small-scale experiments on production of hydrochloric acid by combustion of chlorine gas in hydrogen, the latter also a product of brine electrolysis.† (In the U.S., where the LeBlanc System never established itself, the possibility of manufacturing hydrochloric acid from electrolytically-produced chlorine and hydrogen had been explored as early as 1905.) From 1912 to 1920 production of so-called " synthetic " hydrochloric acid in the U.K. remained on a small scale. With the extinction of the LeBlanc process, the United Alkali Company began to produce hydrochloric acid from electrolytic gases; commercial production of the synthetic acid began in 1922. Various types of " burner " were invented for the purpose of hydrogen chloride production; those developed in the early 1920's are in use today. The principal hydrochloric acid process improvements have been in the absorption systems for dissolving the hydrogen chloride in water. A most important feature of the synthetic process was that it liberated the economics as well as the technology of manu-

* This term is only properly applied to solutions of hydrogen chloride (HCl) in water.

† 0·028 ton of hydrogen is co-produced with each ton of chlorine.

facture of the acid from those of saltcake. The importance of this separation may be realized when it is considered that if hydrochloric acid had continued to be produced solely by the old saltcake route, output of saltcake would have exceeded demand for that chemical by three or four times.

Most of the saltcake hydrochloric acid now made in the U.K. is produced in continuously-operating mechanical furnaces. Since the end of World War I, accompanying the increasing manufacture of chlorine, there has been a great increase in production of synthetic hydrochloric acid. By the mid-1950's almost 40% of the acid made in this country was synthetic. In the post-World War II period the rapidly increasing production of chloro-organic derivatives (by processes involving chlorination of hydrocarbons), has generated a new source of by-product hydrochloric acid. The proportion of hydrochloric acid now produced synthetically is determined by the demand for saltcake and the quantity of the acid recovered from production of chloro-organics.

The principal application of hydrochloric acid in the present century continues to be in the metal industries, where it is used in the pickling process, i.e., the removal of oxide scale from metallic surfaces prior to their being galvanized, electroplated, etc. For a number of technical and economic reasons, although metal pickling is the chief industrial application of hydrochloric acid, its use in this connection has increased at a slower rate than has that of sulphuric acid for the same purpose. In 1932, manufacture of gelatine from bones began in the U.K.; this process consumes between 7 and 8 tons of commercial hydrochloric acid* for each ton of bones treated. Over the last two decades there has been a steadily increasing use of hydrochloric acid in the manufacture of organic chemicals, e.g., ethyl chloride and vinyl chloride. In these chemical processes hydrogen chloride is generally used as such, being taken directly from " burners ", in which chlorine is burned in hydrogen, or from an organic chlorination process in which it is being made as a by-product.

* Most commercial hydrochloric acid contains 28–30% of actual hydrogen chloride.

Nitric Acid

The few figures that have been published for nitric acid outputs indicate that, in this century, its expansion has been rapid. In 1935, U.K. nitric acid production was about 70,000 tons; by the early 1950's it exceeded 300,000 tons a year. No statistics of U.K. nitric acid production or consumption have been published since 1953. Recently there has been a steep increase in nitric acid demand and it appears probable that it will soon reach an annual figure of one million tons.

The modern (Ostwald) process for making nitric acid, i.e., the catalysed air oxidation of ammonia, was introduced in the U.K. in 1918, and by about 1925 it had largely superseded the classical route from sodium nitrate. In 1960 nitric acid was being produced by this process at seven sites in the U.K. It is of interest to observe, in passing, that, as early as 1908, the United Alkali Company was considering the question of importing the Ostwald process for nitric acid; they decided against it on the grounds that it would be unprofitable and that the acid it produced was of lower concentration than that in general demand! Despite the proved advantage of the Ostwald process, the older route from Chile nitre continued in declining use up to World War II. In his 1963 Report the Chief Alkali Inspector comments: "The cast-iron cylinder, Woulfe bottle, earthenware tower system was still operated in southern England when that war (World War II) began."

Owing to almost complete absence of statistical information, it is not possible to give a quantitative use-pattern for nitric acid. The modern uses may be classified under three heads: salt formation, including production of nitrate fertilizers, which represent by far the greatest single demand for the acid; production of organic nitrates and nitro-derivatives, including nitrocellulose and nitro-glycerine; oxidation of organic compounds, e.g., nitric acid oxidation is a process step in the manufacture of "Terylene". It has been estimated that in the 1960's, annual U.K. consumption of nitric acid for manufacture of civil and military explosives and for industrial nitro-cellulose is between 45,000 and 50,000 tons.

3. AMMONIA AND AMMONIUM COMPOUNDS

The oldest site of synthetic ammonia production in Britain is at the ICI Castner-Kellner factory at Runcorn in Cheshire, where the pilot plant was erected for the larger operation of the Haber process at Billingham. Following establishment of the Billingham plant, synthetic ammonia works were set up by ICI and others at Flixborough (1937), Mossend (1939), Dowlais (1940), Heysham and Prudhoe (1942), Shell Haven (1959) and Severnside (1963). At Flixborough, in Lincolnshire, the plant is operated by Nitrogen Fertilisers Ltd, a subsidiary of Fisons Ltd. The factory at Prudhoe was erected by the Government to produce ammonium sulphate by the direct route from ammonia and anhydrite; after World War II the works was purchased by ICI. Shell Chemical Company's Shell Haven works in Essex established the nexus between the petroleum industry and ammonia manufacture in the U.K. During the past few years hydrogen for ammonia synthesis has been increasingly produced from the light fraction from petroleum refining, and the former water gas route* is declining in use. Since its beginning, because of the comparative simplicity of the systems involved, ammonia synthesis has been subjected to increasing measures of automatic control. At Billingham, by 1926, the boiler plant and water gas plant had been brought under automatic operation. Modern Haber ammonia plants provide one of the first examples of automation in chemical industry.

The industrial ammonium compounds are derived by direct reaction of ammonia with acids; their manufacture was established in every case before 1900. Ammonium sulphate, the principal ammonium compound, in 1916 was produced in 641 works, using gasworks ammonia; by 1963 there were only 50 works, producing three times the total 1916 output. This is an important instance of the changing scale and increasing concentration of heavy chemical operations. Only about 20% of U.K.

---

* Steam is passed over red-hot coke, yielding a mixture of hydrogen and carbon monoxide and dioxide.

ammonium sulphate is now produced from gasworks ammonia. Production of ammonium nitrate from synthetic ammonia and nitric acid made by oxidation of synthetic ammonia began at Billingham in 1927. Ammonium chloride (sal ammoniac) is one of the oldest heavy chemicals, having been produced by many small concerns since the end of the eighteenth century, particularly in the Midlands, where it finds application in the metal industries; it is now manufactured mainly from synthetic ammonia. Although, properly, urea is regarded as an *organic* compound, its modern manufacture is associated with Haber ammonia synthesis. For a number of years urea has been produced at ICI Billingham factory from liquid synthetic ammonia and gaseous carbon dioxide.

4. CALCIUM CARBIDE

The end of the long hiatus in indigenous production of calcium carbide was in sight in 1935–37, when attempts were made by chemical concerns to obtain Government support for re-establishing its manufacture in the U.K. This *démarche* was, however, not successful, and on the eve of World War II the British Oxygen Company acquired the Odda Smelteverk, in Norway, to secure independence of supply. Following the outbreak of war the Government erected a carbide works at Kenfig in South Wales, close to sources of coke and limestone. The Kenfig plant, with an annual capacity of 120,000 tons, was operated by British Industrial Solvents Ltd (a subsidiary of Distillers Company Ltd) on behalf of the Government until 1953, when the factory was leased by the Distillers Company Ltd from the Ministry of Supply, with the consent of other U.K. carbide users. Lime for the Kenfig carbide plant is obtained from the Tythegston limestone quarries some five miles distant. The greater part of the carbide produced goes to vinyl chloride manufacture by British Geon Ltd, a concern founded in 1945 and owned by Distillers Company Ltd (55%) and the Goodrich Chemical Company of America.

In order to supply the wartime demand for acetylene to manufacture PVC and chloroethylene solvents, as well as for use as a welding gas, a carbide plant was erected by ICI at its Castner-Kellner works in Runcorn; this came into operation in 1943. Two further ICI carbide plants were erected in 1958 and 1961. In the latter of these years Carbide Industries Ltd began operating a 50,000 tons a year carbide plant at Maydown in Northern Ireland to produce acetylene for a nearby neoprene (synthetic rubber) factory. At present (1965), for the first time in the history of the chemical industry, U.K. production of calcium carbide approximates to national demand. In 1961, U.K. carbide production was 254,000 tons, some three-quarters of which was used to produce acetylene for manufacture of organic chemicals. Carbide has never been used in this country, as on the Continent, for the production of cyanamide fertilizer.

In a number of European countries and America, during the past fifteen years, a number of processes have been developed for production of acetylene by cracking and partial combustion of vaporized or gaseous hydrocarbons. This route has acquired particular importance with the increasing rapprochement between the chemical and petroleum industries and discoveries in various countries of hitherto unsuspected sources of natural gas (methane). In 1964, at their Castner-Kellner factory, ICI began production of acetylene from petroleum hydrocarbons by a German process (BASF); this is the first hydrocarbon–acetylene process to be introduced into the U.K.

5. PHOSPHORUS CHEMICALS

Until the outbreak of World War II, Albright & Wilson's resumed phosphorus production was carried on only at the firm's Widnes factory. In 1939, although the Widnes plant was producing three and a half times the output of Oldbury during World War I, this was inadequate to meet wartime requirements. With Albright & Wilson's technical and administrative assistance the Government erected a £500,000 battery of phosphorus furnaces at the Oldbury site. In 1942, shortage of shipping space limited imports of phosphate rock and Albright & Wilson, at the request of the Ministry of Supply, converted the Oldbury furnaces to produce ferrosilicon. After the war the plant, taken

FIG. 9. CALCIUM CARBIDE FURNACE

over by Albright & Wilson, was reconverted for phosphorus manufacture. During the war half the phosphorus produced in the U.K. went to military applications and half to manufacture of food phosphates and phosphoric acid, the demand for the former having trebled. Following the Dunkirk debacle, when invasion appeared imminent, hundreds of thousands of "Molotoff cocktails" with a self-igniting phosphorus–benzene filling were produced. In the course of World War II Albright & Wilson filled 124 million weapons with phosphorus compositions.

Up to World War I, although the company had diversified its manufacture into both the inorganic and organic fields, e.g., cyanide and carbon tetrachloride production, it had not done so on any substantial scale. From 1918 onwards the firm developed its interest in foodstuff phosphates. In 1932 the company entered the water treatment field with Calgon (sodium hexa-meta phosphate), a product which, in 1935, received the British Industries Fair award as the most interesting chemical of the year. By 1934, 36% of Albright & Wilson's profits came from production of food phosphates. In 1952 the firm established a factory at Kirkby, near Liverpool, to produce phosphoric acid and tripolyphosphates, the latter an important constituent of many modern detergents. At Portishead, near Bristol, in 1953, production began at the company's new phosphorus plant. A great part of the phosphorus output of this last plant is conveyed by the firm's own rail-tank wagons to Kirkby for conversion to detergent phosphates by the so-called "dry" process (oxidation of phosphorus).

Up to about 1950, Albright & Wilson had a U.K. monopoly in the manufacture of phosphorus chemicals; at that time Solway Chemicals Ltd, established in 1943 as a subsidiary of Marchon Products Ltd,* began production of detergent phosphates at Whitehaven in Cumberland, by the "wet" process (i.e., by the action of sulphuric acid on phosphate rock). For some years close commercial relations existed between Albright & Wilson and Solway Chemicals. In 1955 Albright & Wilson purchased Marchon Products and its various subsidiaries for £2·6 million, and thus restored its dominant position in the field of phosphorus chemicals. This 1955 merger brought Albright & Wilson about one-tenth of the synthetic detergent trade of the U.K. In addition to their employment in the foodstuffs industry, inorganic phosphates are used in the making of rust-proofing agents,

* Marchon Products Ltd was founded in 1939 by F. Marzillier and F. A. Schon, from whose surnames its name was coined; originally the firm was a chemical-merchanting concern.

catalysts, soft drinks, activated carbon, dentifrices, metal-cleaning agents, and in the pharmaceutical, dyeing, textile-processing, paper, oil, and tanning industries.

6. SULPHUR CHEMICALS

Apart from sulphuric acid, sulphur is the key raw material in the manufacture of a considerable range of inorganic chemicals; only comparatively few of those, however, are produced in " heavy " tonnages. The history of the production of the sodium sulphites (sulphite, bisulphite, hydrosulphite) and of thiosulphate—all having as their starting materials sulphur dioxide and alkali—is obscure. C. T. Kingzett in his textbook on the alkali trade, published in 1877, describes sodium hyposulphite (hydrosulphite) as a " speciality of the Tyne district " and gives the output for 1866 as 400 tons of total value £7200. The United Alkali Company, having sulphur dioxide available at its sulphuric acid plants and alkali from its LeBlanc and ammonia–soda works, began making various sulphites in the early 1900's. Probably other firms, producing these raw materials, also manufactured the chemicals early in this century. From about 1918 Brotherton & Company produced hydrosulphites at their Mersey Chemical Works at Bromborough in Cheshire, the former factory of the Badische Anilin and Soda Fabrik (BASF). For many years sodium sulphite was a by-product of the manufacture of phenol from benzene *via* benzene sulphonic acid; in recent years, with the development of the cumene process for phenol, this source of sulphite has now almost entirely ceased. The sulphites are used as mild bleaching agents and sodium hydrosulphite has a special application in dyeing with the so-called vat dyes. Since sodium thiosulphate (" hypo ") was used to remove excess silver from Daguerre's photographs (invented 1839), this photographic chemical must have been in laboratory-scale production during the second half of the nineteenth century, but the history of this early manufacture does not appear to be on record. *Macmillan's Magazine*, in 1871, observed that " the sixpenny photograph is doing more for the poor than all the philanthropists in the world ". George Eastman's " Kodak ", which, in the 1890's made photography a popular pastime, made sodium thiosulphate a heavy chemical.

With the decline of the LeBlanc soda industry in the early 1900's, the United Alkali Company had a very large and increasing idle capacity for saltcake manufacture. In order to create an economic outlet for saltcake, the company began production of sodium sulphide. The process was the simple one of using the LeBlanc black ash revolvers, but omitting limestone from the charge, thus arresting reaction at the stage of saltcake reduction by the coal: $Na_2SO_4 + 2C \rightarrow Na_2S + 2CO_2$. Extraction of the sulphide was carried out in the vats and the liquor concentrated in the cast-iron pots which had formerly been used in processing LeBlanc soda. This adaptation resulted in survival, if not of the LeBlanc soda process, of its principal techniques until 1964, when the last two sodium sulphide revolvers at Widnes were shut down. These revolvers which continued in service to produce sulphide instead of soda, had been commissioned in the 1880's and were named " Jumbo " and " Alice ", after the once famous elephants in P. T. Barnum's " Greatest Show on Earth ". Most of the sodium sulphide produced went to the tanning industry for de-hairing, and to the manufacture of dyes. Output of sulphide increased from 5400 tons in 1912 to 19,200 tons in 1920. After the post-war depression demand for sulphide greatly increased. In 1964, ICI introduced production of sodium sulphide by a process which does not use saltcake as the starting material.

Carbon disulphide was produced in small quantities by various concerns in Widnes between 1885 and 1906. Shortly before 1900, the United Alkali Company, in order to enter the profitable market for cyanide, developed the Raschen process for that manufacture, starting from ammonium thiocyanate, produced in turn from carbon disulphide and ammonia. To produce the necessary ammonium thiocyanate, the company, in 1900, began manufacture of carbon disulphide by the established method of reacting sulphur vapour with charcoal at high temperature. In that year 580 tons were produced of which 569 tons were used to make cyanide, the remainder being sold. When the Raschen cyanide process ceased to be used in 1909, the United Alkali Company continued to produce carbon disulphide. The largest single application of the chemical is in the manufacture of viscose rayon and, to control their supply for this purpose, Courtaulds Ltd began manufacture of carbon disulphide in 1916. Prior to 1952, carbon disulphide was not recovered by the rayon industry anywhere, and about one-third of a pound was lost for every pound of viscose made. In recent years recovery of the disulphide has become established, consequently consumption by the rayon industry now represents the amount required to replace disulphide not yet recovered. Carbon disulphide is used in production of transparent paper, carbon tetrachloride, various rubber chemicals, dyes and flotation agents (for mining).

7. CYANIDES

The cyanides in present-day tonnage production in the U.K. are sodium cyanide, hydrocyanic acid (hydrogen cyanide), and the cyanides of potassium, zinc and copper; these are all products of ICI and their manufacture is associated with the production of ammonia at Billingham.

The Castner process for sodium cyanide was an entirely U.K. development, although its inventor was an American working in this country. In 1894, Hamilton Y. Castner, managing director of the Aluminium Company of Oldbury, was granted patents for the manufacture of sodium cyanide by reaction of metallic sodium with carbon and nitrogen (or ammonia) at about 650°C: $2Na + 2C + 2NH_3 \rightarrow 2NaCN + 3H_2$. This process attained industrial status only after its inventor's death in 1899. Its great simplicity and high efficiency enabled it to advance into a field already occupied by at least seven other cyanide processes. It almost immediately displaced the Beilby process operated at Glasgow by the Cassel Gold Extracting Company.

For some years the Castner-Kellner Alkali Company (the descendant of the Aluminium Company of Oldbury) supplied the metallic sodium requirements of the Cassel Company, then, in 1907, the companies established a jointly-owned sodium factory at Wallsend-on-Tyne. By this time the annual U.K. output of sodium cyanide was 7000 tons—about half of the then total world production. In the 1960's ICI has replaced the Castner cyanide process by the simpler procedure of reacting hydrocyanic acid with caustic soda liquor ($HCN + NaOH \rightarrow NaCN + H_2O$). This is one of the most important changes in the manufacture of a key heavy inorganic chemical in this country in recent years.

While, on the eve of World War I, world production of sodium cyanide exceeded 18,000 tons a year, hydrocyanic acid had still no place in the chemical industry. During that war some hundreds of tons of hydrocyanic acid were produced in the U.K. for use in war gas mixtures, and sodium cyanide capacity was increased for that purpose ($2NaCN + H_2SO_4 \rightarrow Na_2SO_4 + 2HCN$). In 1927 the Cassel Cyanide Company, which had long been associated with the Castner-Kellner Alkali Company, latterly a subsidiary of Brunner, Mond & Company, became part of ICI, and its operations were carried on by the General Chemicals Division of that combine. Production of cyanide was, in 1930, transferred from Glasgow to Billingham, where sodium was then being produced, the Wallsend plant having been shut down. In 1931 ICI began to make small tonnages of hydrocyanic acid, as well as zinc, copper and potassium cyanides.

In the late 1930's, gold-mining interests in South Africa, having regard to the then menacing political state of Europe, began to consider measures to ensure their cyanide supplies in the event of war. Various ICI missions, co-operating with the joint ICI–De Beers Company (later African Explosives & Chemical Industries Ltd), investigated the problem and in 1937, a pilot plant for cyanide was erected in South Africa. The process used was the two-stage one of making hydrocyanic acid and reacting that with caustic soda solution. Hydrocyanic acid was produced by high-temperature catalytic reaction of methane with ammonia and air.* Following the successful operation of this process in South Africa, the outbreak of World War II and certain patent and other difficulties delayed until 1949 the setting up of ICI's hydrocyanic acid plant at Billingham.

Sodium cyanide is predominantly an export chemical. The ultimate application of the greater part of exported cyanide is in the gold-mining and metal industries; at least half goes to the latter for electroplating, heat-treatment and case-hardening† processes. In the U.K., some 40% of cyanide consumed is used by electroplating works, 35% for the manufacture of other chemicals, and 25% by the metal industries. The use of cyanide or hydrocyanic acid for vermin extermination began to decline in the early 1950's. In recent years 90% of the hydrocyanic acid produced by ICI has gone to the manufacture of the plastic "Perspex".

8. PEROXYGEN CHEMICALS

There are four principal peroxygen chemicals—hydrogen peroxide, sodium peroxide, sodium perborate and sodium percarbonate. These all contain the oxidizing structure —O . O—, which gives them their industrial application as bleaching agents. At the present time, three U.K. firms are engaged in their manufacture: ICI (sodium peroxide and perborate, and percarbonate), Laporte Ltd (hydrogen peroxide and sodium perborate), Alcock (Peroxide) Ltd‡ (hydrogen peroxide only).

As has been shown, hydrogen peroxide manufacture in the U.K. was begun by Bernard Laporte. The firm, known today as Laporte Industries Ltd, is the product of a series of mergers which began early in the present century; manufacture of

---

* $CH_4 + NH_3 + 1\frac{1}{2}O_2 \text{ (air)} \xrightarrow[\text{(catalyst)}]{1000°C} HCN + 3H_2O$.

† The technique of using sodium cyanide for case-hardening steel originated in Germany.

‡ Founded in 1929 by a former managing director of Laportes.

peroxygen chemicals remains, however, one of its principal operations. For the first thirty years of its existence its chief business was production of hydrogen peroxide by the barium peroxide process, a process which, incidentally, occasioned the firm's embarking on the manufacture of a series of barium chemicals. In 1916, Laportes commenced producing barium peroxide for hydrogen peroxide manufacture; formerly it had imported this raw material. Laportes, in 1964, formed, jointly with Imperial Smelting Corporation, Barium Chemicals Ltd to make barium and zinc products.

In 1932, Laportes introduced manufacture of hydrogen peroxide by an electrolytic process, first at Luton and, after World War II, at Warrington, where sodium perborate (*see below*) manufacture was also started. Electrolytic production of hydrogen peroxide originated on the Continent during the first decade of this century in Austria and Germany. At Warrington in 1958, Laportes established one of the largest peroxide plants in the world, using the autoxidation process: shortly afterwards the electrolytic hydrogen peroxide plant was shut down. The autoxidation process came comparatively late to this country. In 1932, the Americans, Walton and Filson, discovered that autoxidation of hydrazobenzene in alcoholic solution with air yields hydrogen peroxide in practically quantitative amount. Reactions of this type suggested the basis of a new hydrogen peroxide process, and before World War II IG Farbenindustrie was experimenting with hydrazobenzene autoxidation. In 1944, Dupont, in the U.S., were operating a small-scale autoxidative plant employing anthraquinone derivatives as the oxygen vehicle; by 1953 that firm was operating this process on an industrial scale.

In 1890, Hamilton Y. Castner, his sodium aluminium process having been suddenly superseded by the electrolytic processes of Hall and Héroult, was in search of new applications for metallic sodium. Aware of the growing use of hydrogen peroxide, he decided to produce sodium peroxide as a much more stable and concentrated peroxygen bleaching agent than the hydrogen peroxide then being made. With the directness and simplicity which characterized his chemical inventions, he reacted molten sodium with air by exposing the metal in a series of shallow trays carried on wheeled " tramway carriages " moving through a stream of air in a tubular furnace. By controlling the rate of advance of the train of carriages, the sodium was substantially burned to the peroxide (Na_2O_2). Castner's peroxide process was operated on a small scale at Oldbury for some years, then, after 1905, at Runcorn by the Castner-Kellner Alkali Company. The sodium peroxide process as developed by Castner was adopted in Germany and America. In 1927–28 a two-stage process was developed by Roessler and Hasslacher in the U.S., whereby sodium was first converted to monoxide in a rotating kiln, then further oxidized to peroxide in a second rotating furnace. This process obviated the high labour consumption in manhandling, charging and cleaning the " tramway carriages ", of the original procedure; it was subsequently adopted in Germany and the U.K. A large modern plant for sodium peroxide was erected by ICI at its Castner-Kellner factory in 1952–53.

Sodium perborate ($NaBO_2 . H_2O_2 . 3H_2O$) is not in fact a *perborate*, but sodium metaborate associated as shown, with a molecule of hydrogen peroxide of crystallization; that is, the salt functions as a hydrogen peroxide carrier. About 1908, some continental firms began to incorporate sodium peroxide with soap powders; it was soon found, however, that the peroxide was too alkaline ($Na_2O_2 + H_2O \rightarrow 2NaOH + \frac{1}{2}O_2$), and it was replaced by sodium perborate. To meet this new demand, the Castner-Kellner Alkali Company in 1915, began manufacture of perborate using sodium peroxide as the source of peroxygen, rather than hydrogen peroxide. Later Laportes used hydrogen peroxide for their perborate manufacture. In the original process, used by the Castner-Kellner Company, borax was reacted with sodium peroxide and hydrochloric acid; this is generally known as the " acid " process.*

* $Na_2B_4O_7 + 4Na_2O_2 + 6HCl + 13H_2O \rightarrow 4NaBO_2.H_2O_2.3H_2O + 6NaCl$.

In the late 1920's, a perborate process was developed in the U.S.; this was in effect a hybrid of the sodium peroxide-using and hydrogen peroxide-using routes; for that reason it is sometimes called the "duplex" process. When ICI in 1950 erected its new plants at Runcorn for production of sodium peroxide and perborate these formed an integrated and flexible system for manufacture of perborate by the "duplex", acid, or hydrogen peroxide process and also of sodium percarbonate from sodium carbonate and hydrogen peroxide. Manufacture of percarbonate began at Runcorn in 1941; it was intended as a wartime substitute for perborate which, because of drastic reductions in borax imports, was in short supply; after the war, however, demand for percarbonate continued. Like perborate, the so-called percarbonate is not a true per-salt but sodium carbonate associated with three molecules of hydrogen peroxide of crystallization.

9. INDUSTRIAL GASES

The term "industrial gases" is generally applied to oxygen, nitrogen, hydrogen, acetylene and carbon dioxide, when these are produced for technological applications other than as starting materials in chemical manufacture. Oxygen, hydrogen, acetylene and carbon dioxide are used in the chemical factory for production, respectively, of acetylene (by hydrocarbon oxidation), hardening unsaturated oils, chloro-ethylene solvents, and urea. For these chemical applications the gases are generally produced at the consuming site, although, as for example, in the case of hydrogen, not always. Industrial gases find their main applications in the metal and engineering industries. Apart from the initiative taken by the Brins Oxygen Company before 1900, production of industrial gases in the U.K., as elsewhere, is a twentieth-century development, and has been determined by the evolution of metallurgical techniques.

The U.K. industrial gases trade has throughout its modern history been dominated by the British Oxygen Company (earlier

The Brins Oxygen Company). Industrial gases are also produced in the U.K. by Saturn Gases Ltd (a subsidiary of Air Products (Great Britain) Ltd), ICI, the Distillers Company Ltd, Murgatroyd's Salt and Chemical Manufacturing Company Ltd, and Staveley Iron & Chemical Company Ltd. In 1906 The British Oxygen Company ceased producing oxygen by the original barium peroxide route and adopted the Linde process for liquefaction of air followed by its fractional distillation. Liquid air distillation is now the method universally employed for production of oxygen, nitrogen and the rare gas fraction (neon, argon, etc.). In the years immediately following World War I the U.K. annual consumption of industrial oxygen was of the order of 300 million cubic feet. British Oxygen Company laid down in Sheffield an oxygen grid, consisting of 15 miles of pipeline, to distribute the gas to their customers in that district, and, in 1932, introduced the Heylandt process for production, storage and transport of liquid oxygen. In the 1950's, the scale of production of oxygen at British Oxygen works, and at some chemical and steelworks sites, rose to the "tonnage" level. Several tonnage oxygen plants now operating in the U.K. have capacities exceeding 500 tons a day. Nitrogen, co-produced at tonnage oxygen plants, is mainly used as a blanketing gas to exclude air from metallurgical and chemical operations. The rare gases, neon, argon, helium, krypton, xenon became available commercially as co-products of liquid air distillation in the 1930's. Increase in demand for these inert gases has been mainly due to their use as fillings for electric light bulbs and discharge tubes, and for argon-arc welding (after 1940). In recent years the British Oxygen Company has marketed liquefied helium and neon for certain cryogenic applications.

Hydrogen, co-produced with chlorine and caustic soda in brine electrolysis, was originally used as a fuel gas. At the works of the Castner-Kellner Alkali Company this byproduct hydrogen was used, in 1920, to make ammonia in the first Haber plant to be operated in this country. In World War I some hydrogen from that factory was used for filling military balloons and airships; byproduct hydrogen from the plants of ICI and Staveley Iron and Chemical Company was used extensively for barrage balloons during World War II. Byproduct hydrogen from brine electrolysis and produced by electrolysis of water, compressed into steel cylinders, has been in wide use in this country for oxy-hydrogen welding and cutting since about 1920. Since the 1920's a large part of electrolytic byproduct hydrogen has been used for hydrochloric acid synthesis, and for oil hardening.

In 1901, Normann, a German chemist, discovered, while working in the oil mills of Leprince and Siveke, a process for converting unsaturated oils to hard fats by hydrogenation in the presence of a catalyst. German manufacturers did not adopt the process at that time and, in 1905, Joseph Crosfield & Sons of Warrington invited Normann to enter their employment and purchased the British patent for the hydrogenation process. Crosfields, in an attempt to secure a monopoly of the technique, bought up the patent rights in other countries. After protracted development work in their laboratories, they began production of hardened oil for soap in 1909. Apparently discouraged by the difficulties of the process, Crosfields sold their rights in it to Jurgens for the production of edible fats; Normann later joined Jurgens and for a number of years advised various U.K. firms on oil hydrogenation. Later patents in connection with the process became the subject of a prolonged legal action by Brunner, Mond & Company (then owners of Crosfields) and Lever Brothers of Port Sunlight. Although, as a consequence of that action, Normann's patent was found invalid because it was held not to give adequate information for carrying out the process, oil hydrogenation had become successful practice and, as the demand for soap and foodstuff fats rapidly increased after 1914, came into widespread use.

Until calcium carbide manufacture was re-established in the U.K. during World War II, acetylene for the metallurgical industries was produced from imported carbide, about one-quarter of imports going to that use. By 1900 several companies in the industrial countries were exploiting Claude and Hess's discovery* of the high solubility of acetylene in acetone. In the U.K. dissolved acetylene was first marketed by Meteolite Ltd. Today, industrial acetylene in the U.K. is generally supplied under pressure in steel cylinders containing 200 cubic feet of the gas dissolved in acetone and stabilized by a filling consisting of a mixture of kieselguhr and charcoal. During World War I use of oxyacetylene welding for production of warlike equipment greatly increased the use of both oxygen and acetylene. World War II produced an even greater increase in demand for welding gases, the peak acetylene requirement reaching 315 million cubic feet a week, or some ten times the highest rate of consumption during World War I. In the post-war period, owing to its increased use for chemical manufacture, acetylene was in short supply to the engineering industries. To mitigate the difficulty arising from this situation, the Ministry of Supply attempted to promote use of propane instead of acetylene for welding. In recent years there has been a great diversification of welding techniques—thirty-seven are currently in use in this country. Electric arc welding is now of greater importance than gas welding, either with acetylene or hydrogen. Although oxyacetylene welding continues to be of importance, and is more widely used than oxyhydrogen, its field has been narrowed to making metal parts of less than $\frac{3}{8}$-inch thickness, and to cutting and dismantling operations.

A proposal to produce aerated drinks by compressing carbon dioxide into solution is attributed to Joseph Priestley. During the nineteenth century and in the early years of this century, aerated water manufacturers produced the carbonic dioxide they required by reacting whiting or chalk with sulphuric acid. By 1929 the Distillers Company Ltd were recovering the gas from their fermentation processes and marketing it in solid form as a refrigerant (" dry ice "). Today, that company's Carbon Dioxide Division produces liquid and solid carbon dioxide at seven sites, producing the gas from coke combustion, limestone, coal,

* Made in 1896.

and oil, as well as a fermentation byproduct from its distilleries. In the late 1930's carbon dioxide, recovered from the hydrogen purification process at Billingham synthetic ammonia works, was being produced by ICI, both as the compressed gas and as the solid (" Drikold "). During the past thirty years few entirely new applications have been found for carbon dioxide, either as solid or gas. Amongst new uses is the so-called " shrouded arc " welding process. The Calder Hall atomic energy plant, which began supplying electricity to the grid in October 1956, uses gaseous carbon dioxide as a heat-transfer medium. Each reactor has a charge of 26 tons of the gas, which is circulated by blowers at a rate of a ton a second between the pile and the boilers. The great tonnages of impure carbon dioxide annually co-produced in " burning " limestone for the most part are allowed to pass to waste into the atmosphere. Some carbon dioxide, produced captively in lime kilns at the consuming sites, is used in the ammonia–soda process and in the sugar-refining industry.

10. METALS ASSOCIATED WITH HEAVY CHEMICAL MANUFACTURE

The frontier between heavy chemical industry and certain branches of non-ferrous metal production has long been much disputed. Historically, this ambiguity has arisen from the fact that the techniques of isolating the newer metals originated within the chemical industry. Thus, aluminium was first produced cheaply on a tonnage scale in this country by reaction of H. Y. Castner's metallic sodium with aluminium chloride in 1888; with the development of Hall and Hérault's electrolytic processes, connection between the aluminium and the chemical industries was severed: chemical procedures were no longer a major feature of its isolation.

The carbonyl process for isolating nickel is based on a purely chemical discovery made in 1889 by Otto Langer, in Ludwig Mond's private laboratory, when investigating difficulties experienced during work on one of Mond's processes for recovering the chlorine lost in the Solvay ammonia–soda process. Mond took the process out of the chemical industry in 1900, by forming the Mond Nickel Company, in order to exploit Langer's discovery independently of his chemical industrial interests.

Magnesium was first isolated commercially in Germany from the chloride present in the Stassfurt salt beds; its extraction from sea water* was first achieved by British chemists in 1937; the process was a complicated chemical one. Later, during World War II, magnesium production began at Clifton Junction, near Bolton, by electrolysis and also by carbon reduction of the oxide—both conventional " metallurgical " procedures. The technology of magnesium production was developed in the U.S. between World Wars I and II, notably by the Dow Chemical Company.

Sodium has continued throughout its industrial history to be regarded as a heavy chemical. With the introduction, in 1937, of the Downs sodium cell, modern sodium manufacture in this country began. Up to 1954, ICI had a monopoly of sodium production in the U.K.; in that year the Associated Ethyl Company (now The Associated Octel Co Ltd) began production of its own sodium requirement for TEL manufacture. During World War II, ICI, having a part interest in the British Ethyl Corporation, participated in TEL production in the U.K. and supplied sodium for that purpose.† Until 1954, ICI sodium was used by the Associated Ethyl Company, which had acquired ownership of the British Ethyl Corporation after the war. In the U.K., for many years, the use-pattern of sodium was dominated by its employment in the manufacture of sodium cyanide by the Castner process, up to 70% of the output going to this application. The Castner cyanide process has recently been displaced by production of cyanide by reaction of hydrocyanic acid with caustic soda liquor, thus eliminating the former largest single use

---

* One cubic mile of sea-water contains six million tons of magnesium. Steetley's plant at Hartlepool produced 4,000 tons of magnesia in 1939.

† The TEL plants at this period were operated by ICI Alkali Division on behalf of the Ministry of Supply.

of sodium metal; TEL manufacture now constitutes its principal application in the U.K., as it has done in the U.S. since the mid-1920's. Physical, as opposed to chemical, applications of sodium have never attained large tonnages and have not been added to for a number of years; they include its use in sodium-lighting, metal de-scaling, heat-transfer (e.g., in atomic piles).

The large-scale isolation of titanium in a sufficiently pure form for technological application, although the existence of the element had been known since 1790, was made possible by modern techniques in 1940. Small quantities of the metal of 99% purity had been obtained in 1910, but it was found to be hard and brittle and appeared to offer no promise as a constructional material. In 1925, two Dutch chemists produced a very small amount of titanium by high-temperature decomposition of its tetraiodide; the metal, isolated in this way, was shown to be soft and malleable. This was the first indication that titanium, if a process could be developed for its large-scale isolation, might be an important addition to the range of engineering metals. In 1930, W. J. Kroll, a native of Luxembourg and a metallurgist of varied experience, addressed himself to the problem of discovering a practicable titanium process. By 1937, Kroll was in sight of success, having developed a method of reducing titanium tetrachloride by means of metallic magnesium. Eventually after a number of failures to interest U.S. firms in his process, Kroll convinced the Bureau of Mines of the potentiality of his method. World War II greatly stimulated interest in metals of high strength-to-density ratio, and, in 1944, a plant was established at Boulder City to produce 100 lb of titanium a week.

In 1951, ICI set up a small titanium plant at Widnes, using Kroll's process; somewhat later, in the same chemical town, another Kroll plant was installed by McKechnie Bros., a firm long-established in the copper sulphate trade. After experimental work with their Kroll plant, ICI developed its own titanium process based on the reduction of the tetrachloride by metallic sodium, a procedure recalling the Deville-Castner process for aluminium. By 1955, ICI had erected a 1500 tons a year plant, using sodium reduction, at its new Wilton factory in the neighbourhood of Billingham, where sodium was available from the company's Cassel works. After overcoming the considerable difficulties in converting the metallic " sponge ", obtained from the reduction stage, into useable consolidated metal, ICI titanium began to find a range of applications in the engineering and chemical industries, as well as in the construction of aircraft, the latter an application sponsored mainly by the Government. It now appears that in the U.K., as in America and elsewhere, titanium has not so far moved to the place amongst modern light metals that it was expected to assume. ICI capacity for production of the metal is in being and titanium output has recently been resumed; the McKechnie Kroll plant has ceased operating. The slow development of titanium may to a great extent be owing to the unexpected part played in recent years by aluminium and its alloys in displacing ferrous metals from traditional applications which might otherwise have been taken over by titanium.

11. SILICON CHEMICALS

The detergent properties of sodium silicate were first exploited industrially by William Gossage of Widnes in the early 1850's. At this time Joseph Crosfield & Sons of nearby Warrington also began adding silicate to soap. Sodium silicate production requires as its raw material silica (sand) and sodium carbonate, the latter relating it to the Solvay alkali industry. Crosfields produce two washing powders in very large tonnage—" Persil "* and "Rinso"—which are mixtures of sodium silicate, with sodium perborate. These manufactures form an important nexus with production of peroxygen chemicals. In 1962, Crosfields greatly extended and modernized their plants at Warrington. At the present time about one million packets of silicate-containing washing powders a day are produced at the firm's Warrington factory. Crosfields chemical business is founded on silicate, of which they

* So named because it contains sodium *per*borate and sodium *sil*icate.

are today the largest producers in the U.K., manufacturing at their Warrington and Bow (London) factories about 100,000 tons a year, of which 20% is exported to over 80 countries. ICI are also producers of sodium, aluminium and calcium silicates (for water treatment), and of precipitated silica.

Sodium silicate, in a mixture with caustic soda, is used for cleaning beer and milk bottles, and floors and plant in food factories. Thousands of tons of silicate are used annually as an adhesive in the packaging industry; it is also used for bonding cores and moulds in foundries,* and in soil consolidation. Potassium silicate, of which Crosfields are the only U.K. makers, is used in the process for coating cathode ray tubes for television apparatus. Silica gel, produced by action of sulphuric acid on sodium silicate, is used in production of water softeners, as a filler in plastic compositions, and as a flatting agent in paints; silica and silicates are also employed as additives to common salt and other substances to promote their free flow. Since 1951, Crosfields have been producing a silicate-based catalyst for petroleum cracking. The manufacture of this catalyst was developed in the U.S.; the Houdry process, in which it is employed, was invented in 1927. Today the firm is annually supplying some 15,000 tons of the catalyst to the petroleum refining industry. Although, as a complex of alumina and silica, the catalyst is not a simple chemical, the mode and scale of its manufacture align it with the heavy inorganic silicon chemicals.

12. FLUORINE CHEMICALS

In 1959 some 350 academic and industrial scientists from twenty countries attended an international symposium at Birmingham University on the subject of fluorine chemistry. Thirty years earlier such a conference would have had practically

---

* Particularly in the so-called "carbon dioxide process", in which foundry sand, impregnated with silicate, is gassed with carbon dioxide. The reaction which takes place causes sudden hardening of the sand.

nothing of industrial interest to discuss and, even in the academic field, only a series of small preparations of fluoro-organic derivatives and various inorganic fluorides could have concerned it. Although its existence, in hydrofluoric acid, had been deduced by 1810, it was not till 76 years later that elementary fluorine was isolated and described. Until then, and for decades afterwards, hydrofluoric acid (HF) was the fluorine chemical, enabling the too reactive halogen to be brought into inorganic and organic combination in controlled manner. Fluorine chemistry was without importance in industry until 1930. In that year, in order to meet the need for a non-flammable, low toxicity refrigerant, small quantities of difluorodichloro-methane were produced in the U.S. That development first attracted the attention of ICI to the potentialities of industrial fluorine chemistry.

By 1939 the former General Chemicals Division of ICI had developed a fairly successful type of electrolytic cell for fluorine production. At the outbreak of World War II all work by ICI on fluorine cells ceased. In 1940, as part of the so-called Manhattan Project, it was decided by the American scientists engaged in it to produce uranium hexafluoride as a means of isolating the uranium isotope they required for the atomic bomb. ICI was invited to co-operate in the production of the required hexafluoride, which, as the only volatile uranium derivative, appeared to offer the most promising means of effecting the desired separation of uranium isotopes. During the remaining period of the war ICI carried out intensive work on fluorine production and chemistry. In the immediately post-war years the company continued to improve the design of its fluorine cells. By the early 1950's a large-scale fluorine plant was in operation at Runcorn, each cell producing several pounds an hour of elementary fluorine. Between 1950 and 1959 the capacity of the cells was increased five-fold.

Hydrofluoric acid, the fluorine analogue of hydrochloric acid, was reported by Margraff (1768) and Scheele (1771) as being evolved when sulphuric acid is reacted with fluorspar (mineral calcium fluoride). The anhydrous compound (HF) was first produced by Frémy in 1856. From about 1830 onwards aqueous hydrofluoric acid was produced in small commercial quantities in several countries for etching glass. Anhydrous hydrogen fluoride was first produced industrially in 1931 in the U.S. for use in the manufacture of refrigerants; it is now by far the most important inorganic fluorine chemical. In 1934, ICI and Imperial Smelting Corporation in 1946 began production of anhydrous hydrofluoric acid, the latter at Avonmouth. At the present time these companies produce large annual tonnages of hydrogen fluoride, mainly for captive use in the manufacture of refrigerants and aerosols. A few small concerns produce the aqueous acid as a glass-etching agent, although this process is now largely effected by directing a high-velocity jet of sand or other abrasive powder on to the glass. It has been estimated that, in 1957, production of hydrofluoric acid, anhydrous and aqueous, in the U.K. amounted to some 10,000 tons (as HF). All the processes for making hydrofluoric acid use fluorspar as the raw material; the grade of mineral used contains more than 97% calcium fluoride and is obtained from deposits in Durham and Derbyshire, where the mineral is known as "Blue John" or "Derbyshire spar".

Apart from fluorine and hydrogen fluoride, the only other inorganic fluorine chemicals at present produced in this country on a substantial scale are boron trifluoride and sodium silicofluoride, the former by Imperial Smelting Corporation and the latter by ICI at Billingham. Boron trifluoride has some application as a catalyst in organic reaction processes. Sodium silicofluoride, recovered as a byproduct in the processing of phosphate rock, has several minor applications, including use as an insecticide. In 1934, Imperial Smelting Corporation, in partnership with British Aluminium Ltd, began production of aluminium fluoride, an essential material in the modern electrolytic production of aluminium; in face of strong foreign competition manufacture of that fluoride was discontinued.

13. OTHER HEAVY INORGANIC CHEMICALS

Lime in the Chemical Industry

Lime burning and slaking are amongst the very oldest heavy chemical processes; both processes were known in ancient Egypt and are described by the elder Pliny. Over 50 million tons of limestone are annually consumed in the U.K. Traditionally the processing of this vast quantity of mineral is regarded as lying outside the chemical industry; it is to be noted, however, that about 20% of lime made in this country is produced by ICI at its limestone quarries in the Peak District of Derbyshire. While the greater part of the annual output of quick and slaked lime goes to the building industry and agriculture, very large tonnages are consumed in chemical manufacture, for example, in production of caustic soda and calcium carbide.

Chromium Chemicals

In 1927 the Eaglescliffe Chemical Company, in order to utilize its surplus sulphuric acid capacity, began manufacture of sodium dichromate, a chemical which finds application in the textile, dyeing, metallurgical and pigment industries. During World War II the same firm produced chromium oxide for use in producing special metals required in aircraft construction; sodium chromate was also manufactured for smoke-producing formulations. By that time Eaglescliffe Chemical Company was the largest U.K. maker. In 1951 the firm acquired Potter & Company of Bolton and, in 1953, J. & J. White Ltd of Glasgow, thus consolidating the U.K. chromium chemical trade under one control as British Chrome and Chemicals Ltd. The only firm outside this merger was the Lancashire Chemical Works Ltd, a small concern founded in 1936 to make chrome products for the leather tanning trade.

White Lead and Titanium Oxide

In modern times the white pigment industry is dominated by Associated Lead Manufacturers Ltd,* British Titan Products

* A chemical subsidiary of Goodlass Wall & Lead Industries Ltd.

Company and Laportes. The first of these represents a continuity of white lead manufacture in the U.K. dating from the end of the eighteenth century. At Bootle, Associated Lead Manufacturers produce white lead by an electrolytic process in a plant which is the largest of its kind in Europe. For many years in the present century white lead has also been produced by Mersey White Lead Company at Warrington. British Titan Products was jointly established in 1933 by ICI, Imperial Smelting Corporation, Goodlass Wall & Lead Industries, the Titan Company and R. W. Greef. National Titanium Pigments Ltd was formed in 1927; in 1932 Laportes acquired an interest in the firm which, in 1952, became their subsidiary, Laporte Titanium Ltd. Production of titanium dioxide from the ore ilmenite is now one of the largest applications of sulphuric acid in the U.K.

Bromine

Herbert Dow, the American chemical industrialist, long entertained the conviction that the sea could be made a source of bromine in industrial quantity. From 1925 onwards, the Dow Chemical Company carried on investigations and eventually developed a process which they exploited jointly with the Ethyl Company, forming the Ethyl-Dow Chemical Company. Bromine is present in sea-water in amounts of less than 70 parts per million. The question of concentrating by evaporation did not arise; in the process that was evolved the bromine is liberated by treating the raw sea-water with sulphuric acid and chlorine and blowing it out of solution by air. When, during World War II, TEL petrol additive was made in the U.K., the bromine necessary for the associated ethylene dibromide* was produced at Hayle in Cornwall by the Ethyl-Dow process; later, a similar plant was erected at Amlwch in North Wales. Both these bromine works are now operated by The Associated Octel Co Ltd. Although

* The "ethyl fluid", consisting mainly of tetraethyl lead, generally contains ethylene dichloride as well as the dibromide.

statistics are not published, it may be estimated that U.K. output of bromine is of the order of 10,000 tons annually; all, except a minor tonnage going to small-scale production of a number of organic chemicals, is used in ethylene dibromide manufacture.

Aluminium Sulphate and Alum

After manufacturing alum in the Manchester area and elsewhere for over seventy years the firm of Peter Spence (p. 47), in 1919, established a large factory in Widnes; several small firms, also at that chemical centre, where sulphuric acid was conveniently available, began manufacture of aluminium sulphate early in this century and continue in operation today, although now merged with the British Aluminium Company. In 1960, Laportes, who already had interests in two aluminium sulphate firms, acquired Peter Spence & Sons Ltd. Laportes have bauxitic clay mines in Ayrshire; these supply the raw material for their alum and aluminium sulphate production. While there have been improvements in detail, the alum-producing process has remained essentially unchanged since the 1880's. Aluminium sulphate is now chiefly used in the paper industry and for water purification.

Carbon Black

The rapid increase in the number of motor vehicles, particularly after World War II, has greatly increased the use of rubber and consequently the demand for lamp black, which is the principal rubber filler. Lamp black is annually produced in large tonnages by combustion of various hydrocarbons in a controlled supply of air. Until 1936, lamp black for rubber manufacture in the U.K. was wholly imported. In that year Palatine Gas Corporation Ltd (now Philblack Ltd) began production at Bristol. Cabot Carbon Ltd, formed in 1948 as a subsidiary of the Cabot Corporation of Boston, U.S., produces carbon black at Stanlow and Duckinfield in Cheshire. The first plant erected by this firm had a capacity of 8000 tons a year. By 1958, U.K. output of carbon black was about 100,000 tons—equivalent to 4 lb per head of the population.

DYES AND ORGANIC PIGMENTS

The modern epoch of the U.K. dye and organic pigment industry dates from the early 1920's, when the industry, benefiting by experience and the protection afforded by the Dyestuffs (Import Regulation) Act (p. 108), began to acquire new purpose and vigour. These early years of the new era were uneasy: political and financial influences outside the industry were adverse to its expansion and, in particular to large-scale expenditure on research—a principal necessity to its further development. In 1924, proposals were canvassed for a more or less close association of the British Dyestuffs Corporation with the resurgent IG Farbenindustrie; no doubt fortunately, these negotiations were effectively discouraged by the Government. The merging of British Dyestuffs Corporation in ICI in 1926 was of cardinal significance for the future of the industry, linking as it did dye manufacture in the U.K. with the heavy inorganic chemical industry, which for every ton of dyestuff produced supplies between three and five tons of the raw material chemicals. Its integration with ICI enabled the principal sector of the dyestuffs industry not only to finance research but to develop the non-dyestuff potentialities which emerged as a byproduct of that research, namely, rubber chemicals, textile finishes, and pharmaceuticals. During the forty years following World War I, of the seven major discoveries made in the dyestuffs and organic pigments field no fewer than five were made in this country. In its modern era the U.K. dyestuffs industry developed a world-wide technical service to its customers—a similar service had played a role of overwhelming importance in creating the world dominance of the great German dyestuff trusts and its absence had done much to limit the pre-1914 British dye trade, particularly in the export market. In the home market British dyemakers came into their kingdom: in 1930 only 20% of dyes consumed in this country were imported; nine years later that portion had fallen to a mere 5%. Over the decade 1926–36 the tonnage output of dyes in this country approximately doubled. While in 1964, U.K. exports of synthetic organic dyestuffs were valued at £20·4 million, imports were still close on £9 million, indicating the continuous rapid progress in the foreign development of new dyes.

In modern times the dyestuffs industry has continued to be located mainly in the textile-producing regions of northern England; it is carried on by some fourteen firms of which ICI (Dyestuffs Division) is by far the largest. The principal primary organic raw materials of the industry are constituted by a small group of comparatively simple aromatic hydrocarbons of which the principal members are benzene, toluene, naphthalene, and anthracene. These primaries are, in the U.K., still derived mainly from coal tar distillation, although in recent years some have become available from indigenous petroleum refining. A number of firms (about half-a-dozen) specialize in making such so-called "intermediates" as aniline and nitro-derivatives from the primary aromatics, but the larger dye firms generally produce their own intermediates; thus, for example, ICI produces its requirement of phthalic anhydride by catalytic oxidation of naphthalene. Manufacture of organic pigments is a department of dyestuff production; recent developments have tended to blur any frontier there was between the two classes of products, e.g., the "Monastral" pigments (see below) and the brightly-coloured complex phosphorus, molybdenum and tungsten derivatives of soluble dyestuffs developed in the post-World War II period. Formerly organic pigments were mainly lakes of organic dyes, i.e., dyes precipitated with aluminium hydroxide.

It has been said that "The modern dye industry is built upon the coal tar industry as its source of raw material, and upon the

Kekulé benzene theory as its scientific basis ".* At the present time between three and four thousand dyes and related organic pigments are in production; of these a major portion depend for their colour property on the diazo chromophore (—N=N—), discovered in 1858 by Peter Griess, while working as Hofmann's assistant at the Royal College of Chemistry in London (p. 67). It was the most momentous single discovery so far in the history of dye chemistry. Not only the number but the variety and structural complexity of its products make the dyestuffs branch of chemical industry the least susceptible of comprehensive descriptive definition; only the modern pharmaceutical industry compares with it in this respect. The systematic chemical names of most dyes are too recondite to be used outside the specialized literature of the relevant field of organic chemistry. For purposes of communication and commerce dyes are generally referred to by names which may indicate the general chemical class, the colour, a significant property and, in many instances, the place of origin, or the manufacturer's name: no standardized nomenclature, however, exists and many dyestuffs names are almost as fanciful as those of race-horses. Typical commercial dyestuff names are: Indanthrene Blue, Fast Green, Caledon Jade Green, Ciba Scarlet G, and Monastral Fast Blue.

The dyestuffs industry measures the outputs of its products in pounds rather than in tons; most of its processes are of the batch type. Unlike the heavy branches (inorganic and organic) of chemical industry the history of dye manufacture is hardly at all the history of its processes; dye production has not been greatly affected by the rise of chemical engineering: its manufacturing technology is much more the product of the development of the aromatic branch of organic chemistry than of the application of the principles of physical chemistry. In this respect the manufacture of dyes is in contrast to the *techniques of dyeing*. Most of the techniques of industrial dye-making are those of laboratory organic chemistry, the flasks, etc., of the bench-chemist being

* *Fundamental Processes of Dye Chemistry*, Fierz-David and Blangey, Interscience Publishers Inc., 1949.

replaced by stainless steel,* enamelled, glass or rubber-lined mild steel vessels. The technical history of the industry is that of a quest for new molecular structures having the essential characteristics of a dyestuff—colour, the capability of attaching itself to materials sufficiently to resist extensive removal under normal conditions of use, and fastness against the fading action of light. To present a techno-historical picture of the modern British dyestuffs industry in terms of what may be called the " molecular " history would far out-run the scope of the present study and lose the main theme in a maze of aromatic hexagons. In what follows the principal British contributions to the technology from 1920 to the present will be outlined.

It has been estimated that the 3500 or so dyestuffs now in commercial use in the world dye industry are the residuum of the experimental synthesis in the laboratory of about a million coloured organic compounds. Major discoveries in the field of dyestuff chemistry in modern times can be made only as a result of sustained and costly research; it is to be expected, therefore, that the principal advances will be made by the largest firms only: this is clearly shown by the recent history of the industry in the U.K.

1. *The Search for Light-fastness*

Through the centuries artists and dyers have sorrowed over the fugitive nature of some of the most splendid colouring matters, particularly those of organic nature, when exposed to the fading action of direct sunlight. Many synthetic dyes were no more exempt from that action than those of natural origin. Amongst natural dyes indigo, apart from the beauty and range of its colour, was prized for its fastness. Perkin's mauve, its discoverer found, " resisted the action of light remarkably well "; had it done otherwise the factory at Greenford Green—the first synthetic dyeworks in the world—would probably never have been erected, and the beginning of the artificial dye industry

* Stainless steel came into large-scale use in the chemical industry in 1928.

delayed by several years. Many of the aniline dyes that were subsequently produced, although they survived in their brilliance a few appearances under the chandeliers of Victorian ballrooms, were notably lacking in fastness to the light of day. Although it was a well-established fact of experience that dyestuffs differed widely in fastness, no systematic attempt appears to have been made to classify them empirically on this basis before 1900. It is not improbable that some dye-makers were in no haste to reduce occurrence of this in-built obsolescence of their products!

In 1902, James Morton, who carried on business as the Morton Sundour Company at Carlisle, was shocked by the rapid fading of the dyed fabrics his company produced when these were exposed in a shop-window in Regent Street. For some years he systematically exposed large numbers of dyed test-strips of textile to the action of sunlight and was thus enabled gradually to build up a selection of those displaying outstanding fastness. Morton found that these belonged to a class known as vat dyes, of which indigo is an example; indanthrene blue, which appeared on the market in 1901, is another. Morton's discovery established the first relationship between molecular structure and the property of fastness: it pointed the way to further research. A vat dye is capable of existing in two states—a reduced, soluble leuco (" white ") form, and an oxidized insoluble form. In the dyeing process the reduced dye is taken up by the textile from the solution in the vat; subsequent exposure to the air results in the insoluble colour being re-formed and firmly fixed in the fibres. Using his selected fast dyes, Morton, in 1906, produced dyed textiles which he sold under a *Sundour* guarantee against their fading. At that period Morton was obtaining his dyes, like most other British dyers, from Germany. At the outbreak of World War I, this source of supply having ceased, Morton founded Solway Dyes Ltd, to produce dyes of high light-fastness. By early 1915, Morton was producing his first batches of blue and yellow vat dyes at his Carlisle factory. In 1919 he established a new concern, Scottish Dyes Ltd, at Grangemouth to extend his dye-making enterprise.

In 1910–20 Davies, Fraser-Thomson, and Thomas, research chemists employed by Scottish Dyes Ltd, while investigating compounds containing the anthraquinone ring-system, discovered that one of these (12,12′-dimethoxydibenzanthrone) was a green dye of excellent brilliancy, fastness and dyeing properties. Morton announced the discovery at the first Annual Meeting of Scottish Dyes Ltd in 1921. This new vat dye was marketed as Caledon Jade Green—the first really fast green dyestuff.* Its discovery was followed by intensive investigation of analogous compounds based on the anthraquinone structure; thus followed from a series of comparatively simple variations such fast dyes as Caledon Dark Blue, Caledon Ming Blue, and the Caledon Greens, all of which possess high fastness to light and other influences. In addition to the financial results of the development of Caledon Jade Green, its discovery had considerable effect in raising the prestige of the U.K. dye industry at a period still overshadowed by the persisting tradition of German pre-1914 supremacy in chemical industry.

In 1924 a further advance in the field of vat dyes was made by Scottish Dyes chemists, who developed a process for production of solubilized dyes of that type by converting them to copper, zinc, and iron derivatives, thus originating the Soledon group of dyes. Soledon dyes were the first of the vat type that could be applied to animal, as opposed to cellulosic fibres, without the damage that would occur if these were exposed to the strongly alkaline vat normally used in dyeing. When, in 1925, Scottish Dyes Ltd was merged with British Dyestuffs Corporation the technical strength of the post-1920 U.K. dyestuffs industry was significantly increased. Subsequently, in 1931–33, after the ICI merger, further advances were made by development of anthraquinone and azo dyes (Carbolan dyes), capable of dyeing wool, and which combine fastness to light and to washing.

* In 1912 BASF chemists obtained a green dye from 12,12′-dihydroxydibenzanthrone by condensation with *p*-toluidine, but found that it was not fast to light.

2. *Dyes for the Man-made Fibres*

The man-made fibres (cellulose acetate, nylon, "Terylene", etc.) have had an important influence both on the techniques of dyeing and on the development of dyestuffs. Many of the new fibres do not swell in water and a swelling agent has to be added to the dye bath to assist the dyeing process—a procedure which has the disadvantage that the swelling agent has to be removed at the end of the operation. When production of cellulose acetate was begun in the U.K. (p. 80) the difficulty in dyeing the fibre very considerably delayed its development. The acetate fibre displayed little or no affinity for dyes in established use for cotton dyeing; those dyes with which it could be coloured were too readily removed by water-washing. The development of procedures and dyes for dyeing acetate fibre was almost wholly a British achievement. Certain insoluble azo dyes, particularly those containing an amino-($.NH_2$) group, if used in a finely-divided state, were found to dye acetate fibre. The first dyes specifically developed for dyeing the new textile were produced in 1922 by British Dyestuffs Corporation; these were the Ionamine dyes, made from the insoluble aminoazo dyes by rendering them soluble by introduction of a methyl-sulphonic group. In the dye bath the Ionamine dyes slowly decompose, liberating the aminoazo compound in a form readily taken up by the fibre. Using these specialized dyes it was possible to dye acetate textiles yellow, orange and red; other colours were subsequently produced by modifications of the same molecular structural theme. In 1923 another type of acetate fibre dyes was simultaneously discovered by British Dyestuffs Corporation and British Celanese Company. These latter dyes were of the anthraquinone class and were used in association with dispersing agents in the dye bath, i.e., chemicals which maintain them in suspension in a state of very fine division. The use of anthraquinone dyes in this way greatly extended the colour range of cellulose diacetate textiles and gave dyeings of much greater fastness to light than could be obtained with dyes of the aminoazo type. With the development of an adequate series of dyes, commercial development of cellulose diacetate textiles advanced rapidly. More recently, it was found that " Terylene " fibre is also capable of absorbing dispersed anthraquinone dyes, but at a slow rate only. At temperatures above 100°C dyeing by that method becomes practicable. Techniques for dyeing nylon and polyacrylonitrile fibres with dispersed anthraquinone dyes were also pioneered in the U.K.; indeed, these developments constituted a major contribution to the rise of the fundamentally new field of man-made materials.

3. *A New Chromophore*

In 1928, in the course of the manufacture of phthalimide from phthalic anhydride and ammonia at the works of Scottish Dyes, then merged with ICI, a bluish discoloration of the product was observed. This unexpected development of colour was investigated and found to be due to traces of an intensely blue substance. A year before this accidental discovery, Professor Henri de Diesbach of Fribourg had described an organic compound of copper which had a strong blue colour and a remarkable resistance to chemical attack. The nature of the chromophore in Diesbach's blue and that occurring in the phthalimide process at Grangemouth was eventually elucidated by R. P. Linstead and his co-workers in 1934 at the Imperial College of Science and Technology. The blue impurity in the phthalimide was shown to be an iron-containing derivative of an organic compound having a 16-membered ring of alternate carbon and nitrogen atoms; Diesbach's blue was the copper derivative of the same organic compound. At Grangemouth, the iron was derived from the vessels in which the phthalimide process was carried out. Linstead noted the structural resemblance between the metal derivatives of the new compound and the red haemin pigment of blood, which also contains a 16-membered carbon–nitrogen ring encircling a central metal (iron) atom. The name phthalocyanine was given to the new chromophoric compound, which was found to form derivatives with sodium, magnesium, aluminium, lead and nickel, as well as with iron and copper, resulting in production of a range of colours. Lead phthalocyanine is green and the metal-free compound has a greener blue shade than copper phthalocyanine. Various processes were developed by ICI for the manufacture of phthalocyanines and, in 1934, Monastral Fast Blue B.S., the copper derivative, was put on the market.* Other commercial phthalocyanine pigments followed, e.g., Monastral Fast Green G.S. and Durazol Fast Paper Blue 10G.S.

It is to be noted that the phthalocyanines are *pigments*, not dyes, although in highly dispersed form they can be used to impart an extremely fast colour to textile fibres. The extreme fastness and good colour of the phthalocyanines made it manifestly desirable to attempt their conversion to forms suitable for dyeing textiles. Resort to the usual expedient of introducing sulphonic groups to render them soluble resulted in dyes of low fastness to washing. In 1947, Haddock and Wood of ICI Dyestuffs Division found that ammonium derivatives of copper phthalocyanine had a high affinity for cellulose. The first dye produced in that way from copper phthalocyanine was Alcian Blue 8G.S., marketed by ICI. Production of dyes from the phthalocyanines finally justifies extension of the term " dyes " to include organic pigments.

4. *Reactive Dyes*

In 1895, C. F. Cross and E. J. Bevan, whose names are permanently associated with the introduction of cellulose into the field of chemical industry (p. 79), attempted to link cotton fibre (in the form of soda cellulose) with a substituted benzene ring which they subsequently diazotized. Their object was to produce, by this integration of a chromophore with the cellulose, *a self-coloured fibre*, which might be expected to maintain its colour under most of the conditions to which textiles are normally subjected.

* In the same year IG Farbenindustrie marketed the independently discovered copper phthalocyanine as Heliogen Blue B.

Cross and Bevan did succeed in producing a coloured cellulose derivative; in the process, however, the fibre was so modified as to be of no practical use. Other workers followed this initiative, but with little more success. In 1930–33 in Basle, CIBA investigated the esterification of cotton with 2,4,6-trichloro-1,3,5-triazine, one of the chlorine atoms of which was subsequently replaced by a coloured molecular group.

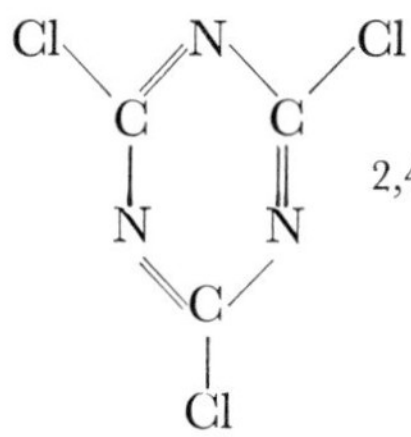

2,4,6-trichloro-1,3,5-triazine

Too severe degradation of the cellulose again occurred, because of the use of strong caustic soda solution to produce the initial soda cellulose with which the triazine was reacted. In the early 1950's, ICI Dyestuffs Division attempted a similar cellulose esterification, using the closely related 2,4-dichloro-1,3,5-triazine, but under very mild alkaline conditions, so that no damage to the cellulose molecule occurred. It was found that, although the extent of reaction was limited, it was sufficient to give a satisfactory degree of colour. As expected, the dyeing was extremely fast. Following this successful outcome, ICI marketed the first commercial series of reactive dyes for cellulose in 1956, under the trade name Procion. This latest and revolutionary discovery within the U.K. dyestuffs industry has been followed by intense research activity on the same theme abroad. A year after the appearance of Procion dyes, CIBA put its similar Cibacron dyes on the market.

BIOLOGICALLY-ACTIVE CHEMICALS

Biologically-active chemicals constitute a wide range of chemically different substances which modify the behaviour of living organisms in a desired manner by participating in their biochemistry; thus they are to be differentiated from crop fertilizers with a purely nutrient function towards normally-growing plants, or grossly destructive agents, such as sulphuric acid, which destroy plant tissue. In the present century, with the increase in knowledge of organic and bio-chemistry, an entirely new branch of chemical industry has arisen, concerned with the synthesis as well as the isolation from natural processes of bio-functional chemicals, almost all of which were unknown some forty years ago. For the purpose of the present review the biologically active chemicals may be conveniently classified as synthetic pharmaceuticals, antibiotics, hormones, new anaesthetics, vitamins, and pest- and weed-control agents.

In many instances, particularly where a synthesis has been effected, whether or not that synthesis is the basis of the actual manufacturing process, the chemical structure of the bio-functional chemical is known. As in the case of the modern dyestuffs, the chemical formulae of the bio-functional chemicals have little or no meaning except to the specialist in the relevant organic chemistry, and a barrier of necessary neologisms, or trade names, stands between their manufacturers and their users. The general medical practitioner, prescribing the often picturesquely-named new pharmaceuticals, usually has no more knowledge of their chemical nature than does the dyer of the synthetic colours he applies, although, of course, the doctor may be expected to be acquainted with the principles underlying their action. There are interesting historical parallels between the emergence of the modern U.K. dyestuffs and biologically-active chemical industries: in both there was an abrupt transition from a trade engaged in extraction of naturally-occurring substances to a scientifically-based industry producing hitherto unknown chemicals, and both emerged after a more or less prolonged prior development abroad, mainly in Germany. It has been estimated that over 70% of the prescriptions issued by the medical profession today could not have been written in 1935, the year in which the first sulphonamide drug was produced industrially. As will be shown, there was, too, a more than analogical relationship between modern dyestuffs and pharmaceuticals manufacture. Like the dyestuffs industry, the bio-functional chemicals industry in the U.K. was greatly benefited by the Safeguarding of Industries Act of 1921 and its key industry duties, which followed closely on the Dyestuffs (Import Regulation) Act.

With the partial exception of pest- and weed-control agents, bio-functional chemicals in the U.K. are the products of what is broadly described as the pharmaceutical industry. At the present time there are between 400 and 500 manufacturing establishments, with a notable concentration in the greater London region; important pharmaceutical factories are located at Edinburgh, Manchester, Macclesfield, Liverpool and Nottingham. The great majority of the firms are small and specialize along a few lines. In 1954 eight firms were producing about one-third of the total U.K. output of pharmaceuticals. Most of the larger firms have research departments. Expenditure in the U.K. on pharmaceutical research increased from £2·8 million in 1953 to £6·25 million in 1959 (this includes research on antibiotics). In most instances the manufacturing firm not only produces the pharmaceutical chemicals but pellets, encapsulates, packages, or formulates them in prescribable or applicable forms for disposal through its wholesaling organization. The industry is one of the largest consumers of glass bottles: ten years ago it took about one-quarter of the total output of the U.K. glass-container industry; in more recent years there has been an increasing use of plastics in the packaging of pharmaceuticals. All manufacturers of pharmaceuticals must be licensed and the quality, strength and labelling of their products are regulated by Government enactments. Governmental control has been greatly extended since the recent unfortunate experiences of the side-effects of thalidomide. All new pharmaceuticals must be submitted for investigation and approval by a state-appointed body (the Dunlop Committee) before they can be put on general issue to the medical profession. The industry has also been much in the public eye in connection with the costs of its products to the National Health Service and the whole position is now under investigation by another Government committee.

The pharmaceutical industry, more than any other branch of British chemical manufacture, has been deeply penetrated by foreign investment—mostly American. In 1891, Parke, Davis & Company was the first foreign pharmaceutical concern to establish itself in this country. By 1960, of sixty-four U.K. pharmaceutical concerns making ethical (i.e., prescription) drugs no fewer than seventeen were subsidiaries of U.S. parent concerns. A recent estimate is that over half of the U.K. prescription drug market is supplied by these American-controlled companies. Many of these American subsidiaries manufacture their starting chemicals in the U.K. Four Swiss pharmaceutical companies have important establishments in this country. An important aspect of the presence of these foreign-based enterprises is that they import into this country the results of research carried out abroad; indeed the need to have the results of continual research has probably been an important factor in the formation of mergers. Important as the foreign component in the modern British pharmaceutical industry is, it hardly justifies the assertion, made by a well-known journalist in an important newspaper,* that " Since the war the British pharmaceutical industry has been virtually taken over; internationalized ".

Between 1935 and 1948 the annual value of pharmaceuticals produced in the U.K. increased from £19 million to £73 million. In the latter year 49,000 persons were employed in the industry. By 1959 the value of the output had increased to £160 million and the number of workers to 66,500. The industry is an important exporter: between 1913 and 1964 the value of its exported

* Brian Inglis in *The Sunday Times*, 28th March 1965, p. 21.

products increased from £2·3 million to £59·2 million. In 1960, one-quarter of all U.K. pharmaceuticals was exported. Statistics of actual quantities are not published, but indications have been given that outputs of, for example, sulphonamide drugs run to several hundreds of tons a year.

1. SYNTHETIC PHARMACEUTICALS

The first steps towards modern chemotherapy were taken in Germany. Paul Ehrlich observed that certain dyes, injected into animals, stained only specific tissues, e.g., the nerve fibres. This observation ultimately gave him the idea that dyes might similarly stain and even kill the micro-organisms of disease. Ehrlich tested 500 different dyes on 2,000 mice inoculated with trypanosomes; the experiments were unsuccessful. His next trials were made with new compounds of arsenic, with similar chemical structures to known dyes: the first 605 such compounds were no more effective than the dyes, but the 606th compound killed the trypanosomes without killing the test animals. The new compound, salvarsan, made in 1910, which came to be known as " 606 ", had an —As=As— group in its structure, analogous to the —N=N— chromophore in the diazo dyes: it was, in effect, a lethal dye—to the micro-organisms, but not to their host; its effectiveness originated in the fact that it had an affinity for the proteinaceous matter of the bacteria, similar to that of a dye for the fibres of a wool textile. The importance of this discovery was immediately established by the fact that it was effective against the spirochetes of syphilis. Ehrlich's achievement destroyed the foundation of the traditional belief held by the medical profession that once disease germs entered the body they were beyond the reach of chemicals; it also marked the beginning of the new science of chemotherapy.

In Germany and America the way shown by Ehrlich was followed during the next two decades. Heidelberger and Jacobs, in 1919, claimed that certain azo derivatives of sulphanilamide displayed antibacterial action, but because of the weakness of the effect *in vitro*, they did not follow this discovery further. An obscure German journal, in 1933, published an account of the curing of a case of septicaemia by means of a sulphonamide derivative. In 1935, Domagk, an employee of Bayer, in Germany, found that the red azo dye, prontosil, was extremely effective against the streptococci which cause puerpural and scarlet fevers. Prontosil has a chemical structure analogous in general " shape " with that of salvarsan. The puzzling observation was made that *outside* the living body prontosil was ineffective against bacteria. A French chemist, Fourneau, suggested that, in the body, the compound was broken down to liberate sulphanilamide, and that this was the effective therapeutic agent,

$$H_2N\langle\bigcirc\rangle SO_2NH_2 \qquad \text{(sulphanilamide)}$$

Fourneau's view was found, in 1936, to be correct, and interest became focussed on the pharmaceutical potentialities of this comparatively simple organic compound. At this time developments began in Britain.

In 1937–38, a small research team, formed by May & Baker Ltd and led by Dr. A. J. Ewins, synthesized a series of sulphanilamide derivatives. The twelfth compound they produced was found to be effective against pneumonia and was patented (B.P. 512 145) in the names of Dr. Ewins and Dr. M. A. Phillips, the latter a member of the research team. Clinical tests of the new pharmaceutical had been carried out at the Middlesex Hospital by Professor (Sir) Lionel Whitby. When first produced the compound was given the firm's appropriate serial identification number, namely 693; thereafter, it became generally known as " M & B 693 ", or, simply, " M & B ". During the six years that followed, a further fifty compounds of similar type were synthezised. A few showed therapeutic action but none was as effective as 693, which, incidentally, was used successfully in the case of Winston Churchill when he contracted pneumonia in North Africa in 1943. The new pharmaceuticals, generally known as the sulphonamide drugs, were from that period in established production by several firms in the U.K. Chemotherapy was defined by the Medical Research Council in its 1936–37 Report " as a form of medical treatment which consists in the administration of chemical compounds synthesized in the laboratory (not yet in the factory) and found to have specific reactions on the infective organisms causing particular diseases in man and animals ". In the early 1930's the Dyestuffs Division of ICI in Manchester was already producing a number of synthetic drugs. Acriflavine, a bright yellow dye, was being made by British Dyes Ltd in 1916 in large quantities for use in disinfecting wounds at the war fronts. ICI in 1936 entered the field of pharmaceutical research and production proper with the formation of its Medicinal Chemicals Section within its Dyestuffs Division Research Department. The original research team consisted of eight chemists, and the general project set before them was that of seeking new and more effective synthetic pharmaceuticals. This development had the significance that it brought to the new British pharmaceutical industry the reservoir of synthetic experience in organic chemistry gained by the most important post-World War I dye firm, and with it the financial resources to sustain an intense and protracted programme of research. By 1942 the activities of the Medicinal Chemicals Section had so greatly extended that a separate company, Imperial Chemicals (Pharmaceuticals) Ltd, was formed; in 1957 this became ICI Pharmaceuticals Division, with a research department of 500 workers at Alderley Park and a factory at Macclesfield. The Dyestuffs Division continues (1965) to manufacture some of the firm's long-established pharmaceutical chemicals. During the period 1936–61 ICI made and tested some 45,000 organic compounds and discovered thereby sixteen new human and veterinary drugs, including sulphamezathine, discovered in 1942, now the most generally employed all-purpose sulphonamide drug in the U.K.

Britain, as a former imperial power and, later, as the centre of a world-wide Commonwealth, has a long history of interest and trade in anti-malarial drugs. When, in the course of

World War II, access was lost to the quinine plantations, resort was had to two synthetic antimalarials, earlier developed in Germany; these were brought into production in the U.K. in 1940. These drugs were Atebrin (Mepacrine) and Plasmoquin, effective respectively against the asexual and sexual forms of the malaria parasite. Both drugs were associated with undesirable side-effects and disadvantages. In 1942, ICI began an investigation to discover an alternative to the German antimalarials, by then in wide use by the Allied armies. A study was made of the chemical structures of known antimalarials; from this it appeared that the desired bio-function was associated with a chlorinated benzene ring connected to a side-chain containing a basic group. In the early syntheses the structural components of a sulphonamide drug and of Mepacrine were used. A large number of the compounds made exhibited some effect on a " laboratory " form of malaria. In all, some 1700 compounds were synthesized and tested before the outstanding antimalarial action of proguanil, later named Paludrine, was discovered in 1946. A few years later, in part following the guidance given by the structure of Paludrine, the Burroughs Wellcome group developed the antimalarial pyrimethamine. Hibitane, an antibacterial agent, discovered by ICI in 1954, has structural components similar to those of Paludrine, and has been said by F. L. Rose* to owe " its birth entirely to the chemistry of Paludrine ".

The post-World War II years have seen a great expansion of the manufacture of synthetic pharmaceuticals in the U.K.; numerous drugs have been developed and are in established production for treatment of leprosy, sleeping-sickness, leukaemia, hypertension, coronary disease and tuberculosis, and for use as antibacterial and antiseptic agents. Veterinary drugs may be exemplified by Antrycide for protecting livestock in Africa against the trypanosome disease carried by the tsetse fly; this pharmaceutical is annually produced to the extent of several hundreds of tons.

---

* " Twenty-five Years of Medicinal Chemistry ", Scheele Centenary Lecture, Stockholm, 9th December 1961.

2. ANTIBIOTICS

Pasteur was, perhaps, the first to suggest that micro-organisms might be made to war against one another to the possible benefit of mankind. In 1901 Emmerich and Low at the University of Munich observed the germ-killing action of a substance generated by the bacterium *Pseudomonas aeruginosa* and employed it in treating patients. Alexander Fleming, who carried on research in vaccine therapy at St. Mary's Hospital in London, discovered in 1922 a substance to which he gave the name " lysozyme ", in the human body; this had the property of killing certain bacteria without damaging cell tissue. It may reasonably be speculated that this discovery prepared Fleming's mind for the one he was to make six years later. In 1928, Fleming, now Professor of Bacteriology at the University of London, observed that, when a spore of the mould *Penicillium notatum* accidentally settled in a medium in which he was cultivating bacteria, the bacteria in the neighbourhood of the mould were destroyed. He immediately recognized the therapeutic potentiality of this discovery and ascertained the types of bacteria affected. Fleming continued to use the mould in his laboratory, but at that time his work had no wider outcome. The fact that the medical world took no account of his discovery was later attributed by him to the surviving belief that the micro-organisms of disease, once within the tissues of the body, were beyond the reach of therapeutic agents. When, in the mid-1930's, the development of the sulphonamide drugs destroyed this obstructive tradition, a more receptive climate supervened.

About 1939, following laboratory work by Professor Raistrick of the London College of Hygiene and Tropical Medicine, on the extraction of the active substance (penicillin) produced by the *Penicillium* mould, a research group at Oxford, led by (Lord) Howard Florey and Ernst Chain, embarked on investigation of the problem. While World War II increased the urgency of developing an extraction method, it increased the practical difficulties of the investigators. In the process originally devised the mould was cultivated on the *surface* of culture medium contained in large numbers of milk bottles. The firm of Glaxo designed special glass vessels for the process; other firms attempted to grow the mould in shallow trays, or to produce the penicillin liquor by trickling the sterile culture medium down a tower containing a packing inoculated with spores of the mould. Having developed methods of extracting the penicillin, concentrating and storing it, Florey, in 1941, crossed to the U.S. In America, in the North Regional Research Laboratory in Illinois, a group of workers evolved a process for cultivating the mould in *depth* in a medium of maize steep liquor and milk sugar (lactose). This conversion of the process from one in two dimensions to one in volume increased production rate about a hundredfold. Output had also been augmented by selecting a more productive strain of mould than *Penicillium notatum*. Evolution of the penicillin process called for the establishment in the factory of sterile zones—a novel condition, at that time, in chemical industry. Metallic ions in the culture medium were found to affect the penicillin, but—fortunately—stainless steel, by then well-established in the chemical industry, proved to be a suitable material of construction for the plant. By 1944–45 production of penicillin was on a tonnage basis in both the U.S. and U.K. In this country, the American deep-cultivation method was adopted from the outset of industrial manufacture. Glaxo was producing penicillin in this country before the end of the war and, in 1946, established a large deep-culture plant at Barnard Castle. Between 1943 and 1947, British production of penicillin increased from 300 to 400,000 million units a month. The University of Cambridge, in 1948, introduced a one-year course in micro-biology, primarily to train staff for the new departure in chemical technology.

From the end of World War II to 1959 new developments in the field of antibiotics took place mainly in America. Dr. Selman A. Waksman, who is credited with coining the term " antibiotics ", a soil bacteriologist, became interested in 1939 in the antibacterial action of micro-organisms in the soil.* By 1943 he had worked out a technique for testing antibiotics; in that year the antibiotic streptomycin, active against the tubercle bacillus, was extracted from the culture of a soil micro-organism. There followed the discovery of streptomycin an intensive investigation of soil samples from all over the world. Within a few years hundreds of antibiotics were isolated, but only about twenty have reached industrial production, e.g., aureomycin, Terramycin, chloromycetin. Manufacture of the later antibiotics has been introduced, mainly by U.S. firms' subsidiaries, into the U.K.

By 1946 the chemical structure of the penicillin molecule had been elucidated by the work of academic chemists in this country and in the U.S. Already limitations to the effectiveness of penicillin had begun to make their appearance. Strains of the bacterium *Staphylococcus aureus* were showing resistance to the antibiotic, because of their acquired ability to generate the enzyme penicillinase, which rapidly destroys penicillin. By adding a series of phenylacetic acid derivatives to the culture process it was found possible to produce penicillins of modified chemical structure. By 1950, several hundred different penicillins had been made in this way; these semi-synthetic variants, however, were not greatly different in their therapeutic activity from the original form of antibiotic and what was more important, showed no significant resistance to penicillinase. It had become apparent that some more drastic modification of the penicillin molecule was necessary.

In January 1959, the Beecham Research Laboratories announced† the isolation of 6-aminopenicillanic acid from the process culture medium. This research organization was founded in 1945 by the Beecham Group Ltd in order to give that firm a more effective entry into the post-war developments in the field of synthetic drugs. The Microbiological Unit was formed

* In 1921, Lieske had shown that soil organisms are capable of destroying some bacteria. Russian researchers in 1937 investigated the anti-bacterial action of the actinomycetes.

† *Nature*, **183**, 258 (1959).

in 1955 to research in antibiotics: their objective was to prepare antibiotics of the penicillin type by chemical modification of a master structure. This approach differed radically from the American one of searching the soils of the world for entirely new natural antibiotics. The discovery of 6-aminopenicillanic acid, the structural " core " of the penicillin molecule, was the master antibiotic sought. Although of itself this substance was without antibacterial action, new and powerfully effective penicillins were prepared from it by linking it with side-chains of various structure. By playing what has been called " molecular roulette " in this way, Beecham Laboratories have so far put on the market four new penicillins—Broxil, Penbritin, Orbenin and Celbenin—the latter two of which are not destroyed by staphylococcal penicillinase and are therefore effective against the bacteria generating it. Isolation and application of 6-aminopenicillanic acid constitute the single most important advance in antibiotic drugs since the development of the original penicillin discovery and of natural antibiotics produced by soil organisms, dating from almost two decades earlier.

3. HORMONES

Claude Bernard, the French experimental physiologist, a pioneer in the application of modern scientific method to the study of medicine, established about a century ago that certain important functions of the human body were regulated by what he called " centres of internal secretion ". These centres were recognized as the ductless glands and their active secretions became known as hormones (from a Greek word meaning " to urge on ").* Hormones may be defined as organic substances produced in minute amount in one part of the body and which, when transported by the blood stream to other parts, produce a profound effect on the physiological functions. Chemically, the hormones do not belong to one class of substance, but a number of them belong to the sterene group which has a molecular

* Animals and plants also generate hormonic catalysts appropriate to their vital functions.

skeleton considerably favoured by Nature in her synthetic operations. The British contribution to the field of hormone chemicals, although limited, has been notable.

Adrenaline, the simplest of the hormones, has the power of causing contraction of the blood vessels and is concerned in the transmission of nervous effects. This substance was the first of the hormonic compounds to be isolated. Originally it was extracted from the suprarenal glands of animals. It was first marketed by Parke, Davis & Company. The chemical structure was established in the 1920's. Adrenaline is now produced on a large scale synthetically.

The discovery and isolation of insulin, the hormonic substance generated in the Islets of Langerhans, a group of specialized cells in the pancreas, has the function of controlling sugar metabolism in the body; its absence due to injury or disease results in the development of diabetes. Insulin was first isolated in Canada in 1921 by F. G. Banting and C. H. Best in the laboratories of the University of Toronto; its spectacular effect on diabetic patients was demonstrated in January 1922. Manufacture of insulin in the U.K. was undertaken shortly afterwards, by the firm of Allen & Hanbury (now a subsidiary of Glaxo) in co-operation with The British Drug Houses Ltd. Insulin has a complicated protein structure, elucidated in 1954 by F. Sanger and co-workers in the Biochemistry Department at Cambridge. So far, a synthesis has not been achieved and it continues to be produced in large quantities from the pancreas of oxen, sheep and pigs by an eight-stage extraction process.

Thyroxine, the hormone generated by the thyroid gland, has the function of controlling the general metabolic rate; its absence results in dwarfism and cretinism. The structure and synthesis of thyroxine were worked out in the U.K. in 1926 by Harington and Barger.

The isolation of a hormonic substance, later to be known as cortisone, from the cortex of the suprarenal glands, was first effected in 1935, simultaneously in America and Switzerland. It was subsequently found that the adrenals are stimulated to produce cortisone by ACTH (adrenocorticotrophic hormone)* generated by the pituitary gland at the base of the brain. Discovery of the beneficial effect of cortisone on arthritis promoted interest in its production. The hormone occurs in too small quantity in the adrenal cortex of animals for it to be obtained from that source. In 1948 a partial synthesis was effected in America, the starting point being a constituent of ox bile; this process involved 36 steps, and 40 cattle were required to yield 100 milligrams of cortisone—enough to treat one patient for one day. It was estimated (in 1950) that one week's treatment with cortisone produced in that way would cost £1500! Search continued for new starting materials from which to produce cortisone by partial synthesis. In 1955, Glaxo in the U.K. produced the hormone from hecogenin, a substance obtained from the juices discarded in the manufacture of sisal fibre; the process comprised thirteen stages. By 1956 a practicable route had been established for the industrial production of cortisone.

The primary and secondary sexual functions are hormonically controlled. Male sexual characteristics are associated with testosterone, generated by the testes; the female hormone, oestradiol, is generated by the ovaries; both hormones occur in both sexes, their effect being determined by their quantitative presence. Like cortisone, the sex hormones are based structurally on the sterene molecular skeleton; generally they are referred to as the steroid hormones. Mammary cancer has been successfully treated by testosterone, and cancer of the prostate gland by stilboestrol. Stilboestrol is a female hormone produced entirely artificially by a route worked out by the Courtauld Institute of Biochemistry in collaboration with Sir Robert Robinson. The commercial production of stilboestrol was immediately undertaken by several firms, notably The British Drug Houses, which now, amongst other synthetic analogues of natural hormones, produces

* ACTH has been shown to have a molecule consisting of 39 amino-acids units linked together in a straight chain. It was first produced commercially in 1948 by Armour Pharmaceutical Company.

Secrosteron, a progestational agent.* From work on these hormone analogues The British Drug Houses went on to investigate other steroid compounds which might act as ovulation inhibitors ("birth pills") and discovered a number of hitherto unknown steroids, one of which proved to be powerfully effective in preventing ovulation.

By 1959, corticosteroid hormone products, cheapened by the discovery of new starting materials and improvements in manufacturing processes, represented an expenditure of some £6 million in National Health Service prescriptions.

4. NEW ANAESTHETICS

A total anaesthetic may be defined as a substance capable of producing a reversible state of unconsciousness in human beings and animals. For many years after their first use in the nineteenth century, and during the first three decades of the present century, chloroform, nitrous oxide, ether, and ethyl chloride dominated the field of surgical anaesthetics. Meanwhile, the fact that various chloroparaffins (e.g., methylene chloride), ethylene, various alcohols and ketones, cyclopropane, acetylene, and even the rare gases argon and xenon could induce anaesthesia had been discovered: none of these, however, had new properties justifying its displacement of the classical inhalants. With the development of the barbiturate drugs in the 1930's, a compound of that class, pentothal (the so-called "truth drug"), came into wide use as an injection anaesthetic. Thus, despite investigations showing that the property of producing narcosis was associated with many fairly simple chemicals, the practice of anaesthesia continued to be notably conservative. This conservatism is partly to be explained by the facts that the physiological mechanism of anaesthesia is not understood and that most anaesthetics are associated in their action with side-effects, which in the case of a new candidate may require long and costly

* I.e., a drug which prepares the reproductive organs for pregnancy.

experiment to assess. As a result of work in the U.K. two new general anaesthetics have come into use since 1940.

In 1934, ICI put on the market a pharmaceutical grade of trichloroethylene, a chemical which had been produced by that firm and its forerunners since 1908, and which has wide application in the engineering and other industries as a solvent (p. 232). This pharmaceutical trichloroethylene was first recommended for cleaning dirty wounds and burns. In 1941, under the name of Trilene, after extensive trials, trichloroethylene was marketed as a general anaesthetic; this very pure, thymol-stabilized* grade continues to be used in the U.K., particularly for childbirth, during which it may be self-administered to relieve the pains of labour. While trichloroethylene has established itself in this country, it has not been generally accepted in the U.S. The Council on Pharmacy and Chemistry of the American Medical Association, in 1936, commented that the evidence " does not justify the acceptance of trichloroethylene for use as a general anaesthetic ".

About 1950 the General Chemicals Division of ICI was considering possible outlets for organic fluorine derivatives for which it had developed new production techniques. It was known that certain simple fluoro-organics were very much less toxic than their chlorine analogues; they were also more resistant to oxidative breakdown on exposure to air. On the basis of a theory, developed mainly by J. Ferguson, on the mechanism of anaesthetic action, the General Chemicals Division, in collaboration with ICI Pharmaceutical Division, embarked upon a " target " research to discover a possible fluorine-containing anaesthetic which might be expected to have outstanding properties. In February 1952, in the laboratory at Widnes, the new compound 2-chloro-2-bromo-1,1,1-trifluoroethane, a volatile liquid boiling at 50°C, was synthesized; later, under the name Fluothane, this substance became the sought-for new anaesthetic. After investigations with test animals, Fluothane, in 1956, was given

* Against the oxidative action of the air.

clinical trials in Manchester Royal Infirmary and at Crumpsall Hospital. A report on these trials read: " After the first few cases it immediately became obvious we were dealing with a drug totally different from all other anaesthetics. The ease and smoothness of induction and maintenance, the excellent relaxation, the effortless respiration and perfect oxygenation . . . and the prompt recovery of consciousness were all incomparably better than with any other agents." After further trials Fluothane was put on the market in 1957 and, since that date, has been used in many parts of the world in between 20 and 30 million surgical and other operations.

5. VITAMINS AND FOOD CHEMICALS

The manufacture of foodstuffs remains today the only major matter-transforming industry lying almost wholly outside the ambit of the chemical industry. In relation to the food industries chemical manufacture is still restricted to the ancillary function of supplying additives and modifying agents (emulsifiers, colouring matters, flavours, etc.). The field of food chemistry, despite its obvious and paramount importance for mankind, has been the site of only one revolutionary discovery in the present century—the discovery of the role and nature of vitamins. Advances have been made in the use of chemical and physical techniques for the control of food manufacturing processes, e.g., in brewing and margarine production. The chemistry of milk, despite intensive study, has still to be elucidated in its fundamental aspects, and little work has been done on the chemistry of eggs. Although, during World War II, there were developments of dehydration processes for fruit and vegetables, canning processes have not progressed significantly except in their mechanics, since the investigations made towards the end of last century by the Massachusetts Institute of Technology. If tonnage manufacture of semi- or wholly-synthetic new foodstuffs were to be achieved by chemical or biochemical processes, then it may reasonably be supposed that a foodstuffs branch of the chemical industry would emerge, as, for example, the plastics industry has arisen on the basis of the chemical manufacture of entirely novel materials.

In 1912, Gowland Hopkins found that rats, fed on theoretically sufficient quantities of *pure* foods, rapidly showed signs of inadequate nourishment, symptoms which could be removed by subsequently feeding the animals with fresh milk. Later it was established that the remarkable effects produced by the milk were due to traces of a substance now known as vitamin A. In the years that followed, a whole range of hitherto unsuspected dietary necessary chemicals—vitamins—was found to be present in trace amounts in the common articles of food. By 1935, the existence of eight vitamins had been recognized and concentrated extracts of them prepared; the chemical identity of several of them had been established. The vitamins, which belong to widely differing classes of chemicals, have all, so far as they are known, now been isolated in a pure state and most of them have been made synthetically; nicotinic acid (vitamin B4) was actually synthesized before 1900, when its vitaminic function was unknown. Vitamins are now produced—some on a tonnage scale*—as pharmaceuticals to prevent or treat the disorders occasioned by their absence, such as photophobia, pellagra, certain forms of anaemia, scurvy, infertility.

The first vitamin preparation produced commercially in the U.K. was Ostelin, marketed by Glaxo in 1924. In 1948 the anti-pernicious anaemia factor (vitamin B12) was isolated and a product containing it, Cytamen, was manufactured by Glaxo at their Barnard Castle factory, using a deep fermentation process. Vitamin products are also made in the U.K. by Cyanamid of Great Britain Ltd, and the former Distillers Company (Biochemicals) Ltd (now a subsidiary of the American firm of Eli Lilly). A Unilever subsidiary, Advita Ltd, formed in 1946, specializes in production of synthetic vitamins for the margarine

* U.K. output of vitamin C is of the order of 4000 tons a year. It is produced by a 6-stage synthesis starting from glucose.

industry and as additions to animal feeding stuffs; these vitamins are processed at their factory at Silvertown, London, where a laboratory for vitamin control and stabilization is also maintained. Statistics of vitamin production in the U.K. are lacking, but the scale and importance of the manufacture is indicated by exports of vitamins from this country, which in the period 1956–60 had an average annual value of £2·8 million.

6. PEST- AND WEED-CONTROL AGENTS

At the present time some 150 bio-active chemicals for agricultural applications are in production in the U.K. Despite the rapid developments in the field of pest- and weed-control chemicals during the past twenty-five years, the estimated annual loss of crops, by the action of pests, weeds and disease, in this country is estimated at about £140 million. The modern pest-control industry had its origins in the discovery, in 1851, by Grison, a gardener at the Palace of Versailles, that a preparation of sulphur and lime was effective against vine mildew. Lime sulphur (also known as pentasulphide of calcium, or in suspension, as *eau Grison*) found some application in this country in the early 1850's, but the difficulty of its preparation soon discouraged its use. It is of interest that lime sulphur was revived as a pest-control agent in the U.K. in the early decades of the present century by the firms of Lewis Berger, Walter Voss, and the Yalding Manufacturing Company. This revival was an incident in the great increase in the U.K. of interest in pest control by the growers of high-value crops at that time. Berk & Company began marketing mercuric chloride for the destruction of earthworms and, in 1919, this firm introduced the same salt as a seed-dressing. In the early 1900's the long-known extract of quassia* and extracts of derris root (later known to contain the insecticide, rotenone) were first commercialized by the firm of McDougall Brothers. Professor

* Quassia, the West Indian shrub from which the extract was obtained, was the subject of a dissertation by Linnaeus in 1761. It was known in the eighteenth century that cabinets of quassia wood were immune from attack by insects.

William McDougall, while on a visit to the Governor of Sarawak, had seen derris employed by the natives for stupefying fish, facilitating their capture but not affecting their use as food. During the first twenty years of this century derris became of major importance as an insecticide in the U.K. Between 1920 and 1940 a number of pest-control firms were established and the trade became highly competitive. An export trade was developed and exports of liquid derris to America introduced the insecticide to that country. The leading firm at that period was Cooper, McDougall and Robertson Ltd. Introduction of copper sulphate as an agricultural chemical by one of the forerunners of that firm (Cooper) has already been noted (p. 56). Before 1930 the range of pest-control agents produced in the U.K. included, besides copper salts, fungicidal copper dusts, derris dust, liquid pyrethrum,* and tar oil. In 1922 the technique of applying pest-control chemicals was advanced by the introduction of crop spraying from aircraft.

In 1931, ICI entered the pest-control industry, adapting German-mercurial seed-dressings to agricultural conditions in the U.K., and developed at this time a fine copper dust for crop spraying. Collaborating with the Shirley Cotton Institute, ICI introduced salicylanilide as a fungicide. The firm of Boots, in 1931, began manufacture of a finely-divided copper oxychloride in aqueous suspension as a substitute for Bordeaux Mixture. In 1937, Plant Protection Limited was formed jointly by ICI and Cooper, McDougall & Robertson Ltd, thus combining the chemical manufacturing resources of the former with the agricultural experience and marketing facilities of the latter. In the mid-1930's Dr. Ripper of Vienna migrated to this country and brought with him the specialized technique of using hot nicotine against aphid on strawberries; in 1939, Dr. Ripper formed Pest Control Ltd, a concern acquired by Fisons Ltd in 1954. As a subsidiary of Fisons, Pest Control Ltd has established, at its

---

* Produced from the flowers of the chrysanthemum, cinerariaefolium cultivated in Dalmatia, Japan, and Africa. In 1959 the African crop of pyrethrum flowers was 7000 tons.

factory at Harston in Cambridgeshire, the largest hormone weed-killer (MCPA, see p. 186) plant in Western Europe; amongst its other manufactures are systemic insecticides, and selective weed-control agents for pea, broad-leafed, and cereal crops.

The first important discovery in the field of synthetic insecticides was made by the Swiss firm of Geigy in 1939. For many years Geigy had produced the moth-proofing agent Mitin FF; in the 1930's the firm decided to search for agents effective against a wide range of insects. After investigation of the natural insecticides pyrethrum and derris, which were discovered to lose their effect under the influence of light, the properties of a number of simple chloro-organic derivatives were studied; one of those, a compound known for over sixty years, dichloro-diphenyl-trichloroethane (DDT) was found to have a powerful and persistent insecticidal action. Geigy, in 1942, informed the British Legation in Switzerland of their discovery—a circumstance of great importance, since the Japanese invasion of Malaya had cut the Western Allies off from supplies of derris. DDT manufacture was immediately begun in the U.K. and other allied countries; thus, during World War II, the armies of the contending powers were protected from pest-borne disease by an insecticide discovered in a neutral country.

The discovery of DDT as an insecticide may have stimulated investigation of the insecticidal potentialities of other fairly simple chloro-organic compounds. In the course of such a search in 1941–42, ICI discovered the outstanding potency of crude benzene hexachloride (BHC) against grain weevils and locusts. Shortly afterwards this insecticidal activity was found to be associated with the gamma-isomer of the hexachloride. Processes were thereafter developed by ICI and other firms for extracting the gamma-isomer in high concentration. Although gamma-BHC formulations are still produced in the U.K., neither the crude BHC nor the gamma concentrate are now made in this country. Germany is the chief European producer. DDT remains the most widely used insecticide for public health and crop protection. Numerous insecticides are now available;

many of these new chemicals are specific rather than general in their activity.

In the 1930's it was established that the growth of plants is controlled by hormone substances (auxins) generated in the growing tip. This discovery was followed by attempts in many laboratories, including those of ICI, to produce artificial plant hormones. ICI workers at Jealotts Hill in 1936 found that the comparatively simple organic compound, α-naphthyl-acetic acid was effective as a growth promoter, but that, when the dosage was increased, some plants were destroyed by its action. It was observed that charlock, plantain and other weeds could be destroyed without damaging the wheat crop in which they had been growing. Search for other substances that could exert this selective action was continued. In collaboration with the Government-sponsored Agricultural Research Council the selective weed-killing action of 4-chloro-2-methyl-phenoxy-acetic acid (MCPA) and 2,4-dichloro-phenoxy-acetic acid (2,4-D) was discovered. Processes for the manufacture of these substances on a large scale were developed in 1940–45 at ICI's laboratory in Widnes. Although many weed-killers of the hormone and other types have been developed since the mid-1940's, the phenoxy-acetic acid weed-killers, now made in large tonnages in this and other countries, continue to be amongst the most important products of this kind.

Plant Protection Ltd, in 1958–59, discovered the powerful weed-destroying action of derivatives of bipyridyl. Since this discovery was made the ICI laboratory at Widnes has investigated several process variants for production of bipyridyl, the structural core of the molecule of the new herbicide (Paraquat).* Commercial production began in 1961; an improved and computer-controlled plant is now in operation. It is believed that the powerful action of Paraquat is brought about by liberation of lethal free radicals within the plant cells, and that these active molecular fragments interfere with the normal processes of

---

* Marketed by ICI (Plant Protection Ltd) as Reglone and Gramoxone.

photosynthesis. Bipyridyl herbicides are not persistent and are neutralized by contact with the soil; it is therefore considered that they could be used for eliminating grass, crop aftermaths and weeds from land before re-sowing, thus obviating the need for ploughing. If this expectation were realized, it would constitute the most important impact of the chemical industry on agriculture since the introduction of synthetic fertilizers.

NEW MATERIALS

The development in the nineteenth century of semi-synthetic new materials from cellulose and proteins has been reviewed (pp. 76 ff). In the first decade of the present century began a further development—the production of wholly-synthetic new materials—which was to prove the most revolutionary event in technological history since the introduction of the textile machinery inventions in the eighteenth century: more particularly, in chemical technology no other emergence approaches comparable importance. Most of the new materials are characterized by their ability to be moulded or otherwise shaped into solid artefacts; some, however, are fluids or semi-liquids used as adhesives, lacquers, polishes, etc., and to these the term " plastics " applies very loosely or not at all; the term " plastics industry " has wide use in both popular and specialist literature. Some technical writers exclude man-made fibres and synthetic rubbers from the class of plastics—a misleading exclusion, since some plastics, such as Nylon, are used both for moulding artefacts and for production of fibres. The new materials industry is a highly-specialized branch of organic chemical manufacture; its products are not " chemicals " in either the academic or technical sense of that term, since they are not reactive in their function but, in fact, their very inertness, is one of their most essential properties. As will be shown, they are not *pure* compounds. Several general aspects of the new materials development fall to be discussed before the particular British developments are described.

Perhaps the most important historical fact about the production of the wholly-synthetic new materials is that they are *new*, that they are novel creations, structurally resembling to a limited extent some of the productions of nature (e.g., cellulose and protein), but as much strangers in the universe as the anthraquinone vat dyes. In their physical characteristics they display combinations of properties that differentiate them from the traditional materials—metals, wood, glass, ceramics—which they have in part displaced. The new chemically-derived materials have emerged opportunely to meet the modern demands for mass-produced articles and the needs of technology for materials with pre-selected novel combinations of physical characteristics. From the techno-historical point of view the most significant circumstance is that they are the results of a series of *process* discoveries, made increasingly in the light of an evolved theoretical understanding. The new materials have been synthesized from a series of comparatively simple organic chemicals, e.g., formaldehyde, phenol, urea, vinyl chloride, ethylene, etc. Many of the "building block" (primary) chemicals have been known for many years; thus formaldehyde was discovered by A. W. Hofmann in 1868, phenol was observed in coal tar in 1834, urea was synthesized in 1828, vinyl chloride in 1838, and ethylene was recognized as long ago as 1797. As will be shown, formation of solid complexes by some of these chemicals had already been observed in the nineteenth century, but the reactions by which this took place were neither understood nor could they be controlled: formation of polyvinyl chloride was recorded about a century before it became the twentieth century plastic, PVC. The long hiatus between these early qualitative discoveries and the emergence of modern plastics technology is to be explained largely by absence of a theoretical basis for their investigation. The notion of creating by chemical operations entirely new technically useful substances to function like those prepared from natural ones, and even to displace these, was itself novel. A moment of prophetic illumination had come to Robert Hooke who wrote in 1664: " I have often thought that probably there might be a way found out to make *an artificial glutinous composition* much resembling that excrement out of which the silkworm wire-draws his clew ". There can be little doubt that the development of semi-synthetic materials, particularly from cellulose, with what Cross and Bevan saw as the possibility of " endless varieties of form and substance ", alerted chemists to the potentialities of the discoloured gums and solids they had obtained by reacting simple organic compounds with one another.

If rubber and celluloid (p. 78) were the pathfinders of the modern plastics branch of the chemical industry, Leo Hendrik Baekeland's discovery in America in 1906–7, of a process for producing technically useful resins by reaction of phenol with formaldehyde, dated its effective foundation. During the thirty years or so before Baekeland's discovery, chemists in Europe had made and described viscous and hard masses produced by interaction of phenol and formaldehyde and even proposed technical applications for them, without, however, establishing them industrially. By use of specified temperatures and application of pressure, Baekeland was able to control the hardening of his resins and thus to make them amenable to moulding. His phenol-formaldehyde (PF) resins were the first modern plastics; they belonged to the so-called thermo-setting class, i.e., once heated in the moulding process, they cannot be thereafter softened or melted by application of heat. Baekeland's discovery was not made against the background of any fundamental understanding of the chemical structure of the products; it did, however, set the direction of plastics development for almost two decades.

Phenol-formaldehyde resins made their appearance at an opportune time: the electrical and motor car industries were creating a new demand for insulating materials, and in the 1920's the wireless industry extended the demand for insulants. Gramophone records were the first important PF resin mouldings. By 1925, modified PF resins were established in the paint trade. The dark colour of PF resins stimulated search for colourless compositions and led to the development of urea-formaldehyde plastics—the erroneously named amino-plastics—in the late 1920's. It was at that stage that British chemical industry made its first significant entry into the new materials field (see Section 2 below). An important feature of the urea-formaldehyde resins was that they brought colour into plastics, since brightly-coloured pigments could be added to the colourless moulding mixtures. The discovery, by a small group of researchers employed by Tootal, Broadhurst Lee & Company, in 1926, of a method of imparting non-creasing properties to cotton fabrics by formation of urea-formaldehyde condensates *in situ* in the fibres established the first link between the textile and plastics industries. During the years between World Wars I and II, the new materials industry rose to a recognized place in the technology of the industrially advanced countries, including the U.K. At this time the new materials in common use were the semi-synthetics, celluloid, cellulose acetate and casein, alongside the wholly-synthetics, which were still limited to the phenol- and urea-formaldehyde plastics.

Two men, Hermann Staudinger, in Germany, and Wallace Hume Carothers, in America, laid the theoretical foundations of plastics chemistry. Much of Staudinger's work in the 1920's was based on his study of the production of solid materials from styrene. Staudinger concluded that the solid complexes, which could have molecular weights of the order of 20,000, were formed by the end-to-end linking of individual styrene molecules into long chains (now known as the process of polymerization). The materials (plastics) consist of *mixtures* of these giant molecules of chain-lengths approximating to an average, characteristic of the particular polymer and determined by the conditions under which it was made. To these giant linear molecules he gave the name macromolecules; he also proposed a mechanism for their formation, a view which all subsequent experimental results have increasingly supported. Staudinger's work provided an explanation of the difference between thermo-setting and thermo-plastic polymers, namely, that in the former between the molecular chains there are cross-links that resist the action of heat and maintain the material in the solid state.

Carothers, an employee of the U.S. firm of Du Pont, began his study of high-polymer chemistry in 1928, declaring his objective to be the synthesis of " giant molecules of known structure by strictly rational methods ". His principal line of approach was to investigate the condensation with one another of small molecules carrying chemically-reactive groups (e.g., —COOH, $—NH_2$, —OH). " Condensation polymerization, " wrote Carothers, " merely involves the use in a multiple fashion of the typical reactions of common functional groups. " He explored the field of condensation polymers extensively and, in particular, investigated their ability to form fine filaments. Carothers' work consolidated and amplified the advances in knowledge of macromolecular chemistry made by Staudinger. From the mid-1930's the empirical element in plastics manufacture was greatly diminished and there developed in dialectical relationship to one another a theoretical science of high polymers and an applied science of new materials. More recently, accumulating knowledge of the physics of the solid state of matter has had a significant impact on research on the behaviour and application of plastics. One of Carothers' polymers was produced by condensation of hexamethylene diamine and adipic acid: it was later to become known as Nylon 66, the prototype of all subsequent condensation polymers, and the material of the first wholly man-made fibre.

On the eve of World War II, four new materials, polyethylene, Nylon, polyvinyl chloride (PVC) and polystyrene, the history of which in the U.K. is dealt with in the sections following this introduction, were already being made; during the war years these became large-tonnage plastics. The war also greatly accelerated in countries other than Britain development of synthetic rubbers, silicones, polyurethanes, and epoxy resins, all of which became established in the U.K. in the decade following the war. In the U.S. the existence of a vast petroleum industry and the related development of the automobile greatly stimulated interest in synthetic rubbers, and in Germany the National Socialist drive to self-sufficiency (*Autarkie*) had a like effect not only in respect to artificial rubbers but to synthetic materials in general. Neither of these factors was operative in pre-war Britain. Absence of indigenous calcium carbide manufacture undoubtedly discouraged interest by chemical firms in the U.K., prior to World War II, in the potentialities of new materials based on acetylene. Urgent production demands almost wholly inhibited exploratory research in Britain in the plastics field during the war. The outcome of these historical circumstances was that, although some of the most important later developments have originated in the U.K., the science of high polymers, and its application to the manufacture of new materials, developed in the main in Germany and America. Production of many of the important types of plastic began five to ten years later in Britain than in those two countries.

The rise of the plastics industry in the U.K. became, from about 1950 onwards, part of another revolutionary change in the chemical industry—the shift from coal to petroleum as the source of organic raw materials. Already in the 1930's it was becoming evident that reliance on phenol from coal tar would ultimately set limits to the expansion of manufacture of phenol-formaldehyde plastics. Urea and formaldehyde, the other two important plastics raw materials at that time, were also derived from coal *via* coke. In 1939 the sole British source of ethylene—required for the manufacture of polythene—was alcohol produced by fermentation of molasses. Traditional sources could not have supplied, either in the necessary quantities or at low enough cost, the raw material needs of the U.K. later post-war plastics iudustry. If some circumstances of World War II delayed development of the British plastics industry by a few years, others ultimately provided it with the raw material basis for its later expansion.

From 1925 onwards, as a result of a series of process developments, made mainly in the U.S., petroleum had become an important source of a number of primary organic chemicals.

During the war, U.K. firms began to employ petroleum as a source of chemical raw materials; indeed, Britain was the first European country to produce and employ petrochemicals. In the immediately post-war years several of the large petroleum companies, which formerly had their refineries at the overseas oilfields, established refining installations in the U.K. This change in the geography of the petroleum industry was rapidly reflected by changes in organization of chemical manufacture in this country. Some of the larger chemical firms installed plants for producing their own organic raw materials from petroleum fractions supplied by the refineries, while the oil companies became chemically orientated and increasingly involved in chemical manufacture, either by developing their own chemical-producing departments or by associating with established chemical firms in joint enterprises. The inter-relation between the petrochemical and plastics industries has been to the advantage of both: plastics manufacture provided a market for the products of petroleum cracking and cheap petrochemicals promoted expansion of the plastics industry.*

The structure of the new materials industry in the U.K. is one of considerable complexity, both in respect to its technical inter-relations and its capitalization. Its geography extends from Grangemouth in Scotland to Fawley on the Channel coast. Location of the manufacturing sites has been determined only in part by proximity to petroleum refineries; thus, for example, the ICI and British Geon PVC plants are associated with nearby calcium carbide manufacture, and the former firm's polyolefin plants in the Tees region are linked to the giant naphtha steam-reforming complex at Billingham. The Plastics and Fibres Divisions of ICI, employing between them some 14,000 workers, form a very large but not dominant sector of the industry. The Distillers Company Ltd, besides carrying on, through its subsidiary Bakelite Xylonite Ltd (owned jointly with Union Carbide

* The non-plastics significance of petrochemical manufacture is discussed on pp. 245 ff.

Corporation), the connection with the development of celluloid and of phenol-formaldehyde resins in this country, has been extensively active in the post-World War II new materials industry, as have the firms of Monsanto, Courtaulds, Union Carbide and the International Synthetic Rubber Company. Three petroleum companies are concerned directly or indirectly with the industry. Shell supplies olefins from its Carrington and Stanlow plants in Cheshire for the manufacture of polyolefins by its subsidiary, Shell Chemical Company; Esso supplies from its Fawley refinery ethylene to Monsanto and butadiene to the International Synthetic Rubber Company; British Petroleum's subsidiary, British Hydrocarbon Chemicals, in which it is associated with the Distillers Company Ltd, supplies ethylene to Union Carbide and Forth Chemicals Ltd (a British Hydrocarbon subsidiary), as well as using that gas for its own production of polyethylene. In addition to these major chemical and petroleum concerns, about a dozen smaller companies carry on manufacture of various types of plastics.

American participation in the U.K. new materials industry is extensive and has taken place almost wholly in the post-World War II period. This participation has taken the form either of the establishment of independent companies, such as Chemstrand Ltd, formed by the U.S. Monsanto Company in 1955, or of membership of joint concerns with existing U.K. companies; examples of the latter are Midland Silicones (Dow-Corning and Albright & Wilson Ltd), British Geon (B. F. Goodrich and Distillers Company Ltd), and Formica International Ltd (De La Rue and American Cyanamid Company). At least five American companies, other than those already mentioned, have acquired interests in U.K. plastics manufacture. The International Synthetic Rubber Company formed in 1956, is a consortium of the British motor tyre industry with the Italian Pirelli and French Michelin companies. Swiss interest is represented by CIBA (ARL) Ltd, a synthetic resins and adhesives concern which derived from a small research company established at Duxford in 1934.

Production of new materials has become a very large tonnage industry. It has been estimated that in 1933 the total U.K. output of plastics was about 10,000 tons, consisting chiefly of phenol- and urea-formaldehyde resins; ten years later production was 273,000 tons, and by 1961 it had risen to 610,000 tons. Between 1954 and 1961 the output of thermo-setting plastics rose from 143,000 to 214,000 tons, while production of thermoplastics increased from 130,000 to 396,000 tons. Plastics manufacture shows one of the highest growth rates of any major industry in the U.K. Over the decade 1949–59 plastics production increased at a rate of 15% p.a.; this compares with an average increase rate of 3% p.a. for all British manufacturing industry.

In the sections that follow the history of the principal new materials produced in the U.K. is outlined; the sequence of these sections corresponds to that of the dates of the beginning of industrial production of the materials in this country.

1. PHENOL-FORMALDEHYDE

Shortly after 1900 James Swinburne (later Sir James Swinburne, Bt) a consulting engineer who was associated with Joseph Swan in the early development of the carbon filament electric lamp (see p. 78), turned his attention to the problem of obtaining improved electrical insulants. " It is largely the insulation of the cables that limits our pressures and therefore our distance of transmission ", he said in his Presidential Address to the Institution of Electrical Engineers in 1902. He was acquainted with patents granted in 1902 to Adolf Luft for resinous materials obtained by condensation of phenol with formaldehyde and considered these could offer the prospect of obtaining the insulators he sought. Swinburne, in 1904, having acquired the British rights to Luft's patents, and thinking of the resins he intended to make as an advance on the established semi-synthetics, formed the Fireproof Celluloid Syndicate in London. By purely empirical modifications of Luft's procedure he produced a resin which, in appearance at least, bettered its predecessor. Swinburne delayed patenting his improvements, and when he finally had recourse to the Patents Office in 1907, he found himself anticipated by Baekeland. Continuing his work, he fabricated film, phonograph records and electric battery boxes; at the same time he became interested in the use of pheno-formaldehyde resins as solutions for the protection of metal surfaces; lacquers suitable for that purpose had been patented in 1905 by W. H. Story. Because of the technical difficulty he found in producing moulded articles from his resins and failure to find a market for those he did make, Swinburne decided to concentrate on the manufacture of phenol-formaldehyde lacquers. This interest in lacquers was probably not unconnected with the fact that Birmingham (where the Damard Company later had its factory) had been the site of an important lacquer industry since the versatile printer John Baskerville had introduced the art of japanning there as long ago as 1740. In 1910 he wound up the Fireproof Celluloid Syndicate and transferred its assets to a new concern, the meaningfully-titled Damard Lacquer Company. Swinburne's initial failure to interest commerce and industry in his phenol-formaldehyde resin mouldings may have been in large part due to lack of what is now called technical service to customers. Even in technical circles, little appears to have been known about the American development of phenol-formaldehyde. The first lecture in this country on the subject was delivered by Dr. H. Lebach to the Society of Chemical Industry at Birmingham in April 1913; its title was " Bakelite and its Applications ".

Swinburne's lacquers proved more successful than his mouldings. He found his most important market for them in the U.S. and was so encouraged by this that, in 1912, he established his associated Damard Lacquer Company on Long Island, N.Y. World War I caused the American sales to decline and, in 1916, the factory there was shut down. Just prior to the war, the Bakelite G.m.b.H. of Germany, formed to exploit Baekeland's patents in Europe, had established a works at Cowley in Middlesex. In 1916, Swinburne acquired the rights to use Baekeland's patents in the U.K. and the Damard Lacquer Company took over the factory at Cowley. During the remainder of World War I the entire output of that works went to meet the electrical requirements of Government departments. After the war, phenol-formaldehyde mouldings were made there in large quantities; amongst the articles produced were cigarette holders, pipe-stems, personal adornments, tea pot and umbrella handles. The firm also supplied resins for use in the manufacture of automotive brake linings. In 1920–21 the Damard Company transferred its manufacturing operations to a new factory at Greet, Birmingham, and production of fast-curing phenol-formaldehyde moulding powders became its principal activity.

At Darley Dale in Derbyshire, Mouldensite Ltd began in 1921 to manufacture resins according to a process patented by the American J. W. Aylsworth, who is credited with having proposed to Edison the use of phenol-formaldehyde resins for making phonograph records. Redmanol Ltd, London, a subsidiary of Redmanol Chemical Products Company, was, in the 1920's, marketing in the U.K. the products of its U.S. parent concern, which operated the patents of Dr. L. V. Redman, a pioneer of the application of phenol-formaldehyde varnishes and adhesives in the furniture industry. These other British companies were, in 1927, merged with the Damard Lacquer Company in Bakelite Ltd, formed for that purpose. The merger was concerned not only with production of the resins but with their application. A year after its formation Bakelite Ltd began manufacture of phenol-formaldehyde laminated materials at Darley Dale and, in 1928, acquired the Ideal Manufacturing Company, then engaged in manufacture of rod and tube laminates at Birmingham. Finally, in 1930–31, all the operations at Bakelite Ltd were concentrated in new works on a 30-acre site at Tyseley, Birmingham; later factories were established at Aycliffe (near Darlington) and at Ware (Hertfordshire). In the post-World War II era Bakelite Ltd, while continuing its production of phenol-formaldehyde resins, extended its operations to the manufacture and fabrication of thermo-plastics. By 1940, the American company Union Carbide Ltd held 51% of the share capital of Bakelite Ltd, and in 1962, on formation of Bakelite Xylonite Ltd, jointly with the Distillers Company Ltd, the remaining capital was acquired.

2. AMINO-PLASTICS*

Two factors stimulated the second stage in the development of thermo-setting resins: (1) the fact that phenol-formaldehyde resins, as normally produced, ranged in colour from various shades of brown to pitch-black; (2) the prospect of producing a transparent organic " glass " that would have the advantage of comparative unbreakability as compared with the oldest of all plastics—ordinary glass. In 1909, Raschig, in Germany, found that when phenol-formaldehyde condensates at the liquid stage were introduced into moulds, maintained for days or weeks at about 80°C, the castings obtained were clear and colourless. It was more than ten years later, however, before cast resins of this type were commonly produced. For the first time it was possible to impart by addition of pigments or dyes, any colour desired to a wholly synthetic plastic. Attractive though these coloured castings were, the very prolonged " curing " period made their manufacture costly and limited their application to expensive luxury articles.

It had been known since 1884 that urea gave condensation products with formaldehyde when Hans John, an American, in 1918, was granted a patent for production of clear viscous condensates of these two substances. John, with technical insight, proposed the use of his urea-formaldehydes as adhesives and impregnants for textiles; he also obtained hard, glass-like materials by modifying the condensation conditions. In the mid-1920's a number of chemists in Austria and Germany continued the quest for organic glasses of the urea-formaldehyde type without achieving complete industrial success, although they patented a number of processes. This was the position reached when the

* This is the customary name given to the plastics dealt with in this section; they are, in fact, derivatives of *amides*; e.g., urea is the amide of carbonic acid.

British Cyanides Company of Oldbury (p. 62), in 1924, under pressure of imminent bankruptcy, turned from unprofitable cyanide manufacture to production of new chemicals; one of these was thiourea, made from ammonium thiocyanate, used in the company's former cyanide process. E. C. Rossiter, a chemist employed by the firm, proposed that he should investigate the condensation of thiourea with formaldehyde, with the express object of producing an organic glass and thus circumventing patents already held in the field of glass-like condensates of urea. The reaction of formaldehyde with thiourea was found to differ considerably from that with urea and no satisfactory resin was produced. Eventually, Rossiter found that a promising material resulted from simultaneous condensation of both urea and thiourea with formaldehyde.

The British Cyanides Company used their complex condensate to impregnate wood flour or cellulose pulp, which was then dried and ground to produce a successful moulding powder. Their first commercial urea-formaldehyde moulding material was marketed in 1928 under the firm's brand name, Beetle. This British success —the second most important primary development in the field of thermo-setting plastics—was rapidly followed by similar developments in the principal European countries; in America commercial production was undertaken, in collaboration with British Cyanides, by a firm now merged with the American Cyanamid Company. The British company assigned its Beetle trademark to the U.S. producers. The mixed urea-thiourea formaldehyde condensates were translucent but not *transparent*, and with their commercial success as moulding powders the search for glass-like materials in the urea-formaldehyde field was finally abandoned. Beetle-type resins had the disadvantages that the hardening (curing) process was lengthy and staining of the metal moulds (owing to sulphur liberation) occurred; they also absorbed moisture, although to a less extent than the simple urea-formaldehyde condensates. The British Cyanides Company, having made plastics its principal business, became, in 1936, British Industrial Plastics Ltd, which became associated with the

Turner & Newall group in 1961. During the 1930's condensation products of urea continued to be intensively investigated on the Continent and in America. In anticipation of costly patent conflicts, an agreement between U.K. and foreign firms to exchange patent rights was arrived at in 1933; until its termination on the outbreak of World War II this arrangement, which was in effect an exchange of the results of research, greatly accelerated progress in the development of amino-plastics in the U.K., U.S. and continental countries.

Melamine, the triamide of cyanuric acid, was first isolated in 1834 by Liebig and remained without practical interest for a century before Henkel in Germany and CIBA in Basle used it as the basis of a new series of resins, formed by condensing it with formaldehyde. The new resins were similar in many respects to

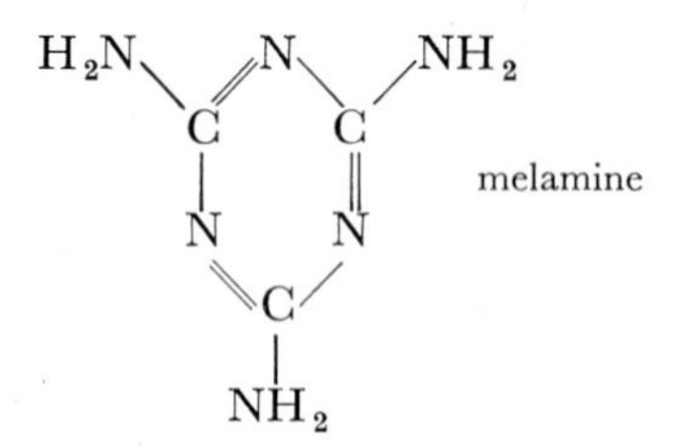

urea-formaldehyde condensates; they had, however, greater resistance to moisture and heat. With the development of the melamine resins the technical history so far of the amino-plastics was complete. Because of their superior moulding and mixing properties (e.g., with asbestos), the melamine resins largely superseded those based on thiourea, except in a few special applications. Melamine resins and melamine laminates have been produced by several firms in the U.K. since the late 1930's. It is estimated that in 1933 production of amino-plastics in the U.K. amounted to between 400 and 500 tons. By 1954 the total British production of amino-resins and moulding powders was 43,000 tons; the output for 1961 was 59,600 tons.

3. METHYL METHACRYLATE (PERSPEX)

In 1929–30, as a result of laboratory work by William Chalmers at McGill University, Montreal, interest was aroused in the polymerization of methacrylates. Chalmers had found that polymers of methacrylic ethyl ester and methacrylic nitrile were hard, clear and transparent, bringing into sight the possibility of the long-sought organic glass. Having obtained Canadian and U.S. patents, Chalmers brought his discovery to the attention of Röhm and Hass in America, and ICI. Both firms recognized its potentialities. An ICI research team concentrated on an investigation of methyl methacrylate, this ester having been found to have a higher softening temperature than the ethyl ester. In 1931, ICI was granted a patent for production of moulded articles from polymerized methyl methacrylate. Further

$$CH_2{=}\underset{\displaystyle CH_3}{\underset{|}{C}}{-}COOCH_3 \qquad \text{methyl methacrylate}$$

progress then depended on development of a sufficiently low cost process for making the monomeric methyl methacrylate. This process was invented by J. W. C. Crawford of ICI in 1932, the starting materials being acetone, hydrocyanic acid, methyl alcohol and sulphuric acid. Commercial production began in 1934. ICI, by virtue of the advantage of this process, was first to achieve economic production of the polymer, which it subsequently marketed as Perspex, the first successful organic glass. In 1936, ICI granted to the American firm of Du Pont a licence to use the Crawford process in the U.S.

Before the outbreak of World War II, ICI had perfected the technique of casting Perspex sheets, and during the war output of the plastic was used almost entirely by the aircraft industry, as a glazing material. It was at that period that, owing to shortage of natural rubber, methyl methacrylate polymer began to be used in the manufacture of false teeth, rapidly replacing vulcanized rubber in that important prosthetic. In 1946 output of polymerized methyl methacrylate in the U.K. was about 4000 tons; later production figures do not appear to have been published. Today Perspex finds one of its largest single outlets in the production of internally-illuminated signs.

4. POLYSTYRENE

Styrene (phenylethylene) was first reported about 1830 as a product of the steam distillation of the resin storax, an exudation of the tree *Liquidambar orientale*, native to Asia Minor; it was given the name " styrol " (still sometimes used) by E. Simon, who prepared it from that resin in 1839. Simon noted its transformation, on prolonged standing, to a viscous liquid which eventually solidified; he assumed, erroneously, that this was an oxidation product. Hofmann and Blyth, at the Royal College of Chemistry in London in 1845, showed that the solid had the same empirical formula as styrene itself and named it " metastyrene ", i.e., although the notion of polymerization had not yet emerged, they unwittingly demonstrated its polymeric character. The nature of the solid was finally elucidated in the early 1920's by Staudinger, who proposed replacing the name " metastyrene " by " polystyrene "; indeed, Staudinger derived many of his views on the nature of polymerization from his study of the behaviour of styrene.

The Naugatuck Chemical Company, in 1925, and, in 1930, the Dow Chemical Company in the U.S. and IG Farbenindustrie in Germany embarked on industrial production of styrene. At Ludwigshafen, about 1930, Mark and Wulff had worked out a process for its manufacture, starting from benzene and ethylene. Three properties of styrene motivated its industrial development: it could be polymerized to an organic glass of brilliant clarity, it was an excellent electrical insulant, and it co-polymerized with butadiene to produce a promising synthetic rubber. The last-named became of great importance in Germany after 1933 as the Third Reich developed its drive towards technological self-sufficiency (see p. 192).

Small quantities of styrene and polystyrene were made in the U.K. before World War II. During the war Boake, Roberts & Company produced polystyrene for electrical applications, using as the starting material phenyl-ethyl alcohol, a constituent of artificial rose perfume. This wartime route could not serve as the basis of economic manufacture, but it had the temporary advantage that it yielded extremely pure styrene, thus obviating the need for the then technically difficult process of its fractional distillation. After the war, the great capacity for styrene and polystyrene production in the U.S. discouraged their manufacture in the U.K., and several years elapsed before production gathered momentum in this country, when styrene became linked with the developing petrochemical industry* and, later, with the rising synthetic rubber industry. By 1959, Monsanto, British Resin Products, BX Plastics, and Shell were manufacturing polystyrene and had a total annual capacity of 37,500 tons; despite this, such was the home and export demand for the plastic, that 4500 tons of polystyrene were imported in 1958. In 1961 the U.K. production capacity had increased to 60,000 tons. About three-quarters of British polystyrene output is consumed by the refrigerator, packaging, electrical, radio, gramophone, and houseware industries. In recent years, toughened and foamed grades of polystyrene have greatly extended its range of utility and raised the material to the status of being the third plastic in order of tonnage in this country.

5. POLYETHYLENE (POLYTHENE)

The discovery of polyethylene is the most remarkable incident in the history of new materials in the U.K., and one of the most important in the world plastics industry. Both the discovery and primary development took place in this country without any

---

* Although ethylene for styrene production was a petroleum chemical in the U.K. from the early 1950's, benzene continued to be wholly coal-derived until 1961.

kind of antecedent in Germany or the U.S., as was so in the case of every other important plastic now manufactured in Britain. Polyethylene was discovered *outside* the new materials industry as the unsought bonus from a general research on the effects of high pressure on chemical reactions, consequently it had no " logical " place in the historical trend of plastics development in the 1930's—its appearance could have been delayed by a decade or longer. Conant and Cohen in the U.S. had already investigated the effect of very high pressures on the polymerization of isoprene, dimethylbutadiene, and styrene without any industrial outcome. Researchers in the high polymer field had generally employed pressure only sufficient to keep volatile monomers in the liquid state or, in the case of the formaldehyde resins, to effect the curing stage: the notion of polymerizing a normally gaseous substance to a solid was, in these circumstances, unlikely to present itself as a research target. Liquid and gaseous polymers of ethylene had been known for some time; these, however, had not been produced by high pressures and had attracted no industrial attention. In any event, the use of very high pressures in chemical processes appeared to lie beyond the resources of chemical engineering in the early 1930's; the highest pressure in use at that time was 1000 atmospheres, employed in the synthesis of ammonia by the Claude process.

By 1930 the production of alkali by the Solvay process, the principal operation carried on by ICI Alkali Division, offered little as a subject of research, nor did the related inorganic heavy chemical processes at Winnington ; the Division, therefore, sought topics outside its own industrial field to employ the resources of the research organization it inherited from its forerunner, Brunner, Mond & Company. F. A. Freeth, F.R.S., who entered the Research Department of Brunner, Mond & Company in 1907 and twenty years later became Joint Research Manager of ICI, has been described as the catalyst in the chain of human reactions that led to polythene, although, in fact, he had no direct connection with its discovery. Freeth first became interested in the techniques for study of chemical reactions under critical conditions of temperature and pressure as a result of his visit to the laboratory of Kamerlingh Onnes in Leiden in 1919. In 1930 Freeth recommended that ICI should undertake a research programme on the effects of very high pressures on chemical reactions and equilibria. The scheme was approved, and Professor A. Michels of Amsterdam, who specialized in the study of the physical and chemical effects of " hyper-pressures ", was taken into consultation. Equipment was installed at Billingham and R. O. Gibson and E. W. Fawcett began work on the project in 1932. Gibson had spent several years in Michels' Amsterdam laboratory and Fawcett had worked for some time with W. H. Carothers, the discoverer of nylon.

After a preliminary investigation of the effect of high pressure on the ethylene-carbon monoxide gas system, Gibson and Fawcett, on the suggestion of Professor (Sir) Robert Robinson, attempted the condensation of ethylene with benzaldehyde. In one of their experiments the ethylene decomposed into its elements; in another, in March 1933, a leak developed in the apparatus and a waxy, white solid was found in the reaction tube. The solid contained no oxygen, melted at 113°C, and was found to have the empirical formula $(CH_2)_n$; it was, therefore, apparently a polymer of ethylene—in fact, the first sample of polythene ever produced. In a subsequent experiment with ehtylene alone very small quantities of the waxy polymer were produced, but in further experiments explosions occurred and work was suspended until improved equipment and safer accommodation were available. The 1933 experiments yielded a total of a mere half-gram of polythene. For a time, particularly because of the economic depression at that period, it appeared probable that the pressure programme, still regarded as very " academic " and unlikely to have immediately valuable industrial results, would be entirely abandoned. Suggestions for possible further work did not include investigation of the polymerization of ethylene! When work was ultimately resumed in 1935, it was found that small amounts of oxygen were necessary to promote the ethylene polymerization and that the presence of the catalytic oxygen in the successful

1933 experiment had been due to the fortuitous leak in the reaction apparatus.

By the summer of 1936, following an intensive study of its physical properties, particularly as an electrical insulant, the industrial potentialities of polythene as it came to be called* were fully appreciated. A pilot plant for its manufacture was built and by the end of 1938 a ton of the new material had been produced. The Telegraph Construction and Maintenance Company manufactured a mile of polythene-clad cable and laid it on the sea-bottom between the Isle of Wight and the mainland. This collaborative evaluation of polythene by ICI and the Telegraph Construction Company established the case for erection of the first large-scale production plant, which came on stream on the eve of World War II. During the war the major part of the polyethylene output went to provide insulation in radar equipment; to increase production, information on polythene manufacture was given to the DuPont Company and Union Carbide and Carbon Corporation, who speedily erected factories in the U.S.

At the end of the war considerable production capacity for polythene was in existence, but up to that date only the electrical applications of the material had been developed. In 1948 only 1% of U.K. polythene was being used in the manufacture of domestic holloware, beer pipes, and similar articles. Amongst the wares made at that time was a washing-up basin which was produced in various attractive colours; it is claimed that this single item did much to raise plastics in general in consumer esteem and to awaken public consciousness to the fact that, properly fabricated, the new materials were not cheap and nasty substitutes for the traditional ones. By 1957, 40% of U.K.-produced polythene was being used in the manufacture of articles for the home. Demand for polyethylene was greatly increased from about 1955 onwards by its use in film form as a wrapping for perishable goods in supermarkets and self-service cafés.

* Originally "polythene" was applied to the ICI product only; it now appears to be the generic name of high pressure ethylene polymers.

In 1952–53, Karl Ziegler, engaged in a more or less academic investigation at the Max Planck Institute for Coal Research at Mülheim, West Germany, discovered a reaction between ethylene and aluminium alkyls that enabled him to polymerize ethylene at ordinary pressure. Ziegler's polymers, however, had much shorter molecular chains than ICI's polythene and were of little use as plastic materials. Continuing his research, Ziegler discovered by pure chance that the presence of certain metallic compounds promoted the growth of long-chain ethylene polymers, which, although of higher density and differing in other respects from high-pressure polythene, could be used in most of the applications of that material. Almost simultaneously with Ziegler's discovery of his high density polyethylene process, similar catalytic, ordinary pressure processes were developed in America by the Philips Petroleum Company and by Standard Oil of Indiana. At the present time six firms are producing polyethylene in the U.K.; several of these make the high density material by processes which they operate under licence from the American patentees. Between 1951 and 1956 output-rate of polyethylenes in the U.K. increased more than ten-fold; in 1960 105,000 tons were manufactured and, in terms of tonnage, it had become the principal plastic.

6 NYLON

Nylon was a purely American development; its discovery and primary industrialization took place in the laboratories and factories of the firm of Du Pont. A decade elapsed between Carothers' discovery of the polymer and the coming on stream of the small-scale production unit. The cost of that development has been estimated to have been about £10 million. Nylon yarn was used in experimental production of stockings in 1937; tooth brushes with nylon bristles were manufactured shortly afterwards. Large-scale nylon production began in the U.S. in October 1938. ICI acquired in 1940 the right to operate under the Du Pont British Patents and with Courtauld Ltd formed a joint-subsidiary, British Nylon Spinners Ltd, to manufacture nylon in the U.K. In Britain, as in America, nylon became the spearhead of the man-made fibre industry. During World War II production went almost entirely into meeting requirements for the manufacture of parachutes, glider tow-ropes, tyre-cord for bombers, and life-rafts. The first mouldings of the polymer were made in 1941; these were gears, bearings and other small engineering parts, all of which went to warlike applications.

Manufacture of nylon in the U.K. in the post-war years continued to be an ICI monopoly until 1963, when the Dutch-controlled firm of British Enkalon established its 7 million lb a year nylon plant in Antrim, Northern Ireland. At the present time (1965) Chemstrand (the U.K. subsidiary of Monsanto, U.S.A.) is in process of erecting an 18 million lb a year nylon 66 plant in Ayrshire, to come on stream in 1966. Production of the polymer has been carried on by the ICI Dyestuffs Division at Wilton Works since 1949 from where it is transported to the British Nylon Spinners' factory in South Wales, and spun into yarn and staple fibre. The capacity of the original nylon plant at Wilton was 5000 tons a year. Since 1964 British Nylon Spinners Ltd has been a wholly-owned ICI company. Some nylon from Wilton goes to ICI Plastics Division for production of moulding compounds, monofilaments for fishing tackle, brush bristles, and surgical sutures. Since the introduction of nylon into the U.K., several similar polyamides have been investigated by ICI, e.g., " nylons " have been produced experimentally by condensing oxalic acid with various diamines.

7. POLYVINYL CHLORIDE (PVC)

Vinyl chloride was first prepared in 1835 by the French chemist Regnault, who observed that when the liquid was exposed to sunlight it was transformed into a white, amorphous solid with no chemical properties. Regnault's white solid we now know to have been polyvinyl chloride.

$$CH_2{=}CHCl \quad \text{vinyl chloride}$$

Chemists in Russia and Germany, probably because of the success of Baekeland's phenol-formaldehyde resins, alerted to the potentialities of hitherto neglected chemically-derived inert materials, turned their attention to vinyl chloride and its polymerization. In 1912 Ostromislensky, in Moscow, patented a method of plasticizing the polymer, i.e., of rendering it mouldable by addition of other substances (now known as plasticizers); in the same year Klatte and Rollett, in Germany, produced vinyl chloride by reacting hydrogen chloride with acetylene, and two years later the firm of Griesheim-Elektron patented the use of mercuric chloride as a catalyst for that reaction. Between World Wars I and II, IG Farbenindustrie produced PVC on a very limited scale, and from 1930 ICI carried on some laboratory work in this country, but did not manufacture. There were two reasons why PVC was not produced in Britain until after the outbreak of World War II: viz., the ready availability of natural rubber for electrical insulating applications, and dependence of the U.K. chemical industry on imported calcium carbide as its only acetylene source. Germany, on the other hand, during the inter-war and war period, based much of its heavy organic manufacture on syntheses from acetylene, based in turn on indigenously produced calcium carbide. In America and in Germany, in the inter-war period, the problem of obtaining mouldable materials from vinyl chloride was solved by the expedient of co-polymerizing it with minor amounts of vinyl acetate. There were also advances in plasticizing techniques in the U.S., particularly by the firm of B. F. Goodrich. The technological prerequisites for large-scale production and application of PVC had fully emerged by 1939.

The loss of the rubber-growing regions in the Far East in the early stages of World War II made it pressingly necessary for the Western Allies to produce a material to take the place of natural rubber as an insulant. A carbide plant having been erected by ICI (p. 132), a vinyl chloride unit was brought into operation at Runcorn in 1941; this produced, in that year, some 20 tons of the monomer, which was used in experimental production of PVC. DCL, in 1943, was operating a 75-ton a year pilot plant at Tonbridge in Kent, producing vinyl chloride, not from acetylene, but by cracking ethylene dichloride, produced from ethylene, made in its turn from fermentation alcohol. During the post-war years PVC capacity was extended in 1948 by British Geon (a joint-subsidiary of DCL and B. F. Goodrich (45%)) and, in 1953 and 1958 by ICI. By 1958 PVC constituted 40% of the total U.K. plastics production. ICI and British Geon continue to be the only producers of vinyl chloride and its polymers in this country. The national output of all grades of PVC now probably substantially exceeds 100,000 tons a year.

During World War II the great part of PVC production in the U.K. went to the electrical cable industry, although many of the present large outlets for the polymer were being tentatively developed. In 1941, PVC paste was being used by ICI at its Hyde leathercloth factory for bus seat coverings for London Passenger Transport Board. Various types of packing and protective clothing were also made from PVC at that time. Before the war ended, PVC compositions were being used for motor car upholstery by Morris Motors. Manufacture of safety gloves by a dipping technique was also developed in the later war years. PVC was first extrusion-moulded in the U.K. by Extrudex Ltd (later a subsidiary of Bakelite Xylonite Ltd) in 1954. By 1948, modifications in the nature of the polymer were facilitating extension of PVC applications. ICI, in 1949, developed methods of controlling the molecular weight of the polymers in order to increase their suitability to specific applications. Lower molecular weight polymers have found application in the fabrication of pipes for chemicals and corrosive wastes; these and similar rigid artefacts are made by extruding the *unplasticized* polymer. PVC is now well-established in the manufacture of floor-tiles, toys, conveyor belting, gramophone records, and electrical fittings. The so-called " easy processing " (EP) polymers were introduced in the U.K. by British Geon in 1951.

8. POLYTETRAFLUOROETHYLENE (PTFE)

A considerable number of solid polymers containing fluorine have been prepared, but only two of these (polytetra- and polytri-fluoroethylene) have so far found industrial application; by far the more important is polytetrafluoroethylene. In the course of investigations on the organic chemistry of fluorine in the 1930's, Ruff and Bretschneider, in Germany, produced the gaseous compound tetrafluoroethylene by high temperature reaction of hydrogen fluoride with chloroform. Plunkett, in the

$CF_2{=}CF_2$ tetrafluoroethylene

Kinetic Chemicals laboratory, in the U.S. in 1938, discovered by accident that this fluoro-carbon compound polymerized to an intractable white solid, capable of withstanding a higher temperature than any known organic derivative, and also it was unaffected by the strongest acids and alkalis. The new material obtained was, in chemical structure, polyethylene with all the hydrogen atoms replaced by fluorine. It was soon found that even the most adhesive substances would not adhere to it, and that its coefficient of friction against steel was as low as that with polished, lightly-loaded metals actually lubricated with oil, i.e., it appeared to have potentialities as a *self-lubricating* bearing material. Its very low and constant power factor also suggested that the material would be an effective insulator for high-frequency current conductors.

In 1943 a pilot plant for PTFE was erected by Du Pont and, by 1945, the polymer was in large-scale production. ICI, partly because of its own developments in fluorine chemistry (p. 151), became interested in PTFE in the early 1940's and acquired the rights to manufacture it. In the U.K., where it is known as Fluon, industrial production was established by 1947. The non-sticking property of PTFE has been utilized in the manufacture of easy-to-clean domestic cooking utensils. Its self-lubricating property has been used in the production of many types of " dry " bearing in light engineering; it has been reported that a coating of PTFE on skis enables higher speeds to be achieved than with the usual ski-wax. In certain specialized electrical applications the polymer is now almost indispensable.

Despite the growing range of uses that have been found for it, the high cost of PTFE (£2 per lb) appears likely to restrict it to small-bulk and " two-dimensional " applications.

9. THE ARDIL EPISODE

In 1935, Professor W. T. Astbury, then Director of the Textile Physics Laboratory at the University of Leeds, and Professor A. C. Chibnall suggested to ICI that textile fibres might be produced by dissolving vegetable protein in urea and extruding the solution into a suitable curing bath. At that time ICI was involved in research connected with the development of the Empire's natural resources, in particular with the later Government-sponsored groundnut scheme. Groundnuts were being imported into the U.K. at a rate of about eight million tons a year as a source of arachis oil for margarine manufacture. The residue, after removal of the oil, contained a high proportion of vegetable protein, and it appeared that this would provide a good source of starting material for production of the proposed protein fibre. The experimental work on the project was carried out in the laboratories of ICI Nobel Division at Ardeer from which site the name of the new fibre—Ardil—was coined. By 1938, filaments had been produced which were sufficiently promising in their properties to justify setting up a pilot plant, which, however, was not done at this time. Enough fibre had been produced in the laboratory before the outbreak of World War II for a few suits of clothes to be made from a mixture of equal parts of Ardil and natural wool. The war interrupted further development at that stage.

After World War II, Ardil fibre, produced in a small pilot plant, was offered for examination by the textile industry. The new fibre was reported to be warm and hard-wearing and not only to look like wool but to behave like it, for example, in dyeing processes. Reaction of the trade was such that, in 1947, ICI decided to embark upon industrial-scale manufacture of Ardil. A £2 million plant was erected at Dumfries. The planned output of this factory was twenty million pounds of Ardil a year. Production at Dumfries began in 1951. During the next few years mixtures of Ardil with wool, rayon, and nylon were developed by the textile trade. Advantages of the new fibre were that it was more resistant than wool to mildew and moths; it was also less flammable than wool. Despite its desirable characteristics and the economic fact that Ardil was cheaper than wool, however, it failed to make headway against the competition of the established semi-synthetic fibres and nylon and the threat of Terylene (see below), already in experimental production in 1949. This *technically* successful excursion into the field of regenerated natural materials was finally abandoned in 1957. The Ardil episode was not unique: in Japan and the U.S. regenerated protein fibres had been made from soya beans, and a fibre, known as Vicara, had been produced in America from the corn protein, zein. Like Ardil, these other regenerated materials failed to survive against the onrush of wholly-synthetic fibres based on simple monomers derived from coal and—increasingly—from petroleum.

10. SILICONES

Silicon, the key element in the oldest of all plastics—ordinary glass—is the only " inorganic " element so far to have been introduced into modern organic plastics. Silicon-containing plastics appeared late in the history of organo-silicon chemistry, a field largely developed in Britain by Professor F. S. Kipping of the University of Nottingham. Between 1899 and 1937 Professor Kipping prepared and investigated the silicon analogues of most of the common classes of carbon compounds; at the end of that protracted research he came to the considered conclusion that his work was without technological potentialities: he informed the Royal Society in 1937 that " the prospect of any immediate and important advance in this section of chemistry does not seem to be very hopeful ". Kipping's specialized approach did not predispose him to explore the nature of the intractable " glues " he sometimes obtained in the course of his research;

his work, however, was later to have a significance he could not suspect.

In 1931 the notion that organic glasses might be produced by polymerization of silicon-containing organic monomers appears to have determined the direction of a research carried out in the laboratories of the Corning Glass Works in New York. A group of organic chemists in these laboratories, utilizing the capital of information accumulated by Kipping, prepared a series of polymers in which the main molecular chain was a sequence of alternating carbon and silicon atoms. The polymers were given, perhaps unfortunately, the name " silicones "—the term used by Kipping for his silicon analogues of ordinary ketones. These new materials did not provide the sought-for organic glass, but their high resistance to the action of heat and their outstanding electrical properties linked them almost immediately with other work then being done in the Corning research department. The Corning Company was then engaged in developing woven glass fibre tape as an electrical insulating material; for this a heat and water-resistant impregnant with good electrical properties was required: the silicones were found to be entirely suitable. In the meantime, the Mellon Institute at Pittsburg and the General Electric Company began to investigate the technical possibilities of the new polymers. To facilitate the chemical industrial development of the silicones, Corning Glass Works, in 1942, collaborated with the Dow Chemical Company and, in 1943, the Dow-Corning Corporation was formed to promote their large-scale production and application. By 1945, Dow-Corning had a full-scale plant on stream, producing a range of different silicones, including silicone rubbers.

In 1943 the Ministry of Supply invited the Dow-Corning Corporation to send representatives to the U.K.; this visit was the occasion of preliminary discussions between Dow-Corning and Albright & Wilson Ltd. Members of Albright & Wilson's staff visited the U.S. in 1946, and later Albright & Wilson became selling agents for Dow-Corning Silicones in the U.K. In 1950 the two concerns formed a jointly-owned company—

Midland Silicones Ltd, which, in addition to manufacturing silicones, took over the selling agency for the American parent company. Construction of a plant, in a former magnesium-from-sea-water factory began at Barry in South Wales in mid-1950. Silicones were first produced at the new factory in 1952, in which year output was limited to several tons. As the U.K. producer of methyl chloride, the organic starting material for silicones, and as a firm deeply committed in the field of new materials, ICI began manufacture of the silicone polymers in the works of its Nobel Division at Ardeer in 1955. Albright & Wilson and ICI continue to be the only silicone manufacturers in the U.K. The process used by ICI was developed by the General Electric Company in the U.S., and is operated under licence from that concern; thus both the manufacturing processes used in the U.K. are of American origin.

Although statistics are not published, total production of silicones in the U.K. now probably exceeds 1000 tons a year. High cost (of the order of £1 per lb) has restricted the technological development of silicones mainly to surface ("two-dimensional") applications, for which, however, they are in many ways uniquely suitable. Present applications include their use as impregnants, laminating cements, thermally-resistant coatings, waterproofing agents, mould-release compositions, polishes, lubricants. Silicones can be produced in the form of liquids, semi-solids, and solids. The first silicones made in America during World War II were used as non-flammable damping fluids for aircraft; later, semi-solid silicones were employed as ignition sealing compositions in the sparking-plug wells of aeroplane engines. Silicones share with PTFE, amongst new materials, very marked anti-adhesive properties, which they impart to surfaces to which they are applied.

11. POLYURETHANES

In the course of a research directed towards discovering a new fibre-forming material the use of which would not be dominated by Du Pont nylon patents, Dr. Otto Bayer, in Germany, found that di-isocyanates could undergo addition polymerization. In chemical structure the materials obtained were polyureas. These proved incapable of forming fibres or of being used as plastic resins. When, however, di-isocyanates were co-polymerized with simple hydroxyl-containing compounds such as ethylene glycol, materials having a very wide range of flexibility were obtained; these so-called polyurethanes readily formed fibres, in addition to possessing moulding properties. In Germany by 1941, the new polymers were in industrial production for various fibre-forming and plastic applications; their use as foams, adhesives and textile coatings, had also been demonstrated. Polyurethane fibres were found to be too stiff and resistant to dyeing for use in textile manufacture. During World War II the development of polyurethanes was confined to Germany and only became known outside that country when British and American observers visited German chemical factories after the war. A principal wartime use of polyurethanes in Germany had been in the manufacture of light-weight laminated materials for aircraft construction. It was also learned that Bayer had developed a process for producing foams by addition of water to the isocyanate-glycol reaction, whereby carbonic acid gas was released and inflated the polymer as it formed. Production of polyurethane foams was thereafter rapidly developed in the U.S. In the post-war years, both in Germany and America, new types of polyurethanes, based on aromatic compounds, were produced. Manufacture of polyurethanes began in the U.K. in 1953, using starting materials imported from Germany. By 1956 the necessary isocyanates were being manufactured in large tonnage by ICI; later, other firms entered the field. Polyurethanes are now in large-tonnage production in this country, and are mainly used in the manufacture of foams for insulating and packing purposes.

12. EPOXY RESINS

The first epoxy-type polymer was produced in the laboratory by Lindemann in 1891, but was not recognized by him as such, or as having any possible utility. The research leading to the discovery of the epoxy resins was carried out in Germany shortly before World War II. Their wartime development became known to the Western Allies as the result of publication of a CIBA patent in 1943. Epoxy resins are a group of chemically different materials formed by the application of so-called cross-linking agents to various liquid compounds, e.g., diphenylol propane, the proportions of the reactants and the reaction conditions determining the properties of the resin obtained. In the immediately post-war years resins based on a number of cross-linking systems were developed in the U.S. Production of the resins became associated with petrochemical manufacture in 1947 in America when the Shell Corporation, in collaboration with other firms, began making them from petroleum-derived epichlorhydrin and diphenylol propane. The Shell products, known as Epikote resins, were first produced in the U.K. in 1955, when the Shell Chemical Company's plant at Stanlow in Cheshire came on stream. Initially the epichlorhydrin cross-linking agent was imported from the Shell Chemical Corporation in the U.S. About this time Leicester, Lovell & Company (since 1953 a subsidiary of Borden Chemical Company, U.S.A.) began making epoxy resins in this country. Epoxy resins have found wide application as surface coatings and adhesives; they are also used in the electrical industry and in the manufacture of laminated materials. Present (1965) output of epoxy resins in the U.K. is probably of the order of 5000 tons a year.

13. TERYLENE (POLYETHYLENE TEREPHTHALATE)

It has been claimed for Terylene that it was " the predicted and desired product of systematic research ". This seems to be an all too sweeping assertion. J. R. Whinfield and J. T. Dickson, who first produced the polymer in the course of a research for fibre-producing materials in the laboratory of the Calico Printers' Association Ltd, had certainly prepared minds for their discovery. These researchers, when they began their quest in 1939, knew, from the work of Carothers, that a substance capable of forming industrially-useful fibres must have a high melting point, high resistance to chemical action and solvents, a high molecular weight, micro-crystalline structure, and a high degree of linear symmetry in its constructional unit. Whinfield had worked for some time in the laboratory of Cross, the co-discoverer of viscose, a circumstance which appears to have originated his interest in fibre chemistry. It is not clear from the record whether Whinfield and his co-worker were aware at the outset, that Carothers, in 1930–31, had been in sight of anticipating them in their discovery by preparing polyesters from ethylene glycol and phthalic acid. Investigation of the polyesters of glycol with the phthalic acids was the starting point of Whinfield and Dickson's research, chosen, mainly it seems, on the ground that the aromatic molecules of the phthalic acids would be expected to yield products of high melting point. Carothers had used *ortho*-phthalic acid and obtained a material which did not form fibres. Whinfield and Dickson, proceeding empirically, rang the isomeric changes. Finally, they found that *para*-phthalic acid (more usually known as terephthalic acid), the linearly most symmetrical of the three isomeric phthalic acids, then a comparatively rare chemical, combined with glycol to give a material that was later demonstrated to possess in combination all the desired properties, in particular that of being capable of being cold-drawn into thin, strong fibres. It was evident that a discovery comparable to that of nylon had been made.

$$n\ CH_2OH\ .\ CH_2OH + n\ HOOC\langle\bigcirc\rangle COOH$$

(ethylene glycol) (terephthalic acid)

$$n\ (.\ .\ .\ CH_2CH_2OOC\langle\bigcirc\rangle COOCH_2\ .\ CH_2OOC\langle\bigcirc\rangle\text{—}$$

$$COOCH_2\ .\ CH_2OOC\langle\bigcirc\rangle COOCH_2\ .\ .\ .)$$

[part of the chain of polyethylene terephthalate (Terylene) + nH_2O (eliminated)]

Because the resources of the laboratory of the Calico Printers' Association were inadequate for further development of the new polymer, and because it appeared that it might have some importance in warlike technology, the Ministry of Supply, in 1941, requested the Chemical Research Laboratory of the Department of Scientific and Industrial Research to continue the work and to produce experimental quantities. In 1943 a sample of the new polymer, which became known as Terylene, was passed to ICI for full evaluation as a fibre-forming material. Du Pont, in the U.S., had been aware of the new development from 1944 and had engaged in an independent investigation of terephthalic polyesters; that firm acquired, in 1946, a licence to operate the Calico Printers' Association's patent in America. ICI purchased world manufacturing rights (outside the U.S.) in 1947, and by 1949 had a £4 million pilot Terylene plant in operation.

In 1951 ICI work on Terylene was organized under its Terylene Council, which five years later became ICI Fibres Division, and which now as ICI Fibres Ltd controls the firm's Terylene and nylon operations. Full-scale Terylene production in the U.K. began in January 1955 at ICI's Wilton factory—almost two years after commercial production of Du Pont's Dacron began in the U.S. The Wilton plant, erected at a cost of £15 million, had a capacity of 5000 tons a year, this being judged to be the minimum to ensure an effective commercial impact by the new fibre. At Wilton, mixed xylenes, brought by coastal tankers from the petroleum refineries, are fractionated and the para-isomer converted to terephthalic acid; glycol, the other Terylene raw material, is also produced at the site. In 1963, ICI's second Terylene plant, with a capacity of about 10,000 tons a year, came into operation at Belfast Lough, Northern Ireland. Terylene finds its main application in textile manufacture, generally in mixtures with natural fibres; it is also used in the production of sewing threads, ropes, nets, and fishing lines. ICI in 1951 began development of Terylene film, production of which was put on plant scale in 1953. Unlike nylon, Terylene has not proved suitable for plastic moulding.

14. SYNTHETIC RUBBERS

During the past fifty years a series of polymeric materials have been developed with many of the properties of natural rubber and, in some instances, having advantages over that substance. Because of their elastic properties these synthetic materials are known as elastomers. Their initial and main development took place in Germany and the U.S. During World War I over 2000 tons of so-called " methyl " rubber were produced at Leverkusen in Germany to compensate for wartime shortage of the natural commodity. This first industrially-produced synthetic rubber was not highly successful, because its manufacturers were ignorant of the beneficial effects of admixing carbon black. Despite the discouraging results of this first essay on the synthetic rubber problem, BASF in the early 1930's returned to the attack and achieved success with a co-polymer of butadiene and styrene, which was named Buna S.* Similar developments were in progress in the U.S., where it was found that co-polymers of butadiene and vinyl derivatives gave promising elastomers. In America synthesis of rubbers was from the outset associated with the petroleum industry. The Standard Oil Company, in 1937, produced butyl rubber by co-polymerizing petroleum-derived isobutylene and butadiene. The American work was followed up in Nazi Germany. By 1936, both Buna S and Buna N were in large tonnage production in the Reich. In 1938, Buna S output was 5000 tons. Hitler announced to a Party rally that Germany had solved its rubber problem. During World War II Germany's Buna output reached 40,000 tons a year. A 5000 tons a year Buna N plant was in operation in the U.S. before the outbreak of war. American output of synthetic rubber from butadiene rose, during the war, to about 1 million tons. After the war Buna manufacture was prohibited in Germany; its production on a limited scale was permitted by the Allies from 1951 onwards.

---

*" Buna " was coined from butadiene and Na (sodium) the polymerizing catalyst; the " S " indicates styrene. " Buna N " is used for the butadiene-acrylonitrile co-polymer.

Interest in the synthetic rubber problem in America dates from 1925, when Father Nieuwland, then Professor of Chemistry at Notre Dame, revealed in an address to the American Chemical Society that he had obtained rubber-like materials by polymerization of divinyl acetylene. This information led the firm of Du Pont to acquire the right to exploit Nieuwland's discovery. In the Du Pont laboratories attention was turned from divinyl to monovinyl acetylene as the more promising monomer. W. H. Carothers, the discoverer of nylon, suggested that the polymers of chloroprene, the condensation product of monovinyl acetylene and hydrogen chloride, should be investigated: the result was the discovery of neoprene, a material resembling natural rubber and in certain respects superior to it. Manufacture of this chlorine-containing synthetic rubber was rapidly put on an industrial scale, and, by 1960, annual output in the U.S. had reached 110,000 tons; as early as 1934 neoprene had been used in manufacture of the first synthetic rubber tyre in America.

As late as 1952 there appeared to be no early prospect of a synthetic rubber industry in the U.K.; W. J. S. Naunton of ICI wrote in that year: "The possibilities of a synthetic rubber industry in Britain depend upon the availability of the raw materials, the chief of which are ethylene, acetylene, butylene and benzene." Already these materials were being produced in this country and with the rapid rise of the post-war petroleum industry those formerly in deficient supply became available. Production of synthetic rubber on an industrial scale in the U.K. began in 1958, and, by 1960, national output had reached some 80,000 tons a year, mainly of the butadiene-styrene co-polymers. In 1956 a consortium of the principal rubber companies operating in this country (Dunlop, Firestone, Goodyear, North British Rubber, BTR Industries and Pirelli) was formed under the name of the International Synthetic Rubber Company. This combine began production of styrene-butadiene synthetic rubbers in 1958, and by 1964 had a production capacity of 120,000 tons; at the same time construction was in progress of capacity for manufacture of other butadiene polymers and co-polymers. At its

Wilton Works ICI established a large plant for extracting butadiene from the hydrocarbon fractions produced in its olefin plants. Wilton butadiene is co-polymerized with styrene, acrylonitrile, and with methyl methacrylate, to give, respectively, Butakon S, Butakon A, and Butakon latices. These ICI special purpose synthetic rubbers are used as shoe soling material, and in the manufacture of flexible fuel tanks, hose-pipes, etc.; the Butakon latices go mainly to the leather finishing, paper, and textile industries. Butadiene not converted to rubbers at Wilton is sent by sea or road to other synthetic rubber producers. Esso Petroleum Company entered petrochemical manufacture in 1957 and became a supplier of butadiene to the International Synthetic Rubber Company; in 1963, Esso began the first manufacture of butyl rubber in the U.K. The Dunlop Company's Chemical Products Division, formed in 1957, shortly afterwards began production of synthetic gutta-percha (trans-polyisoprene). Du Pont, in 1960, began production of neoprene at their Maydown factory in Northern Ireland. This factory, which has a production capacity of 20,000 tons a year, is supplied with acetylene—its raw material—from a carbide works erected by Carbide Industries Ltd, a subsidiary of the British Oxygen Company. Complete statistics are not published, but it may be estimated that present (1965) U.K. production of all synthetic rubbers is of the order of a quarter of a million tons a year.

15. POLYACRYLONITRILE

Acrylonitrile (vinyl cyanide) was not produced in the U.K. until 1959, when a demand for it arose for the manufacture of nitrile synthetic rubber, i.e., co-polymers of butadiene with acrylonitrile. The first U.K. acrylonitrile plant was operated by ICI at its Cassel Works, Billingham; the process used involved the catalytic reaction of acetylene with hydrocyanic acid (HCN). It is reported that this process has now been replaced by one based on the reaction in the presence of air of ammonia with propylene—both products of the giant Billingham–Wilton complex of chemical processes. In America and Germany an additional demand for

$$CH_2{=}CH\,.\,CN \quad \text{acrylonitrile}$$

acrylonitrile arose in the early 1950's, when it began to be employed as a fibre-making material; mainly owing to this new use, production of the compound in the U.S. more than doubled between 1955 and 1960. Polyacrylonitrile fibre has appeared on the market under various names, e.g., Acrilan, Courtelle, Creslan, Dynel, Orlon, Pan Fibre, Verel and Zefran. In 1963 the American Monsanto Company's subsidiary, Chemstrand Ltd, began production of Acrilan in Northern Ireland with a plant of 27 million lb a year capacity. At the present time (1965) Northern Ireland produces about one-sixth of all man-made fibres (Acrilan, nylon, rayon, Terylene, and polypropylene) manufactured in the U.K. This concentration of a specialized branch of the chemical industry has arisen principally as a consequence of the highly favourable policy of the Stormont Castle Government towards new industries coming into the province, and the decline of the traditional linen industry has provided a reservoir of skilled workers, habituated to textile factory conditions.

16. POLYPROPYLENE

$$\begin{array}{l} CH{=}CH_2 \\ | \\ CH_3 \end{array}$$

Propylene, structurally regarded, is methyl ethylene. In view of the outstanding products obtained by high and low pressure polymerization of ethylene, it was expected that propylene would give similarly useful new polymers. This was found, however, not to be so; the products of polymerization were either viscous liquids or rubber-like solids of no industrial potential. Professor Giulio Natta of the Milan Polytechnic Institute in 1954, using catalysts similar to those employed by Ziegler to polymerize ethylene, succeeded in making solid high molecular weight polymers from propylene. The secret of Natta's success was soon discovered to be in the fact that his catalysts resulted in an *orderly* projection of the methyl groups from the main carbon chain of the polymer; for that reason the catalysts became known as stereo-specific. In the commercially useless materials previously obtained from propylene the methyl groups are orientated in random fashion. Natta named his orderly polymers " isotactic ". His surprising discovery which was rapidly developed first in America, then in the U.K. and other countries, added polypropylene to the established industrially important polymers. Like nylon, polypropylene is capable of being fabricated as mouldings, films and fibres. In some of its applications it is competitive with polyethylene, particularly that produced by low pressure processes, and is said to have slowed the rate of increase of demand for that product.

The first U.K. polypropylene plant was erected by ICI at its Wilton Works in 1959. At Wilton the high-purity propylene required as the starting material is supplied by the petroleum naphtha cracking plants on a nearby site. Polypropylene produced at Wilton is marketed as Propathene. ICI began, at its Belfast Lough works in Northern Ireland, manufacture of polypropylene fibre (Ulstron) in 1963.

17. OTHER POLYMERIC NEW MATERIALS

Vinyl acetate and its polymer have been produced by Courtaulds at Spondon and by DCL's subsidiary, Hedon Chemicals Ltd, at Hull, since the mid-1950's. The Spondon process was based on acetic acid and acetaldehyde, the Hull process, now operated on a scale of 10,000 tons a year, was developed by the Canadian firm of Shawinigan Chemicals and is based on acetylene and acetic acid. In 1964, DCL and Courtaulds entered into an agreement to manufacture vinyl acetate co-operatively at a new larger plant at Hull.

Although polyvinyl alcohol had been manufactured in other industrial countries, including Canada, since the early 1920's, its production was begun only a few years ago in this country by

British Resin Products, a subsidiary of the Distillers Company Ltd. Polycarbonates, investigated by W. H. Carothers in the 1930's and subsequently manufactured in the U.S. and Germany from the early 1950's, appear to have so far no history in the U.K. Polyformaldehydes, produced by polymerization of formaldehyde from about 1950, were the subject of much research in the U.S. by Du Pont, Eastman Kodak, and General Electric Companies, as well as by Farbenfabriken Bayer in Germany; no parallel research was carried out by the British chemical industry. In 1959 both Du Pont and Bayer began marketing polyformaldehydes which have since found wide application as engineering materials.

HEAVY ORGANIC CHEMICALS

Manufacture of organic chemicals in large-tonnage quantities is, along with the development of the post-World War I dyestuffs industry, the production of new materials (plastics, etc.), synthesis of pharmaceutical and of non-soap detergents, historically one of the most important aspects of the chemical industry in the present century. Expansion of heavy organic chemical manufacture is a phenomenon of the post-World War II years: between 1949 and 1959 their production increased 280%, i.e., at a rate of 11% *per annum* (compound interest). Although, as will be shown, the production of heavy *organic* chemicals originated in certain techno-historical developments in the heavy *inorganic* chemical industry, it later became the supply base of the other chemical industrial developments just mentioned; indeed, it determined the possibility of their growth to their present impressive dimensions; without the existence of an extensive and complex heavy organic chemical industry based, latterly, on petroleum as its source of carbon raw material, the great post-World War II new materials and detergent industries would have reached their limits of expansion some two decades ago, or, more probably, have become dependent on imported starting materials, largely from America, the leading producer of primary chemicals from petroleum.

Heavy organic chemical manufacture arose, not as an effort to exploit industrially the very numerous developments in the field of pure organic chemistry, which long before the turn of the century had proved a very rewarding science in terms of the wealth of new compounds the laboratories had brought into being, but as a consequence of the transition from the old to the modern system of *inorganic* chemical manufacture, particularly the bringing into being of extensive capacity for the electrolytic manufacture of chlorine. The first really large tonnage modern organic chemicals were the simple chloro-hydrocarbons (see Section 1 below). (Ethyl alcohol had, of course, been a fairly large tonnage organic chemical even in the nineteenth century, but its full emergence as a heavy organic chemical took place in the present century (see Section 2 below). In addition to providing an outlet for chlorine, no longer required in very large tonnages for export as bleaching powder, chloro-organics were based on coal, at that time the principal source of chemical carbon. The requirements of the dyestuffs industry for intermediates derived from coal tar aromatic hydrocarbons, and of the extractive, paint, lacquer, and (later) engineering industries for solvents provided the primary markets for the first heavy organic chemicals.

Up to the period of World War II heavy organic manufacture in the U.K. was based on fermentation alcohol, coal (as coke or calcium carbide), and coal tar derivatives. The revolutionary shift to petroleum as a source of organic starting materials was initiated during that war and was almost entirely promoted by the raw material and other shortages it produced. Production of fuel petrol from coal, begun by ICI in the 1930's (Section 4 below), may be seen historically as a bridge between the coal-carbon heavy organic chemical industry and the modern petrochemical development. This petrol process arose primarily, however, as a consequence of the development by that company of techniques for purifying and utilizing hydrogen on a vast scale for ammonia manufacture; in other words, this heavy *organic* departure arose, like that of the chloro-hydrocarbons, as a consequence of changes in the technology of heavy *inorganic* chemical manufacture. It is to be noted that the large tonnage modern production of methanol by reduction of carbon monoxide ($CO + 2H_2 \rightleftarrows CH_3OH$), begun at Billingham in 1925, bore a similar relationship to the existing technology of nitrogen reduction—i.e., to the Haber ammonia process.

While some smaller firms have specialized in the production of organic chemicals on a tonnage scale, in the main, the very large financial outlay necessary, especially for petrochemical manufacture, has made heavy organic production the domain of a few of the very largest chemical companies. ICI came into the field as the continuator of the developments already existing in the pre-merger firms; Distillers Company Ltd followed the technological implications of its interest in the chemical exploitation of fermentation alcohol; Courtaulds, originally interested in a few important organics in relation to rayon manufacture, extended that interest; the British Petroleum Company Ltd and Esso Petroleum represented the entry in the U.K. of the petroleum industry into chemical manufacture. Organizationally there are associative links between those leading heavy organic chemical producers, these being determined largely by the uses to which the primary chemicals are put.

Despite the existence of heavy organic chemical manufacture from the earliest years of this century, recognition of its existence as a distinct branch of the chemical industry was long delayed and hesitant. When the Association of British Chemical Manufacturers produced in 1949 its well-known "Report on the Chemical Industry", the term had come only into recent use. According to that Report: "It applied to compounds mainly synthetic, used as prime industrial raw materials and produced in bulk and is intended to distinguish them from those grouped under the term 'fine chemicals' which are generally produced in comparatively small quantities." Confusion of organic chemical manufacture with fine chemical production was of long standing. By the early 1900's some three thousand organic chemicals were in established production in this country on scales ranging from that of analytical reagents to hundreds and, sometimes, thousands of tons; attempts to divide " fine " from " heavy " provided a classical example of taxonomic controversy. Some authors were driven to define a fine chemical as one produced by the fine chemical industry, thus paralleling the definition of an archdeacon as a priest who performs archidiaconal functions! In his 1939 Cantor Lectures the late Prof. Sir Gilbert T. Morgan equated fine chemicals with dyestuffs. Without entering here into the historical causes of this confusion, it can be said that it rested on the fact that many chemicals were at that time not produced primarily in the state of purity in which even high-tonnage chemicals are manufactured today. At the present time many inorganic and organic chemicals, as formerly, are produced in refined grades for medical, analytical, and research purposes; these, if it is so wished may be described as fine chemicals, but there is now no important technological justification for treating their production as a fundamentally distinct branch of the chemical industry.

The principal applications of heavy organics are as intermediates for dyes and pharmaceuticals, monomers and plasticizers for plastics, solvents in a large range of manufacturing industries, and as starting materials for domestic and industrial detergents. The dyes industry has for many years been a producer of such heavy organic chemicals as nitrobenzene, aniline, naphthylamine and phthalic anhydride, but these have been made captively for its own purposes and have not generally been marketed for use in other branches of technology. Likewise, in the explosive industry, nitro-cellulose and nitro-glycerine had a lengthy history by 1900. In more recent times the plastics industry has to a great extent produced its own starting monomers, e.g., vinyl chloride and acetate, methylmethacrylate, and styrene (see pp. 201 ff.). In the sections which follow an attempt will be made to map the principal features of the heavy organic chemical landscape and to show how these were technohistorically determined.

1. CHLORO-HYDROCARBONS

Chloro-derivatives of the simpler hydrocarbons were amongst the first heavy organics to be produced in the U.K. Their initial production was speculative—no established outlet existed for them, except for the chloro-aromatics which were primary chemicals in the dyestuffs industry. In fact, the early chloro-organics were mainly produced in an endeavour to find an economic outlet for the chlorine no longer required for bleaching powder (p. 119). In 1897 a battery of Semet-Solway coking ovens was set up at the Sullivan Works of the United Alkali Company at Widnes; this produced benzene, light oils and ammonia. The last-named chemical was disposed of as fertilizer sulphate, and the benzene was chlorinated to various chlorobenzenes. Coke from the ovens was used for steam-raising. The whole arrangement was regarded as a potentially profitable extension of the LeBlanc system of alkali production. By 1907 the United Alkali Company was marketing about a dozen " tar products ". In 1914 the coking installation was dismantled, but production of chloro-benzenes from coal tar benzole was continued, and these are still made at that site today. Chloro-benzene first acquired large-tonnage importance during World War I when it was required for making phenol for picric acid manufacture. During World War II chloro-benzene began to be used in production of DDT (p. 185). The dichloro-benzenes are produced on scales of several thousand tons annually, for use as pest control agents and for air purification. Towards 1900, at Runcorn, production of carbon disulphide, an organic derived by direct interaction of coke and sulphur, began (p. 137). This volatile, inflammable liquid found considerable application as a solvent and as a vulcanizing agent in the rubber industry; it was also one of the starting materials of the Raschen cyanide process (p. 62). The history of carbon disulphide manufacture in the U.K. is obscure; it must, however, have been produced from about 1840 when it was used by Jesse Fischer in his fat extraction process. Carbon disulphide manufacture by an electric furnace process was carried on in Widnes till 1964.

In the early 1900's, Albright & Wilson Ltd, the only owners of a chlorine plant in the Midlands, found an outlet for this product by making carbon tetrachloride by its reaction with carbon disulphide ($CS_2 + 2Cl_2 \rightarrow CCl_4 + 2S$). This process, begun on a 250-ton-a-year scale, was extended during World War I to supply Government requirements. In 1926 the process was established as a continuous one, in place of the original batch production, and the sulphur for carbon disulphide manufacture was recovered from the phosphorus sesqui-sulphide plant. Before 1933, when Albright & Wilson transferred their carbon tetrachloride production to Widnes, they were making carbon tetrachloride on a 5000-ton-a-year scale. At Widnes, both the chlorine and carbon disulphide required for the manufacture were obtained from the nearby ICI works producing these chemicals. Thus for half a century, carbon tetrachloride, a typical heavy *organic*, continued to be a product of the heavy *inorganic* chemical industry. Carbon tetrachloride found its first applications as an extraction and cleaning solvent and as a fire extinguishing fluid; it was also used for grain fumigation; its modern importance in the chemical industry lies in its use as the starting material in the production of the chloro-fluoro methanes (p. 151).

After a brief and promising history as an illuminant and establishing itself as a welding gas, the potentialities of acetylene as a chemical raw material were recognized. The original development of chemical syntheses from acetylene took place in Germany and Austria. The first possibility to be investigated was that of chlorination. In 1903–05, the Consortium für Electrochemische Industrie in its laboratories at Nuremberg developed processes for making tetrachloroethane from acetylene and of producing trichloroethylene from tetrachloroethane. By 1908, industrial production of trichloroethylene had been established at the factory of the Bosnische Elektrizitäts AG at Jajce, using chlorine from a nearby electrolytic caustic-chlorine works. There being only tentative outlets for the new chemical, production was initially on a scale of only a few tons a year. In 1908 the Consortium formed with the Castner-Kellner Alkali Company the Weston Chemical Company, with a site near Runcorn in Cheshire. At that time the Castner-Kellner Company was looking for employment for its excess chlorine capacity. The operations of the Weston Company were under the management of Gustav Koller, a German chemist, who in an article in the *Chemical Trade Journal* (3rd July 1909, 7) publicized the products his company was producing and claimed: "The boot polish, metal polish, and furniture cream trades have already availed themselves of this useful material (tetrachloroethane). Printing establishments use it for cleaning rollers, lithographic stones and clichés. Dichloride of ethylene* is an excellent material for dissolving rubber and for producing insulating varnishes. In the manufacture of perfumes from flowers dichloride of ethylene . . . is taking the place of Petroleum ether." Although, at this time, attention appears to have been directed mainly to tetrachloroethane, subsequent development was mainly concerned with its derivative, trichloroethylene. Dichloroethylene has continued to be manufactured on a comparatively small scale up to the present day. The important property of this group of chloro-derivatives is that, while being powerful solvents like carbon disulphide and petroleum ether, they are non-flammable.

During the years 1909–29 manufacture of Westrosol, the Weston Company's brand name for their trichloroethylene, provided a very variable outlet for only a few hundred tons of chlorine annually. In the meantime, as a result of World War I and the merger of 1926, the activities of the Weston Company had come under the control of ICI. A principal reason for the slow commercial development was that no thorough-going effort was made to interest industry in the technical potentialities of the solvent, presumably because it was not seen to be a promising outlet for chlorine. Several simple but highly effective inventions in the mid-1920's revolutionized the technique of degreasing metals. Introduction of these to the engineering trades by an energetic and extensive series of demonstration campaigns created

---

* Koller was here referring to dichloroethylene, $CHCl{=}CHCl$.

from then onwards a very rapidly increasing demand for trichloroethylene. Between 1928 and 1936 there was a ten-fold increase of consumption of trichloroethylene by industry in the U.K. During World War II and subsequently consumption of trichloroethylene and its derivative perchloroethylene ($CCl_2=CCl_2$) has continued to rise at almost the pre-war rate. Today, trichloroethylene, made in the U.K. only by ICI, is one of the largest tonnage heavy chloro-organic chemicals. The manufacturing relationships of the various chloro-solvents derived from acetylene are as follows:

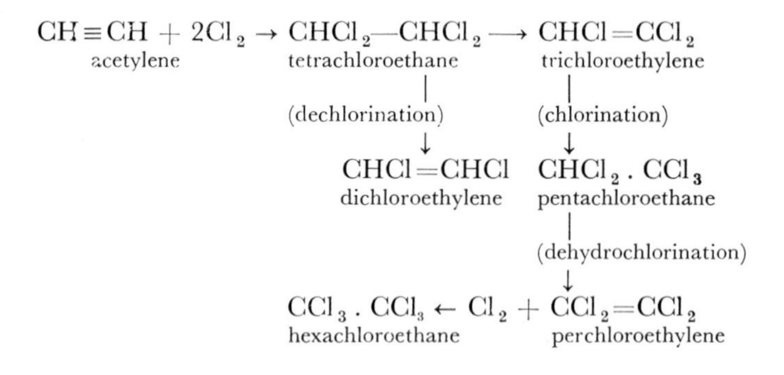

The chlorination of acetylene is carried out in solution in tetrachloroethane in the presence of ferric chloride or antimony sulphide catalyst. Dehydrochlorination of tetrachloroethane and pentachloroethane is effected by means of lime slurry—a waste from the acetylene generators ($CaC_2 + 2H_2O \rightarrow Ca(OH)_2 + C_2H_2$). Chlorination of perchloroethylene to hexachloroethane has been carried out at Runcorn since the mid-1930's. In peacetime, hexachloroethane finds limited application for degassing magnesium, as a constituent of extreme pressure lubricants, and as a moth repellent; during World War II it was produced in considerable tonnages for use in military smoke-screen mixtures.

Apart from carbon tetrachloride and small-tonnage manufacture of methyl chloride in the 1930's, the chloromethanes had little industrial importance in the U.K. until World War II and the immediately post-war years. Methyl chloride, known since 1835, was first produced in this country by ICI in 1930 as a refrigerant. The process used involved reaction of hydrogen chloride with methyl alcohol in the presence of zinc chloride. With the beginning of silicone manufacture and the production of the new chloro-fluoro-refrigerants, demand for the chloromethanes suddenly increased, and ICI established a new plant at Runcorn for their manufacture. Methylene chloride, produced by chlorinating methyl chloride, also acquired importance at this period as a solvent for use in paint-strippers, in aerosols, and in the manufacture of photographic film and synthetic fibres. Since World War II there has been a ten-fold increase in methylene chloride demand in the U.K.

Ethyl chloride was for many years manufactured in this country on a small scale as a local anaesthetic (p. 75); its large-tonnage production by ICI began during World War II, when it was required for TEL production (p. 154). The process employed was similar to that for methyl chloride, but using ethyl alcohol. Ethyl chloride is now produced by the Associated Octel Company Ltd the monopoly makers of TEL in the U.K. Octel manufacture the chloride by an American process in which ethylene is reacted with hydrogen chloride ($CH_2{=}CH_2 + HCl \rightarrow CH_3CH_2Cl$). Hydrogen chloride for this stage is produced by chlorination of ethane-rich gas, a step which also results in formation of ethyl chloride. The ethylene is obtained from the nearby Shell petroleum refinery at Stanlow in Cheshire: ethyl chloride, since 1955, has been a petrochemical. It is estimated that British output of ethyl chloride is now about 40,000 tons annually.

Following Bolley's study in 1856–58 of the viscous products obtained by passing chlorine into molten paraffin wax, little attention was given to these materials until the early 1900's, when the United Alkali Company made an abortive attempt to produce them commercially. ICI, at its Widnes works in 1940, began manufacture of a series of chloro-paraffin waxes of specified viscosities; since that time demand for them has expanded to a scale of many thousands of tons annually. The chloro-paraffin

FIG. 10. A PETROCHEMICAL LANDSCAPE

A general view of the Carrington Works of Petrochemicals Ltd (a wholly-owned subsidiary of Shell Chemical Company Ltd). The scene conveys the complex yet orderly organization of a giant modern chemical factory based on petroleum. In the middle distance the chimney (375 ft) was built to avoid atmospheric contamination.

waxes are mainly used as flame-proofing agents, extreme pressure lubricants, additives to cutting and drawing oils, and as so-called plasticizer " extenders ", i.e., substances which increase the effectiveness of conventional plasticizers and, therefore, result in economy in their use.

As long ago as 1833, Laurent discovered that wax-like materials resulted from reaction of chlorine with naphthalene. The technical potentialities of these chloro-waxes were first realized in the U.S. in the early 1900's, and during World War I waxes of this type were used in Germany as protective coating materials and insulants. Manufacture of chloronaphthalene waxes has been carried on in the U.K. by ICI since 1932. Although chloronaphthalenes are still used on a tonnage scale, the discovery of their high toxicity has considerably limited their application in recent years.

Chlorination of rubber has a long history. In the early years of this century the United Alkali Company marketed a chlorinated rubber under the name of " Duroprene ". This product had some application as a chemically-resistant varnish in the brewing, tanning and tin-plate industries for protection of plant against corrosion. Chlorination of rubber to give a clear viscous liquid was the result of long research. Paralleling developments in the U.S. and Germany, the Widnes Laboratory of ICI General Chemicals Division (now Mond Division) succeeded in 1935 in making a chlorinated rubber having the sought-for properties, as well as in producing solid chloro-derivatives of rubber with technically valuable properties: these products were marketed under the name " Alloprene ", and have been produced in large tonnages for many years. " Alloprene " contains over 60% of chlorine and is, therefore, an important vehicle for marketing that element in a " high value " form.

Chlorinated hydrocarbons now have a unique and important place in the economy of the modern chemical industry: they provide in their manufacture the principal routes by which chlorine reaches the heavy organic chemical market, and constitute by far the most important link (if the chloro-plastic monomers are included) between the inorganic and organic sides of the industry. As has been shown, their development began with the century and was accelerated by the emergence, during World War I, of liquid chlorine as the chlorine commodity in place of bleaching powder. During World War II the chlorohydrocarbons rose to their present dominant position in the chemical industrial field, participating in the post-war development of heavy organics. It may be estimated that at the present time (1965) between 300,000 and 350,000 tons of chlorine is annually used in the U.K. in the manufacture of heavy organic chemicals of all kinds.

2. ETHYL ALCOHOL AND ITS DERIVATIVES

In 1877 six grain whisky distillers merged to form the largest Scotch whisky concern in the U.K. Some half-century later the Distillers Company Ltd (DCL) decided to exploit the potentialities of industrial as well as potable alcohol; by that time DCL was producing half the national turnover of the latter and was in an excellent economic position to finance a new enterprise connected with the former. Earlier this century DCL, an essentially Scottish enterprise, recognizing the growing importance of alcohol in industry, established a number of distilleries in England. DCL's subsidiary, Hull Distillery Company Ltd, established in 1925–26 a distillery at Salt End, Hull. Alongside this Hull works was erected in 1928 the factory of British Industrial Solvents Ltd, another subsidiary, founded to make industrial derivatives of ethyl alcohol. At Hull, the distillery used as its starting material, not grain, but molasses, a sugar residue containing about 50% of fermentable saccharides. In 1927 the Methylating Company Ltd, then a subsidiary of DCL, acquired a small chemical works on the River Wande at Carshalton in Surrey; this works was the first chemical factory to be owned by the Company; it was later operated by British Industrial Solvents Ltd, and is now part of DCL's Carshalton Division.

At Hull, ethyl alcohol was converted to five chemicals: acetone (by action of steam in presence of a catalyst); acetaldehyde (by air oxidation); butaldehyde and butyl alcohol (by intermediate formation of aldol and crotonaldehyde, followed by hydrogenation of the latter), and acetic acid. Ethyl alcohol, acetic acid and butyl alcohol were transported to the Carshalton works where they were used in the manufacture of a range of esters which were marketed as solvents and plasticizers. During this period DCL was also producing acetone and butyl alcohol by deep-fermentation processes, first at a factory in King's Lynn, and later at Bromborough in Cheshire (Commercial Solvents (GB) Ltd). During World War II, DCL manufacture of industrial alcohol and related chemicals was largely under Government control and technological changes in processes were not possible. At that time the U.K. chemical industry could not meet the total demand for acetone (largely required for cordite production) and acetic acid, which had to be imported in large tonnages from the U.S. and Canada.

Already in 1929, DCL, in its research laboratories, was giving consideration to the potentialities of petroleum as a starting material for manufacture of chemicals. At the same time the British Petroleum Company was beginning to investigate the possibilities of a U.K. petrochemical industry. Petroleum appeared attractive as a raw material for alcohol manufacture both because of its great abundance and the stability of its price, compared with that of molasses which fluctuated with the sugar crop and the policy of the sugar-making companies.

Before World War II was at an end DCL had decided to base its chemical manufacture on petroleum. Immediately after World War II, British Petroleum embarked upon a far-reaching reorganization of its Grangemouth refinery; this was taken as a propitious moment for DCL and the petroleum company to form a joint chemical manufacturing concern. In 1947, British Petroleum Chemicals Ltd was established to produce the olefins ethylene and propylene and the derived ethyl and isopropyl alcohols as the basic materials for a range of further syntheses. Production of petrochemical ethanol began in 1951 and, by 1957, DCL had only one distillery in operation for production of industrial alcohol. This development, while providing a most striking illustration of the transition to petroleum as a chemical raw material, is also an instance of an interesting process reversal. While, formerly, ethylene was produced by chemical *dehydration* of fermentation ethyl alcohol, ethyl alcohol itself is now almost wholly made by *hydration* of ethylene from petroleum cracking. The convention has now arisen of referring to the product of fermentation as " ethyl alcohol ", and to that from ethylene hydration as " ethanol "; they are, of course, chemically identical. Today the products of DCL's Hull factory, formerly based on fermentation alcohol, are produced from petrochemicals; they include acetone and acetates for the paint and lacquer, printing ink, cosmetics, and textile industries; other heavy organics produced include acetic acid and its anhydride as well as higher alcohols. A new process for acetic acid, developed in DCL laboratories, came into operation in a £2 million plant at Hull in 1962; this plant is being extended, when it will be the largest acetic acid plant in the U.K.

3. HEAVY ORGANIC CHEMICALS FROM COAL

The spectacular development of manufacture of chemicals from petroleum has, in recent years, tended to diminish in the public eye the continuing importance of the production of chemicals from coal. Heavy organics are obtained from coal by three routes: distillation of tar, by various extraction techniques from coal gas, and from coke ovens. Coal continues, in the U.K., to be the principal source of aromatic heavy organics, such as benzene, toluene, naphthalene, and anthracene. Although some aromatics can be obtained from petroleum, and are so obtained, particularly in the U.S., their production from this source has not proceeded as yet to a corresponding extent in this country, but is now likely greatly to increase. Following Perkin's discovery of the first aniline dye (p. 65), a demand for benzene was created by the rising dyestuffs industry; the discovery of processes for producing synthetic alizarin (1864) brought into being a demand for anthracene recovered from coal tar. Lister's introduction of antiseptic surgery in 1865 established the industrial production of phenolic disinfectants, with a corresponding requirement of phenol and " tar acids ". In the present century demand for phenol was greatly increased by the development of the phenol-formaldehyde plastics (p. 195). In the 1930's creosote became established as the starting material for synthetic petrol (Section 4 below)—an application which has now ceased.

For many years coal carbonization supplied the national requirement of benzene. The benzene-containing fraction from coal products is referred to in the trade as " benzole ", on purification by further fractional distillation this yields less than 50% of pure benzene. Benzole is produced mainly by coke ovens and gasworks. Increasing demand for aniline, for which benzene is the starting material, in the late 1860's made it impossible for the tar distillers to meet the demand. In 1869 a method was patented for recovering benzene from coal gas by extraction with heavy petroleum and tar oils. W. Young, in 1875, devised a scrubbing tower for that purpose and thereby established the modern method of obtaining benzene from that source. Recovery of benzene by extraction from coal gas became important in the early 1880's, when its price rose from three to fifteen shillings a gallon. In 1882, the first range of by-product coke ovens in the U.K. was installed at Crook in County Durham, and two years later Carvès devised his oil-extraction process for obtaining benzene from coke oven gas; from then onwards benzene recovery has been increasingly associated with the coking industry. During World Wars I and II benzene recovery became an operation of national importance. Between World Wars I and II the industry rose to its present extensive development.

Coal carbonization, until the formation of the Gas Council in 1949, was in the hands of over one thousand municipal and private undertakings—one-third of the former and two-thirds of the latter. Tar distillation was, in the main, a private enterprise; as it expanded there was a tendency—at first resisted by the distillers—for tar producers to set up their own distillation

Fig. 11. MODERN CHEMICAL FACTORY
This shows typical vessels for storage of gaseous, liquid and solid products.

plants. When the Association of Tar Distillers—one of the earliest trade organizations of its kind—was formed in 1885, it had twelve founding members; by 1939 this number had risen to sixty-six. At the present time fewer than twenty-five concerns in the U.K. are engaged in production of chemicals from coal, either *via* tar or coking; most of these modern concerns, however, consist of mergers of groups of long-established companies and include Divisions of the National Coal Board. On vesting day, 1st January 1947, the Board found itself owner of 55 coking plants. Today the three sources of coal tar in the U.K. are the Gas Boards, the National Coal Board, and the steel companies' coking ovens, representing an impressive degree of Government control over an important sector of a primary raw material of heavy organic chemical industry!

Coal tar is fractionated into four organic fractions (after removal of the ammonia content); the residue from the distillation process is pitch. Several hundred chemicals have been isolated from the organic fractions, although only a few of these are generally extracted industrially. The most important heavy organics obtained from tar are: benzene, toluene, xylenes, naphthas, naphthalene, anthracene, phenol, cresols, and pyridine with its derivatives the picolines. The tonnage of tar available, which was 2,591,000 in 1914, has decreased steadily, but not very rapidly, in recent years. In 1957 the tonnage was 3,083,000; by 1964 this had declined to 2,518,000. Since 1957, however, the output of the principal heavy organics from coal tar has not greatly changed, except in the case of naphthalene, which has considerably increased:

	1957	1962	1964
naphthalene (all grades)	49,000	68,400	88,000 tons
cresols	70,000	66,000	67,300
phenol	15,000	16,300	19,400

Since the early 1950's phenol has been produced in the U.K. synthetically from benzene by the sulphonation and cumene processes. The latter route employs petroleum-derived propylene in making the cumene (*iso*propylbenzene) ; phenol has, therefore, to that extent become a petrochemical. DCL employs the cumene process to produce the phenol required for synthetic resin manufacture. Statistics of synthetic phenol production are not available.

4. SYNTHETIC PETROL

The term " petrol " was first used by the firm of Carless, Capel & Leonard Ltd and applied to the highly volatile spirit that company produced for the new internal combustion engine in the 1890's. Interest in coal as a source of petrol arose in this country before 1920. Considerable research was done by the National Benzole Association (since 1954 known as the National Benzole and Allied Products Association) on making suitable fuel fractions from coal tar benzole. Benzole as a motor car fuel was marketed by the Benzole Association from the early 1920's onwards; since World War II benzole has been marketed by the Shell Company. During the 1920's chemists in the U.K. began to interest themselves in extending the use of coal as a petrol source, having in view the fact that suitable coal was available at the low cost of less than £1 a ton. Petrol contains about three times as much hydrogen as there is in an average coal; the problem of using coal as a petrol source resolved itself, therefore, into decomposing the coal in the presence of hydrogen under such conditions as to form the more highly-hydrogenated hydrocarbons. Three processes were developed for effecting the synthesis: carbonization, the Fischer-Tropsch, and Bergius processes; the latter two originated in Germany. Carbonization was the method already in use, namely, recovery of benzole from coke ovens and gasworks. This yielded only about 2 gallons of suitable motor fuel per ton of coal carbonized. Even under the most favourable low-temperature carbonizing conditions the petrol yield is only 10% of the coal decomposed.

The Fischer-Tropsch process, which uses expensive catalysts, has as its starting materials hydrogen and carbon monoxide derived from coal. Bergius' process, invented in 1913, involves the direct hydrogenation of coal in powder form under a pressure of 250 atmospheres. The product of the Bergius process is a hydrocarbon oil, easily further hydrogenated to yield a good petrol.

At Billingham Works the technique of high pressure hydrogenation had been studied and developed on a vast scale in connection with ammonia manufacture. It was inevitable, therefore, that when there appeared to be a prospect of a world petrol shortage in the late 1920's, the possibility of applying this hydrogenation technique to coal should have been considered by ICI. It was decided to adopt a variant of the Bergius process. Research began in 1927, just after the ICI merger had made possible the financing of large-scale new enterprises in chemical production. By 1929 a pilot plant was in operation, hydrogenating 10 tons of coal a day. During the 1931–32 slump the recession in fertilizer demand released hydrogen and steam-raising capacity at Billingham. Coal was available at 13 shillings a ton. There was thus every incentive to continue with the petrol-from-coal project; there was, however, little prospect of the synthetic petrol being economically competitive with that obtained from petroleum.

In order to encourage production of motor fuel from coal, the Budget of 1928 imposed a duty of 4*d.* a gallon on imported oil; this preference was guaranteed, in 1934, by the British Hydrocarbon Oil Act. A full-scale petrol plant was then erected at Billingham, and came into operation in 1935. From the outset, coal hydrogenation was carried on in conjunction with hydrogenation of creosote oil, it being found more satisfactory to use the two raw materials rather than coal alone. The creosote used was brought by rail-tankers from the tar distilleries of Lancashire, the Midlands and Scotland. About a quarter of the total came from South Yorkshire and the London area by coastal steamers from the Humber and the Thames. In 1936, 36 million gallons of petrol were produced at Billingham. Output continued at that level until the outbreak of World War II. In 1938 the Falmouth Committee reviewed the coal hydrogenation situation, and, as a result of its recommendations, the Finance Act of 1938 guaranteed a preference of 8*d.* a gallon on synthetic motor oil. When World War II began, coal hydrogenation ceased, but creosote hydrogenation was continued. During the war Billingham produced high-grade aviation spirit from that raw material. The peak annual output achieved was 100,000 tons from 36 million gallons of creosote.

Research at Billingham during World War II discovered that aviation spirit could be upgraded by addition of butyl benzene, a chemical that became known as " Victane ". A fuel containing 12% of " Victane " and only a small amount of TEL had an octane rating of up to 160. From early 1944 " Victane " was produced at Billingham at a rate of 25,000 tons a year. The high-octane spirit produced by its means gave the Spitfire fighter an additional 25 mph at a time when additional speed was required to chase the German flying bombs. After the war, aviation spirit continued to be made at Billingham until 1947, when output was reduced to 25,000 tons a year. During the Berlin Airlift production was temporarily raised to 40,000 tons a year.

Important by-products of coal hydrogenation at Billingham were the gases propane and butane. These were liquefied, the former finding application in the engineering industry as a welding gas, the latter being marketed as a domestic fuel under the name " Calor Gas ".

By the mid-1950's the price of coal was rising and the price of oil falling at such rates that production of motor fuel from coal or coal products was no longer economic. Along with other coal-based processes at Billingham coal hydrogenation was abandoned and the plants dismantled. Petrol, propane and butane are now products of the petroleum-based complex established at ICI's Billingham and Wilton factories (Section 5 below).

5. PETROCHEMICALS

It may be claimed with techno-historical accuracy that development of the use of petroleum as a raw material for organic chemicals arose as a direct result of the steadily extending use of the internal combustion engine. When petroleum stocks were more intensively converted to petrol, an economic outlet had to be found for the by-product hydrocarbons from the conversion processes. Addition of petroleum to the carbonaceous raw materials of the chemical industry has made possible the manufacture of heavy organics on a scale and at a cost which would not have been possible on the basis of the older carbon materials. In particular, the great post-World War II detergent industry (Section 6 below) could not have attained its present impressive proportions without the availability of petroleum-derived primary chemicals. Between 1953 and 1962 production of organic chemicals from the various carbon sources increased as follows:

	1953	1962	
petroleum	200	1500	$\times 10^3$ metric tons
coal tar	230	540	
calcium carbide	80	150	
synthesis gas	70	90	

In 1949 petroleum stocks provided only 6% of the raw material for organic chemical manufacture in the U.K.; ten years later the proportion was 51%, and at the present time (1965) probably between 60 and 70% of organic chemicals in this country are produced from petroleum. Such extensive increases in production involved correspondingly great investment in new plants. It has been estimated that by 1959 a total of some £120 million had gone to the erection of petrochemical plants to produce chemicals many of which had not previously been manufactured in the U.K. By 1962 investment in organic* petrochemical installations had risen to a total of about £200 million. Outside the U.S., Britain has so far made the largest commitment to

* *Inorganic* petrochemicals (sulphur, carbon black and ammonia) amounted to 300,000 metric tons in 1962.

FIG. 12. GENERAL VIEW OF WILTON WORKS

This ICI factory represents an investment to date of £120 million. Integration of the various process operations has determined the lay-out in the "grid-iron" plan of a modern town with streets paralleled by trenches containing the piped services (water, steam, air, etc.) to the plants, and also the pipes conveying their products. At Severnside, ICI has established a similar petrochemical production. Between 1953 and 1960 the number of petrochemical plants in operation in the U.K. increased from 5 to 15.

A petrochemical may be defined as a material produced by a combination of thermal and catalytic conditions from a petroleum fraction or from refinery or natural gas (methane) and used technologically, either as such, or as a primary substance in further chemical manufacture. The primary products of petrochemical manufacture are comparatively simple organic compounds such as ethylene, propylene, butylene and the aromatics benzene, toluene, and the xylenes; synthetic elaboration of these may take place at the petrochemical works or at another chemical industrial site, usually nearby. In the U.K. three principal types of petrochemical processes are in operation: cracking (to produce olefins), the so-called Oxo-process (to produce aldehydes and alcohols), pyrolysis or platforming (followed by solvent extraction processes to obtain aromatics).

The cracking process consists in subjecting petroleum fractions of the 30 to 210°C boiling range to thermal decomposition in the presence of catalysts. It was developed in the U.S. in the course of the expansion of motor car fuel production in that country. The petrochemical industry started in 1918, when Standard Oil of New Jersey produced a few tons of propylene by thermally cracking crude oil fractions. The process, which is well-named " cracking ", breaks down the complex molecules of the petroleum. The cracking stage is followed by one for the separation of the simplified hydrocarbons. It has been established empirically that a given petroleum feed-stock under the same cracking conditions will always yield a similar mixture of products. While the general mechanism of cracking is clearly understood, the physical-chemical steps by which the individual hydrocarbons are produced can not yet be explained: indeed, petrochemical manufacture rests on a well-established *empirical* basis, rather than on the kinetic elucidations of physical chemists. A typical cracker gas contains hydrogen, methane, ethylene, acetylene, ethane, propylene, propane, and hydrocarbons containing four carbon atoms in their structure. By modification of the process conditions, it is possible to increase the yield of particularly desired products. The Oxo-process, discovered in 1938, was developed in the U.S. and Germany. It consists in the reaction of olefins, such as propylene, with carbon monoxide and hydrogen in the presence of cobalt-containing catalysts, whereby aldehydes are produced. By reduction of the product aldehydes, alcohols are obtained which form important starting materials in the manufacture of plasticizers; acids are produced by oxidation of the aldehydes. In the Udex process aromatics are obtained from the products of thermal-catalytic treatment of petroleum fractions by means of extraction with a solvent or solvents such as aqueous diethylene glycol.

In the thermal-catalytic, or catalytic reforming or platforming processes, the vaporized petroleum fraction is passed over a dehydrogenation catalyst in an atmosphere of hydrogen. An important aspect of the petrochemical processes is that they are all of the *continuous* type as opposed to the batch processes by which many organic chemicals were formerly manufactured. By the processes of hydrogenation, oxidation, hydration, chlorination and polymerization of the primary olefins and aromatics obtained by the initial petrochemical processes, several hundred heavy organic chemicals may be obtained; these have their applications in almost every branch of modern technology from agriculture to textiles and from paper manufacture to synthetic rubber.

British Celanese Company (now merged with Courtaulds) was the first concern in the U.K. to produce olefins and chemicals derived from them, by a petrochemical process. The cracker, designed by that Company's own chemical engineers, came into operation in 1942 at Spondon mainly to produce chemicals required for rayon manufacture. After World War II large oil refineries were established in this country in areas where large chemical markets existed. By constructing chemical plants alongside these refineries it was possible to make the most advantageous incursion into the chemical field. The Shell Company's petrochemical plants at Stanlow and Carrington, Monsanto and Union Carbide's plants at Fawley on Southampton Water, and British Hydrocarbon Chemicals factory at Grangemouth in Scotland all provide examples of petrochemical manufacture adjacent to the petroleum refineries supplying their feed-stocks. ICI's petrochemical operations at Billingham and Wilton were, however, sited there primarily because of the local availability of labour and suitability as a centre from which to distribute the products to other parts of the company's manufacturing organization. The chemical industrial complex established by ICI at Severnside will obtain its feed-stocks by pipe-line from Fawley.

As has already been stated (p. 227), petrochemical production, because of the large financial outlays involved, is the domain of the largest chemical manufacturing companies. The firm of Shell began its operations in this field before World War II. Shell, in 1945, formed two subsidiaries, Shell Chemical Manufacturing Ltd and Shell Chemicals Ltd, to deal with its petrochemical operations; these, in 1955, were merged as Shell Chemical Company Ltd and at the same time Petrochemicals Ltd was acquired along with the Weizman and Ziegler patents which that concern was exploiting. In 1947 the Distillers Company Ltd formed a 50–50 partnership with the British Petroleum Company Ltd in establishing British Petroleum Chemicals Ltd with a factory at Grangemouth in Scotland (see p. 280), where a petroleum refinery had operated since early in World War II. British Petroleum Chemicals began manufacture in 1951, producing ethylene and propylene by steam-cracking a light petroleum distillate, and making from these, by hydration, ethanol and isopropanol Since that date the now named British Hydrocarbon Chemicals has erected plants for manufacture of butadiene, cumene, phenol, acetone, ethylene dichloride, isobutylene, and high-density polyethylene. The petrochemical complex at Grangemouth has attracted to the area several concerns requiring the primary chemicals produced by British Hydrocarbon Chemicals; these include Forth Chemicals, Grange Chemicals, Border Chemicals and Bakelite Xylonite Ltd, the last-named producing high-density polyethylene. International Synthetic Rubber Company and Marbon Chemicals Ltd have factories closely adjacent. British Hydrocarbon's second factory was erected, in 1961, at Baglan Bay, near Port Talbot in Glamorgan, where plants produce ethylene, butadiene and ethylene dichloride. The Port Talbot works has attracted to its neighbourhood factories operated by Forth Chemicals, Grace Chemicals and Pfizer Ltd. British Hydrocarbon Chemicals is now one of the largest petrochemical concerns in Europe, producing some half-a-million tons of organic chemicals annually. Forth Chemicals (already mentioned) was formed in 1950 to produce styrene monomer from petrochemical ethylene and benzene. This DCL subsidiary began production in 1953. Its Grangemouth and Baglan Bay plants together produce about 100,000 tons of styrene annually. Grange Chemicals was formed in 1953 to produce detergent alkylate (see Section 6 below). Border Chemicals is a tripartite subsidiary of DCL, ICI and British Petroleum. Founded in 1963, it produces acrylonitrile at Grangemouth, using propylene from British Hydrocarbon Chemicals plant and ammonia from ICI's Billingham works. The process employed was developed by DCL in collaboration with Ugine of France. Acrylonitrile is required for synthetic fibre manufacture and in the production of synthetic rubbers.

ICI began in 1943 to plan an expansion of its production at a new chemical complex in which the plants would supply each other with chemicals and where there would be a unified provision of services. (The individual plants were to be operated by the appropriate manufacturing divisions of the Company.) This project was particularly suited to basing much of the new works' activities on petroleum. A 2000-acre site was acquired on the south side of the estuary of the River Tees at Wilton. Work on the new site began in 1946, and the first plants came into operation three years later; by 1960, the plants, by then twenty-five in number, covered 600 acres. A 10-mile pipeline, passing under the Tees, links Wilton with ICI's Billingham works and enables products to be conveyed between the two sites, which together form the largest chemical complex in Europe and one of the largest in the world. At the present time Wilton has almost 12,000 employees and some £120 million has so far been invested in installations there. Wilton is administered by Wilton Council, but each manufacturing division retains complete technical and commercial control of its own operations within the complex. The chemical heart of operations at Wilton is formed by the three naphtha crackers, supplied with petroleum naphtha brought by tankers to Teesport and piped to the site. Ethylene and propylene are the principal products of the cracking operation, along with hydrogen, methane, butylene and butadiene; residual liquid hydrocarbons provide a premier grade of petrol and fuel oil. Ethylene is used at Wilton for polyethylene (" Alkathene ") manufacture and production of ethylene oxide, which latter is employed for making ethylene glycol and " Lissapol N " (a constituent of detergents). Butadiene from the crackers is used at the site in the manufacture of various synthetic rubbers. Not all the heavy organics made at Wilton are petrochemicals; xylenes, imported from the coal tar distilleries, are fractionated and the para-isomer oxidized to terephthalic acid for " Terylene " manufacture (p. 218). Naphthalene, also from tar distillation, is oxidized to phthalic anhydride, and methanol (from Billingham works) is oxidized to formaldehyde; alpha-naphthylamine* (a dye intermediate) was produced from naphthalene, using nitric acid from Billingham and hydrogen from the electrolytic caustic-chlorine plant on site. In all, some fifteen heavy organic products are manufactured at Wilton, each in very large annual tonnage.

Esso Petroleum Company entered the U.K. petrochemical industry in 1957, initially to supply a specific range of heptenes to ICI; later this was extended to ethylene and butadiene. International Synthetic Rubber Company, Monsanto and Union Carbide established plants alongside the Esso refineries to take petrochemical products by pipeline for their own manufactures,

* Wilton alpha-naphthylamine plant has recently been shut down.

synthetic rubber, polyethylene, and ethylene glycol, respectively. Ethylene is also supplied to ICI's Severnside complex.

Although in production of chemicals from petroleum, Britain has led amongst European countries, manufacture of acetylene from that raw material, despite developments in Germany, Italy and France, has only just begun in the U.K. In 1964, ICI brought into operation at its Castner-Kellner factory the first acetylene-from-petroleum plant in the country.

6. THE NEW DETERGENTS

During an unknown number of centuries domestic and industrial cleansing has been effected by means of the sodium or potassium salts of long-chain fatty acids—i.e., soaps. Only in very recent modern times has there been a departure from this technological tradition. Chemically, most water-soluble detergents belong to the class of paraffin-chain salts of which soaps are examples. They contain one or several chains of carbon atoms linked (usually at one end) to an electrolytically-dissociatable group which is responsible for their effect at surfaces. This structure of hydrocarbon skeleton linked to a surface-active group provides the mechanism for detergent action. The new detergents in most instances depend for their action on possessing these two structural components in their molecules: the modern detergent departure is not so fundamental as it has sometimes been made to appear—in a wide sense the new detergents are soaps.

Modern detergents have emerged as a spectacular consequence of chemical industrial advance into an established field—the soap industry—with which it has become competitive. Exploitation of the new detergents has the socio-historical interest that, apart from patent medicine advertising, it has been the first extensive and persistent attempt to promote a class of chemicals directly to the consuming public. If our era may be described as the Aspirin Age (Ortega y Gasset), it can also be described as the Detergent Age, or as one writer has humorously called it, the Era of *Omo Sapiens*! The principal application of the new detergents is in cleansing and whitening textile goods, exemplifying the continuing relationship of textiles to chemical industry.

The first non-traditional detergent was produced as long ago as 1831, when seventeen-year-old Edmond Frémy, a pupil of Gay-Lussac, sulphated olive and castor oils. During World War I some use was made of sulphated detergents of this type in Germany, to counter the acute fat shortage then prevailing. BASF in 1916 produced Nekal A, the sodium salt of a sulphonated naphthalene carrying a paraffinic chain. Nekal was used in Germany as a wetting agent in textile manufacture but was not usually employed for general washing purposes. It was, however, the first detergent to be produced in industrial quantity and which was entirely synthetic and could be used as a soap substitute.

In the U.K. in the early 1930's, sodium dodecyl sulphate was marketed by Thomas Hedley & Company and Unilever as a constituent of soapless shampoos. " Teepol ", produced by Shell Chemicals Ltd, is made by sulphation of petrochemically-derived olefins and is, chemically, an alkyl sulphate. Between 1938 and 1949 " Teepol " was the detergent constituent of most of the synthetic detergents produced in the U.K.; it has since been displaced by alkyl-aryl-sulphonates, produced by condensation of petrochemical polypropylene with benzene in the presence of aluminium chloride catalysts, and sulphonating the product. Sulphonation of the alkyl-aryl hydrocarbon is often carried out by the detergent-making firm and not at the petrochemical factory. Since 1957 this so-called detergent alkylate has formed the active constituent of most domestic detergent powders. By 1959 capacity for alkylate production in the U.K. was twice that required to meet home demand and much had to be exported against severe competition. " Lissapol N " is made by ICI at its Wilton works from petrochemical ethylene oxide and alkyl phenol from Billingham. It is a liquid detergent used as a wetting and emulsifying agent in many industries; domestically it is employed in dilute form in various detergent formulations, e.g., " Stergene ".

Detergents as marketed are mixtures in which the active organic cleansing compound is generally a minor component.

A typical commercial detergent mixture contains, besides the organic constituent, sodium silicate, phosphate, sulphate and perborate, an optical whitening agent, and traces of a blue or green dyestuff. The detergent trade has thus provided a new market for several classical *inorganic* chemicals as well as for modern heavy organics. The optical whiteners are the producers of the " whiter than white " effect. It was discovered by Krais, in the 1930's, that a number of organic compounds (e.g., diaminostilbene) had this property; their use in commercial detergents became common from about 1955 onwards.

As early as 1953 the problem of foaming caused by detergents contained in domestic and industrial sewage was causing concern, since the blanket of foam produced is capable of destroying the purifying bacteria. It was found that when dodecyl sulphonate had a simple straight paraffinic chain, instead of a branched structure, it became biologically " soft ", i.e., it was capable of being destroyed by bacteria in sewage. Shell Chemicals in 1959 modified their Shell Haven plant to produce biologically "soft" "Dobane JN", straight-chain alkyl benzene. In July 1959 the Ministry of Housing and Local Government announced that a large-scale trial at Luton had shown reduction of detergent in the effluent from the sewage and in the River Lee.

Although the synthetic detergents of the early post-World War II period were inferior by present standards, some 35,000 tons of domestic detergents were sold in the U.K. in 1946, and a further 9000 were used industrially. By 1959 over 200,000 tons of detergents were being used annually, of which the greater part were substitutes for soap, the remainder being employed industrially for cleaning and laundering, and as wetting and dispersing agents. Present consumption of synthetic detergents in the U.K. probably lies between 350,000 and 400,000 tons a year.

7. OTHER ORGANIC CHEMICALS

A considerable number of organic chemicals are made in tonnage amounts by firms specializing in their production. Thus, for example, the Associated Octel Company (formed in 1938 as Associated Ethyl Ltd) specializes in the manufacture of tetraethyl lead (TEL) from petrochemical ethylene and ethane supplied by Shell refineries at Stanlow in Cheshire, and the small firm of Bowmans Chemicals Ltd (founded in 1908) at Widnes has a monopoly of lactic acid and lactates production in the U.K.; the long-established firms of Kemball, Bishop (now part of Pfizers) and J. & E. Sturge Ltd specialize in the manufacture of citric acid by fermentation. A number of concerns, such as W. J. Bush & Company (a subsidiary of Albright & Wilson Ltd), Berk Ltd, Chemical Compounds Ltd, Geigy (Holdings Ltd), and Hickson & Welch Ltd (for particulars of these and other companies see Part B of this work) produce ranges of organic chemicals and intermediates in widely varying quantities; many of these products are of the type which formerly would have been designated "fine chemicals". In the mid-1940's ICI and Imperial Smelting Corporation began manufacture of their ranges of organic fluorine derivatives, mainly as a consequence of their established production of hydrofluoric acid. These chloro-fluoro-methane and ethane derivatives are now produced in "heavy" tonnages and have acquired importance as refrigerants, non-toxic solvents and fire-extinguishing fluids; they have entered the public domain as propellants used in the dispensing of perfumes, insecticides, paints, and even condiments. At present the proportion of these fluoro-derivatives used as aerosol propellants exceeds that used for refrigeration.

The number of organic chemicals, other than dyes and pharmaceuticals, in established manufacture in the U.K. at the present time may be estimated as being of the order of 5000. Of organics produced on a scale of more than 1000 tons a year the number is probably less than 500.

CHAPTER IV

The Economic Importance of the Chemical Industry

GENERAL

The chemical industry is a service industry—the bulk of its production is used by other industries. Only a comparatively small proportion goes direct to the general public; examples are medicines, cosmetics and domestic cleaning materials. Every industry requires chemicals for some purpose or other—as raw materials or intermediates, or for finishing and testing. The main customers are agriculture, textiles, engineering and the building trades. The Organisation for Economic Co-operation and Development (OECD), from data from Western European countries, has estimated that the consumption pattern for chemicals is roughly as follows: chemical industry 25%, agriculture 10%, other industries 40% and direct to the general public 25%. The chemical industry is its own best customer. The products of one manufacturer become the raw materials or the intermediates of another, for example, sulphuric acid for making superphosphate and pigments for paints. The appended chart shows how complex the links are between the different sections of the industry.

The primary business of the industry is to supply the needs of the home market and after that to dispose of surplus production in the export field. This responsibility was laid down by the Government in the years following World War II, when industry in general was instructed to cater for export as its first duty even at the expense of the home consumer. This recognized officially the key significance of the chemical industry in promoting development of the national economy. As a corollary, the prosperity of the industry is intimately bound up with the well-being of its main customers. However competitive the chemical industry itself may be, its development will be largely influenced by the ability of its customers to meet their rivals on at least equal terms, both at home and abroad.

The modern chemical industry is based on scientific research and development. It is highly inventive and creative. The development of improved products and processes is going on all the time. It is an industry of rapid change. In the foregoing chapters it has been shown how from its discoveries new industries have arisen as, for example, plastics and man-made fibres. A constant task is to discover and produce new materials and methods to meet the changing needs of its customers as, for example, dyestuffs to colour the new fibres it has created and more sophisticated products to combat the many pests which threaten food supplies. In recent years change has become so rapid that products and processes may become obsolete or uneconomic long before the expensive plant has reached the end of its normal life. This calls for a high level of capital investment.

STATISTICS

There are masses of statistics relating to the chemical industry, both national and international, published by Government departments and industrial organizations. Global figures about the industry suffer from the defect that these generally differ in their coverage. This makes comparisons unreliable. The recently agreed International Standard Industrial Classification has been used by the U.K. since 1962 and by the OECD in its reports. The coverage for the chemical industry is wide but man-made fibres, an important chemical manufacture, are in the textile group, and are therefore not included in any of the statistics which follow. Even though this international classification has been adopted in principle, it is not always strictly adhered to in practice. In this country in the past, Government departments have used different definitions of the industry for their statistics and these have changed over the years. It would therefore not be profitable—in fact misleading—to attempt to trace the production history of the industry on the basis of these statistics. Only a few sample figures derived from the latest official publications will therefore be quoted to indicate the place of the industry in the economy of the United Kingdom. The Annual Abstract of Statistics, published by the Central Statistical Office, is a most useful compendium of information about the economics of this country, while the monthly Trade and Navigation Accounts, which since the beginning of 1965 have become the Overseas Trade Accounts of the United Kingdom, give up-to-date information regarding imports and exports. It must be mentioned, however, that the " Chemical and Allied Industries " division in the Annual Abstract of Statistics has a substantially wider coverage than the " Chemicals " division in the reports on imports and exports. The periodic Census of Production is the most complete official publication on production, but for reasons of industrial security, the data for most of the important individual chemicals are not shown separately, but are included in group totals.

PRODUCTION DATA

The first Census of Production for British Industry was for the year 1907. Since then, similar surveys have been made at irregular intervals, the last being for 1963. The figures are based on sales, which is considered to be the best indication of the position. In 1907, the sales of the chemical industry, as then defined, at £74 million were 5% of the total sales by all manufacturing industry. The percentage has increased steadily over the years. For 1958, the last census for which figures are available, the figure was 10·5%, based on chemical sales of £1700 million. Based on value of production, the chemical industry is the second largest in the U.K., mechanical engineering being first.

According to the latest OECD report the U.K. chemical industry had a turnover valued at £1930 million in 1963. In Western Europe, this was second to Germany, where the figure was 6% higher. In the U.S.A., the turnover was approximately six times that of the U.K.

The labour force in the U.K. chemical industry is about half-a-million, which is only 6% of the labour force of all manufacturing industry. This low percentage is due in part to the extensive automation in the industry, especially in large modern plants. The industry is not characteristically a large consumer of labour. Of the total number of employees, 34·5% are administrative, technical and clerical, a proportion which has very slightly increased in recent years and, today, is only exceeded by that of the U.S. among the chemical industries of the world.

Chemical enterprise is capital intensive, i.e., it employs a large amount of capital in relation to labour. It employs more fixed capital per employee than any other industry, except steel manufacture and oil refining. In the years indicated, capital expenditure in the industry on new plants and processes has been estimated to have been as follows:

1948 £38m, 1956 £155m, 1960 £150m, 1963 £159m.

In Germany and Italy capital per employee, in 1960, substantially exceeded the British figure:

U.K. £317, W. Germany £430, Italy £642.

The higher ratios in Germany and Italy arise from reconstruction programmes calling for heavy capital expenditures to repair the ravages of World War II.

Productivity in the U.K. chemical industry is lower than in many other countries. The latest OECD report gives the U.K. production index for 1963, based on 100 for 1958, as 141, compared with Italy 207, West Germany 173, European countries combined 167, U.S.A. 149, and Japan 202. In June 1964, the U.K. figure was 163.

It has been estimated by the OECD that the added value per employee in the leading chemical manufacturing countries in 1962 was as follows:

U.S.	£6600
Italy	2700
Germany	2570
France	2280
U.K.	1990

Comparison of these figures does not necessarily indicate which industries have a higher *over-all* productivity, but they do point to which industries are more highly capitalized and which operate modern processes on a large output scale.

The rate of growth of the U.K. chemical industry greatly increased in World War II and post-war years as the following compound interest increases in volume of output show:

1901–1938	2·5%
1938–1948	5·5
1948–1958	6·0
1958–1961	5·9

The greater part of this recent expansion is accounted for by the development of the organic sector of the industry—plastics, detergents, pharmaceuticals and heavy organic chemicals.

During the period 1958–1963, the output index of the U.K. chemical industry (based on 1958 = 100) has steadily increased:

1958	100	(100)	1961	125	(115)
1959	112	(106)	1962	130	(115)
1960	123	(115)	1963	140	(120)

The output indices for U.K. manufacturing industry as a whole are shown in brackets.

Like the chemical industry of America and Europe, the British is dominated by a relatively small group of very large firms. In 1961, of 2451 chemical firms in this country, over 60% employed fewer than a hundred workers, 685 had between one and five hundred, and only 29 had labour forces exceeding 2000.

This picture compares with West Germany where the three IG Farbenindustrie successors in 1960 produced 31% of the country's chemicals; France, where the ten largest companies produced a quarter of the total chemical output, and Italy where the single firm of Montecatini made half that nation's chemicals.

The difference between the 1960 *per capita* consumption of a number of key chemicals in the U.K. and the U.S. affords some indication of the potential market in this country. The consumptions are given in lb/head in the following table:

Per Capita Consumption of Chemicals in U.K. and U.S.A.

	U.K.	U.S.
Sulphuric acid	97	172
Alkalis	77	90
Nitrogen fertilizer	15	27
Ethylene	12	23
Benzene	4	12
Calcium carbide	8	10
Carbon black	4	9
Titanium dioxide (pigments)	4	5
Synthetic fibres	1·3	3·2

EXPORTS

The chemical industry makes an important contribution to the export trade. In 1964, the exports of its products as such were valued at £412·2 million, which was 11·6% of the total for U.K. manufacturing industry. They were exceeded only by machinery, other than electric (£858m) and road motor vehicles (£538m).

The following were the largest chemical exports in 1964:

	£m.
Plastics	74·6
Pharmaceuticals	59·2
Paints and pigments	54·2
Essential oils, perfumes and cleaning materials	36·6
Dyes	20·4
Tetraethyl lead (TEL)	10·8
Other anti-knock mixtures	9·0

Other exports, each valued at over £1 million, were as follows in decreasing order of importance: explosives, fertilizers, rubber accelerators, insecticides, caustic soda, weedkillers, carbon (blacks, etc.), sodium phosphates, soda ash (sodium carbonate), fungicides, creosote oil and cresylic acid.

The above figures represent exports of chemical products as such—in other words direct exports. They take no account of the chemicals which are associated with the exports of other industries—the indirect exports. Examples are dyes on textiles and plastics and paint in motor cars. It is not possible to form an estimate of these, though one authority has suggested at least 10% of the direct export figures. The value is certainly substantial.

IMPORTS

In 1964, the U.K. imported chemicals to the value of £252·5 million, of which £5·9 million were re-exported, leaving a net import of £246·6 million, which represented 12·8% of the retained imports of all manufactured goods. The main import were: plastics (£54·4m), fertilizers (mainly potash) (£14·7m) dyestuffs (£8·8m), pharmaceuticals (£7·1m), rosin (£5·2m), and anti-knock chemicals (£3·7m). The imports include many natural products not available in the U.K., such as potash salts, rosin, borax and mercury, some of which are essential raw materials for chemical manufacture. There are also specialities which are covered by patents. The 1964 imports were inordinately high, due largely to temporary shortages of essential chemicals and to intensive foreign competition. Comparing exports with retained imports, it will be seen that even in the unusual year 1964, there was a favourable chemical balance of £106·3 million, a not inconsiderable contribution to the overseas payments position.

OVERSEAS ACTIVITIES

Situated in the metropolitan country of a former world-wide Empire turned Commonwealth, the larger firms of the British chemical industry have for many years displayed a considerable centrifugal tendency, developing subsidiary companies and forming associated concerns with existing firms in former imperial territories. Many overseas companies have been established also in non-Commonwealth lands, such as South America and even the U.S. These foreign interests were developed in order to compete on an economic level with local chemical production within tariff protected boundaries and with access to indigenous resources. A number of the overseas firms are what would have formerly been styled trading posts and are either partly or wholly concerned with marketing chemicals manufactured in the U.K. In the U.K. in 1949, when data were specially collected by the Association of British Chemical Manufacturers for its report to the Board of Trade, some 40 U.K. chemical companies had 107 subsidiary and associated firms overseas, representing a total capital holding of £45 million. Today, although statistics have not been compiled, the foreign commitments of the U.K. chemical industry must be very much more extensive. In the main, foreign undertakings are carried on only by the largest firms. At the present time, to quote only two examples, ICI has 48 Commonwealth and foreign subsidiaries and 18 associated companies, while Fisons Overseas Ltd controls the operations of 12 subsidiaries and 9 associates. More examples will be found in the section of this history dealing with the organization of the more important firms in the industry.

RESEARCH

Research, as has already been shown, plays a vital part in maintaining the tempo of development of the chemical industry. It was estimated that in 1958 expenditure on research and development was about £22 million, equivalent to 10% of the sum expended by all manufacturing industry. This statistic is impressive, and has now been, doubtless, greatly exceeded. In addition, the industry has also advanced on the basis of imported processes, purchased in ready-to-operate state and the product of research carried out in other countries. Chemical processes, the tangible products of research, have always been the subject of international trading by the world's chemical industries, and with the development of the modern organic chemical industry this tendency has greatly accelerated. This trade in processes has been facilitated by international patent agreements and the stimulus of " invisible exports " in the form of royalties from foreign firms who have acquired licences to operate the patents. Thus it is no longer possible to discern the total contribution of research on a purely national basis. The formation of mergers and the incursion of foreign companies, e.g., the entry of the U.S. concerns into the U.K. pharmaceutical industry, obscure the effects of research by individual laboratories and institutes.

Chemical science may be likened to a rapidly expanding sphere, the surface of which is in contact with the unknown. This area is so vast that no country is able profitably to conduct research over the whole of it. Different nations will concentrate on different sectors and from their researches will spring the specialities they are able to offer to consumers. This is a factor of general applicability. All the great chemical manufacturing countries are themselves importers of specialities from their competitors. No country will for the above reasons ever become chemically independent—or strive to become so.

The U.K. chemical industry, the oldest in the world, and, in consequence, for long the most conservative, is now well on the way to being modernized in its entirety. As it advances into the second half of this century, it becomes clearly an industry in revolution, not only in its raw materials and processes, but in its new dominance in fields of technology in which it was formerly an unimportant ancillary. It is no business of the historian to attempt to predict even its proximate goals, but it can assuredly be expected fully to justify Leo Baekeland's observation that " the whole fabric of modern civilization becomes every day more interwoven with the endless ramifications of applied chemistry ".

Parts B and C

Companies of Importance in the British Chemical Industry and Trade Associations

Part B

Companies of Importance in the British Chemical Industry

The original idea was to include in this part very condensed or " potted " summaries of the organizational history and present day activities of all the companies engaged in chemical manufacture. It was found that the number of firms involved was too large to be covered in the space available in this volume. The Board of Trade Census of Production includes nearly three thousand companies with chemical activities. The paint section of the chemical industry alone is reputed to contain some 600 companies, many of them admittedly quite small. It was, therefore, decided to limit the section to companies regarded as important, not solely on the basis of size, but having regard also to the part which they play in the national economy. The advice of the relevant trade associations and Government Departments was sought and there emerged a list of over 200 companies. To ensure the accuracy of the information, the listed firms were supplied with an *aide-memoire* indicating the points it was hoped to include in the summary. There was an excellent response, which included all the major companies, and, in many cases, the data provided were far more than adequate for the purpose. There were some, however, who did not desire inclusion for one reason or another and some from whom no reply was received, although it was made clear that they were not being asked to supply any confidential information.

While an analysis of the capital structure was considered to be inappropriate in these condensed summaries, it was felt that the reader would be interested in data which would give some idea of the scale of operation in the chemical field. The *aide-memoire* therefore had a query relating to " employed capital ". While accountants sometimes argue as to the precise items to be included in the calculation of employed capital, the term is widely used in commercial and financial circles, since share capital is generally a very inadequate indication of a company's financial commitments. Many of the firms in this booklet are familiar with it, from their membership of the Association of British Chemical Manufacturers,* where employed capital is used as the basis of subscription and is defined as including " in addition to issued capital of all classes, debentures,

* Merged since 1st January, 1966, in the new Chemical Industries Association Ltd. (C.I.A.)

loans and bank overdrafts, also the capital and revenue reserves as shown in the balance sheet, together with trade investments in the chemical and closely allied industries ". This definition is quoted for the benefit of the reader.

The *aide-memoire* asked for the figure showing the amount of capital employed in chemical manufacture in the U.K. within the wide field defined for this series of monographs. As many firms have activities outside the chemical field this necessitated an estimate on their part, so that the figures quoted in this section must be regarded purely as an indication of the scale of chemical operation in this country. Some firms perforce gave variants of this. Others were unable to give any figures as, for example, in the case of private companies, whose financial position is confidential, or of those whose non-chemical activities were too complex to allow them even to attempt an estimate. Firms were also asked to indicate their average number of U.K. chemical employees, but again there were difficulties with mixed manufactures.

The authors are most grateful to the industry for the information which has been supplied and used as the basis of these summaries.

Although the data set out below were collected in the latter part of 1964, note has been taken of changes recorded in technical and lay publications, up to the beginning of October 1965. Nevertheless, organizational and other changes occur with such frequency these days, that this section must inevitably be out of date in certain particulars in a very short time. This will not affect the account of the previous history of the various undertakings. (See Addenda p. 329.)

The companies are arranged in alphabetical order, generally under the title of the main company or group, but all subsidiaries or associates are also shown in the index, while to help the reader, attention is drawn to other relevant summaries by the symbols (q.v.) (*quo vide*) after the firm's name.

A.B.M. Industrial Products Ltd

Formed 1964 as a subsidiary of Associated British Maltsters Ltd, which in 1963 had acquired Norman Evans & Rais Ltd (established 1920), to combine the last-mentioned interests with those of J. M. Collett & Co. Ltd, Gloucester (about 100 years old); The British DiaMalt Co. (an old established firm of malt extract makers) and Sunvi-Torrax Ltd (millers of malt and other special flours). On the chemical side, Norman Evans & Rais Ltd now produce bacterial and fungal enzymes, sequestering agents, wetting agents, emulsifiers, detergents, dispersing agents and metallic soaps, while J. M. Collett & Co. Ltd, who used to make products for the brewing, foodstuffs and pharmaceutical industries, are large producers of sulphites and fine chemicals. Capital £700,000. Employees 400. ABCM member. Woodley, Stockport.

Abbott Laboratories Ltd

This company, established 1937, is a subsidiary of Abbott Laboratories of America, which started in the ethical drug industry in 1886 as the Abbott Alkaloidal Company—it adopted its present name in 1914, and has now branches or subsidiaries in 37 countries. Its U.K. manufactures include barbiturates and cyclamates. U.K. assets about £2m. Employees 300. Queenborough, Kent.

Air Products Ltd

Formed 1907 to acquire Hughes & Lancaster under that name. Subsequently became subsidiary of The Butterley Co. Ltd. In 1957 became U.K. subsidiary of Air Products Inc. (now Air Products and Chemicals Incorporated), U.S.A. Butterley interest bought out in 1961 and the company, which had for some time been Air Products (Great Britain) Ltd, assumed its present name. The original objectives, no longer relevant, were to construct and operate air separation plants; to make industrial gases and equipment for their use, and to make sewage movement and treatment plant. In 1963 acquired Saturn Industrial Gases Ltd and manufacturing rights for all pumps of Tangyes Ltd. The Hughes & Lancaster Department handles sewage treatment. Number of non-trading subsidiaries. Associates in Italy. Now manufactures all industrial gases; industrial gas production plant; cryogenic apparatus; welding equipment; and mobile transportation systems for gases and liquids. Capital £5·85m. Employees 1250. ABCM member. Waverley House, Noel Street, London W1.

The Albright & Wilson Group

In 1844, Arthur Albright began the manufacture of white phosphorus for the match industry. In 1852, he perfected the production of the non-toxic amorphous (red) phosphorus to replace the poisonous white variety. In 1856, he took his fellow Quaker, J. E. Wilson, into partnership. In 1892, Albright & Wilson (A. & W.) became a private limited liability company. In 1896, A. & W. formed the Oldbury Electro-Chemical Co., Niagara, to produce phosphorus and its compounds and later chlorates. (In 1956, the assets of the latter were exchanged for a share-holding in the Hooker Chemical Corporation, Niagara.) In 1902, A. & W. acquired the Electric Reduction Co. of Canada Ltd, also makers of phosphorus. In 1932, Clifford Christopherson & Co. Ltd, chemical agents, was acquired to provide A. & W.'s first sales organization in London. In 1935, Albright & Wilson (Ireland) Ltd was formed to make acid calcium phosphate for the baking industry; the products now include other food phosphates, detergents and mineral supplements for livestock; control was transferred to the Irish directors in 1945. A. & W. (Australia) Pty. Ltd was formed 1939. Thomas Tyrer & Co. Ltd, fine chemical manufacturers, was acquired 1942 and its factory has become part of A. & W. (Mfg.) Ltd. A. & W. Ltd became a public company 1948. In 1950, Midland Silicones Ltd was formed. In 1955 Marchon Products Ltd and its subsidiary, Solway Chemicals Ltd, were acquired and Proban Ltd formed. In 1957, A. & W. Ltd became the holding company of the group and the principal manufacturing and sales activities were transferred to a new company, A. & W. (Mfg.) Ltd. In 1960, A. Boake, Roberts & Co. (Holding) Ltd and its four subsidiaries became members of the group, and W. J. Bush & Co. Ltd and its seven subsidiaries, including Potter & Moore Ltd, in 1961. In 1963, Geka Trading Co. Ltd, Northern Nigeria (which in 1965 became W. J. Bush & Co. (Nigeria) Ltd), and Bush Hellas Fruit Industries, S.A., Greece, joined the group, followed in 1964 by Stafford Allen & Sons Ltd and its three subsidiaries. In the same year, Bush, Boake, Allen Ltd was established to co-ordinate the activities of W. J. Bush, A. Boake, Roberts and Stafford Allen.

The present organization is briefly as follows: the parent company, A. & W. Ltd, co-ordinates the group activities, covering over 35 companies predominantly in the chemical industry. The wholly-owned U.K. subsidiaries are:

1. A. & W. (Mfg.) Ltd, with Clifford Christopherson & Co. (selling) and Electropol Ltd, the last-mentioned owning the world patent rights for the Electropol processes for electro-polishing and -machining stainless steel.

2. Bush, Boake, Allen Ltd, as already mentioned.

3. W. J. Bush & Co. Ltd, founded in 1851, makers *inter alia* of essences, flavours and fine chemicals, with Potter & Moore Ltd, founded 1749, becoming a Bush subsidiary 1886, and making perfumes and toiletries, with companies in Australia, Canada, Greece, India, New Zealand, South Africa and U.S.A. Potter & Moore acquired Meggeson & Co. in 1965.

4. A. Boake, Roberts & Co. (Holding) Ltd, with A. Boake, Roberts & Co. Ltd, established in 1869 and making industrial and fine chemicals and specialized materials for other industries, and with companies in India and South Africa.

5. Stafford Allen & Sons Ltd (established 1833) with diverse interests in the food and pharmaceuticals (drug extracts and galenicals) industries, with the Allen Chlorophyll Co. Ltd (formed 1952), Wyleys Ltd, wholesale Midland chemists, and Warrick Bros., founded 1753 and pioneer makers of medicated pastilles and lozenges.

6. Marchon Products, established 1939, with associates in Italy and producers of raw materials for detergents, shampoos and cosmetics, with Cumbria Trading Co. Ltd, Leo Lines Ltd and Solway Chemicals Ltd, which started production of sulphuric acid from anhydrite, mined underneath the factory, in 1955.

7. A. & W. (Overseas Developments) Ltd, owning ships for the transport of phosphate rock.

8. Associated Chemical Companies Ltd (q.v.), taken over 1965.

Midland Silicones Ltd is a partly-owned subsidiary with Dow-Corning Corporation having a 40% interest, and with associates in Denmark and the Netherlands.

The U.K. associated companies include A. & W. (Ireland) Ltd, already mentioned, with Goodbody Ltd, sales agents; A. & W. Match Phosphorus Co. Ltd, owned jointly with Bryant & May Ltd, for making phosphorus for the match trade; and Proban Ltd, owned jointly with the Bradford Dyers Association Ltd, for textile-finishing processes.

The A. & W. group have a large number of overseas wholly-owned subsidiaries, many of them derived from the firms they have taken over in recent years and other associates. These are too many to list. They operate in Australia, Canada (where the Electric Reduction Co. with its subsidiaries is the largest A. & W. overseas company), Greece, India, Italy, New Zealand, Nigeria, South Africa, and U.S.A.

A. & W. (Australia) Pty. Ltd is a partly-owned subsidiary in which ICI of Australia and New Zealand has a 42·5% interest; it makes a wide range of A. & W. products. There are partly-owned subsidiaries of Midland Silicones Ltd in Denmark and the Netherlands.

The group makes a wide range of products which include phosphorus and its compounds; materials for detergents, perfumery and flavourings; essential oils; pharmaceuticals; silicones; chemicals for plastics, paints, water treatment and agriculture (insecticides and herbicides); chemicals and processes for metal and textile finishing, and many other industrial and fine chemicals. UK employed capital £30m. Employees exceed 11,000. ABCM member. 1 Knightsbridge Green, London SW1.

Alcock (Peroxide) Ltd

Formed 1929 by H. E. Alcock on severance of his connection with Laportes. Main products are hydrogen peroxide, sodium metasilicate and detergents. Capital £400,000. Employees 150. ABCM member. Leicester Road, Luton, Beds.

Alginate Industries Ltd

Formed 1934 as a private company called Cefoil Ltd, and adopted the present name 1945. The object was to extract alginic acid from seaweed and to make products therefrom. Over 100 salts and derivatives of alginic acid are made for industries such as textiles, paper, food and water treatment. There are three subsidiaries, Manucol Products Ltd (formed 1936, now dormant), Alginate Industries (Scotland) Ltd (formed 1946 for raw material supply), and Alginate Industries (Ireland) Ltd (formed 1964 to utilize seaweed on site). There is an associate in Eire. Agents in 70 countries. Capital £700,000. Employees 450. ABCM member. Walter House, Bedford Street, Strand, London WC2.

Frederick Allen & Sons (Chemicals) Ltd

Founded 1826 by Frederick John Allen for the manufacture and sale of chemicals. The Allen family owned the company till 1937, when it passed into the ownership of the Unwin family. It still remains a private company. Makes wide range of chemicals, notably oxalates, acetates, nitrates and chlorides; also distributors of chemicals for home and export markets. Capital £500,000. Employees 100. ABCM member. Phoenix Chemical Works, Upper North Street, London E14.

The Alumina Company Ltd

Formed 1908, and described in its articles as makers and traders in alumina, dyes, drugs, paints, soap, etc., acquired 1915 by The British Aluminium Co. Ltd. Associated with Aluminium Sulphate Ltd, Thames Alum Ltd, Don Alum Ltd, and Alumina Chemicals Ltd (Dublin). Manufactures solid aluminium sulphate. Employees 70. ABCM member. Iron Bridge Chemical Works, Widnes, Lancs.

Amalgamated Oxides (1939) Ltd

A private company formed 1939 to make zinc oxide, with links back to Edwin Coley's process for zinc extraction in 1923. Associated with Morris Ashby Ltd (formed 1868), its selling agents. Associates in Canada and Germany. Products are zinc dust and zinc oxide. Dartford, Kent.

Anchor Chemical Co. Ltd

Started 1892, as small manufacturer of chemicals for the rubber industry. Private limited company 1915. Public company 1938. Now both manufacturers and merchants, mainly for U.S. concerns, of a variety of products for the rubber industry. J. & G. Hardie & Co. Ltd are selling associates in Scotland. World wide agents. Capital £700,000. Employees 300. ABCM member. Clayton, Manchester 11.

Associated Chemical Companies Ltd

Associated Chemical Companies Ltd is the holding company for a number of subsidiaries, manufacturing industrial inorganic and organic chemicals and fertilizers, formed by absorption and amalgamation as set out below. Eaglescliffe Chemical Co. Ltd, formed 1833 (registered 1910) to manufacture fertilizers and other agricultural products, extending into the chromium salts field in 1927. In 1951, E. P. Potter & Co. Ltd (dichromate makers formed in 1893) was acquired. In 1953, it amalgamated with John & James White Ltd, established 1820, original U.K. dichromate manufacturers which had amalgamated 1932 with H. C. Fairlie & Co. Ltd, formed in the 1880's. This brought all the chromium chemical production under single control and British Chrome & Chemicals Ltd was formed as the holding company (1954). Brotherton & Co. Ltd joined in 1957 and the holding company became Associated Chemical Companies Ltd (1958). Brothertons had been established as a private company in 1878 to make ammonia and ammonium sulphate from gas liquor, and subsequently extended its activities into tar distillation, which it relinquished to the local tar co-operative group after World War I. In 1917, Brothertons acquired the Mersey Chemical Works of the Berlin Aniline Co. (later to become part of the German IG) and made textile reducing agents (hydrosulphites) and dyestuffs, but the latter were abandoned in 1956. In 1949, Brothertons had become a public company.

Associated Chemical Companies Ltd now operates through the following subsidiaries:

1. Associated Chemical Companies (Mfg.) Ltd, formed 1962 to unify the production of British Chrome & Chemicals Ltd and Brotherton & Co. Ltd.
2. Associated Chemical Companies (Sales) Ltd, formed 1958 to sell the products of 1.
3. ACC (Fertilisers) Ltd, formed 1962 to bring together the agricultural activities of Eaglescliffe Chemical Co. Ltd (superphosphate and compound fertilizers), Robert Stephenson & Son Ltd (formed 1826 as seed-crushers, later compound fertilizer makers, acquired 1958), William E. Marshall Ltd, incorporating Alliance Feeds Ltd (started 1930 to make animal feedingstuffs and acquired by Robert Stephenson 1952), and The Farmers Company Ltd

(formed 1874 to make fertilizers and animal feeds and absorbed 1961), to which was added Mays Chemical Manure Co. in 1964.

4. ACC (Chrome & Chemicals) Ltd, Canada, formed 1951 to make chromium products.

Employed capital £10m. Employees 2200. ABCM member. P.O. Box 28, Harrogate, Yorks.

Company acquired by Albright & Wilson group (q.v.) in 1965.

The Associated Octel Company Ltd

Formed 1938 as the Associated Ethyl Company Ltd, a private company with a then capital of £10m, in which the following international oil groups have a direct or indirect interest: Shell, BP, Socony-Mobil and Caltex. Present name adopted 1961 to avoid possible confusion with competitive producers in the U.S.A. Its purpose was the manufacture and distribution of lead alkyl anti-knock compounds, in particular tetraethyl lead (TEL).

In the 1936–40 period as a national security measure, a British TEL plant with the associated extraction of the necessary bromine from seawater, had been built under licence, owned by the Ministry of Aircraft Production (MAP) and leased to the British Ethyl Corporation Ltd, a company then owned jointly by the Ethyl Export Corporation (formed 1931 by the American Ethyl Gasoline Corporation to handle all sales outside U.S.A.) and ICI Ltd. ICI Alkali Division operated both plants during World War II. The Associated Ethyl Company Ltd, when formed in 1938, took over the interests of the Ethyl Export Corporation. In 1945 the British Ethyl Corporation (now wholly owned by the Associated Ethyl Co. Ltd), bought the TEL and bromine plants from MAP and the latter took over their operation as from 1st January 1948. There are overseas associates in France and Germany. The anti-knock " Octel " compounds are based on lead tetraethyl (TEL) and lead tetramethyl (TML). Employees 2100. ABCM member. 20 Berkeley Square, London W1.

Ault & Wiborg Ltd

The Ault & Wiborg Company, U.S.A. (founded 1878), opened a U.K. branch in 1899 to make letterpress and litho inks. From this, the London subsidiary company was established 1909. In 1934, it became a British-owned public company, but maintains close technical contacts with the International Printing Ink Corporation. In 1939, it acquired the old-established Willesden Varnish Company. Since World War II, it has purchased Wembley Paint Company, Associated British Cellulose Ltd, A. Learner & Co. Ltd, Allied Paints & Chemicals Ltd, Wilson Blackadder & Co. Ltd, Stanford Wylie & Fraser Ltd, William R. Todd & Son Ltd, General Industrial Paints Ltd, and William Harland & Son Ltd. This has resulted in a big interest in paint and plastics, as well as the printing ink industry. There are 31 companies in the group with world-wide connections. Employed capital £4m. Standen Road, Southfields, London SW18.

BDH Group Ltd

Incorporated 1908 as The British Drug Houses Ltd by the amalgamation of four pharmaceutical companies, viz., Barron, Harveys & Co. (started about

1790), Davy Hill & Hodgkinsons Ltd (founded 1750), Hearon, Squire & Francis Ltd (established 1714), and Geo. Curling Wyman & Co. (formed 1788). Became public company 1926. In 1959 it entered the wholesale pharmaceutical business and has since acquired half-a-dozen wholesaling firms to provide a distribution system throughout the country. Its U.K. subsidiaries are Bradley & Bliss Ltd (founded 1818, incorporated 1909, acquired 1964), Claude Duval Ltd (incorporated 1950, acquired 1964), J. R. Gibbs Ltd (formed 1937, acquired 1959 with subsidiary Ferris & Co. Ltd, incorporated 1904), BDH (Knights) Ltd (incorporated 1936 as Knights Oil & Chemical Co. Ltd, acquired and name changed 1963; with subsidiary Janda Chemicals Ltd, formed 1945), BDH (London Wholesale) Ltd (formed 1964), BDH (Middletons) Ltd (established 1901 as Middleton & Co. Ltd, acquired 1963, name changed 1964), BDH (Nutritionals) Ltd (formed 1960 as Mead Johnson Ltd, acquired and name changed 1962), Rowland James Ltd (incorporated 1927, acquired 1959), BDH (Woolley & Arnfield) Ltd (established 1895 as James Woolley, Sons & Co. Ltd, became Woolley & Arnfield Ltd 1962, acquired 1962, present name 1963), and AB Insulin Ltd, owned jointly with Allen & Hanburys Ltd, incorporated 1961. In 1957, there was an agreement for collaboration in research with Mead Johnson & Co. of America, and in 1961, a financial association. It has subsidiaries in Australia, Canada, Eire, India, Malaysia, New Zealand and South Africa. In January 1965, the British Drug Houses Ltd changed its name to BDH Group Ltd and hived off the London Wholesale Depot as a separate company—BDH (London Wholesale) Ltd. Its other activities were transferred to BDH (Nutritionals) Ltd, which changed its name to the British Drug Houses Ltd, so that all the activities, formerly carried on by the British Drug Houses Ltd, except wholesaling, are still conducted by a company having the same name. All the U.K. and overseas operating companies are controlled by BDH Group Ltd as a holding company. It manufactures a wide range of pharmaceutical products and medical specialities and some 7000 laboratory chemicals at its division at Poole. Joint company formed with Glaxo group (q.v.) in 1965 to combine their pharmaceutical wholesale interests. Employed chemical (excluding wholesaling) capital £3m. Employees 2700. ABCM member. Graham Street, London N1.

BP Chemical Company Ltd

Founded 1962 as a subsidiary of The British Petroleum Co. Ltd (BP), which was registered 1909 as The Anglo-Persian Oil Co. Ltd; in 1935 the name was changed to Anglo-Iranian Oil Co. Ltd, and to its present title in 1954. The U.K. chemical associates are: British Hydrocarbon Chemicals Ltd (q.v.) (50% BP, 50% DCL), Forth Chemicals Ltd (q.v.) (two-thirds BHC, one-third Monsanto Chemicals), Grange Chemicals Ltd (q.v.) (two-thirds BHC, one-third California Chemical Co.), BP California Ltd (50% BP, 50% California Chemical Co.) formed 1961 to make *o*- and *p*-xylene, Border Chemicals Ltd (q.v.) (one-third each BP, DCL and ICI) and Associated Octel Co. Ltd (q.v.) (36·7% BP, 36·7% Shell, 21·3% Caltex and 5·3% Mobil). B.P. acquired in 1965 from Socony Mobil the U.K. plastics interests and Mobil Chemicals (q.v.) and its associates, Erinoid Ltd, Kleestron Ltd, Wokingham

Plastics Ltd, O. & M. Kleeman Ltd and Imperial Chemical Sales Ltd. BP Plastics Ltd is the name of this new holding.

There are overseas chemical associates in Australia, France, Germany and Kuwait.

BP itself makes pure normal paraffins for the chemical industry while its associates, administered by the BP Chemical Co., make intermediates for the plastics, synthetic rubber, man-made fibre, fertilizer, detergent and general chemical industries, as well as finished products. Capital £18m. Employees (including associates) 5500. Britannic House, Finsbury Circus, London EC2.

Bakelite Xylonite Ltd (BXL)

This was originally British Electro Metallurgical Co. Ltd (formed 1937); name changed to Union Carbide 1954, and to Bakelite Xylonite Ltd 1963. Jointly owned by the Distillers Co. Ltd and Union Carbide Corporation of U.S.A. The objective was to pool the resources of the two parent companies in the field of plastics production to secure increased efficiency. Associated U.K. companies are:

1. *Bakelite Ltd.* Formed 1904 to make phenol-formaldehyde resin, first studied by Baekland in the U.S. in 1902, as the Fireproof Celluloid Syndicate Ltd, which in 1910 was dissolved and replaced by the Damard Lacquer Co. Ltd. Bakelite Ltd, formed by Baekland 1926, acquired the assets of Damard 1927, and united with it those of Mouldensite Ltd and Redmanol Ltd, in the same field. The plastics interests of Bakelite and its associate, Union Carbide Ltd, were merged in Bakelite Xylonite in 1963 (*vide supra*).

2. *British Xylonite Co. Ltd (BX).* Incorporated 1877 to make xylonite, the British celluloid. Cascelloid of Leicester acquired 1931. In 1939, the British Xylonite Group was created by establishing companies for materials and manufacture (BX Plastics Ltd) and for fabrication (Halex Ltd).

Later that year, Distillers Co. Ltd obtained half-interest in BX Plastics Ltd, which acquired 1943 the Expanded Rubber Co. Ltd, transferred to the parent company 1948. BX Plastics Ltd and Ilford Ltd, formed 1946 a new company, Bexford Ltd, to make film base for photographic and other purposes. In 1961, Distillers Co. Ltd acquired the whole holding of the British Xylonite Co. Ltd and sold it to Bakelite Xylonite Ltd 1963 for shares in the latter.

3. *Extrudex Ltd.* Founded 1954, as the first company commercially to extrude polyvinyl chloride in the U.K. Transferred to the Aycliffe site owned by Bakelite Ltd 1964.

4. *Darton Manufacturing Co. Ltd.* Founded during World War II to fabricate polythene; acquired by BX Plastics 1958. Amalgamated with other polythene interests into the Films Division of BXL, January 1965.

5. *Jackson Polythene Ltd.* Polythene fabricators acquired by BXL 1964 and amalgamated into its Film Division, January 1965.

Overseas associates in Australia, Canada and U.S.A.

Its fields of activity cover synthetic resins, both thermosetting and thermoplastic, with their intermediate raw materials; chemicals such as synthetic camphor, nitro-cellulose and organo-tin compounds; all types of plastics and products, such as laminates, man-made textiles and materials for the building industry; and manufacturing services such as injection moulding, laminating and fabricating. Employed capital £20m. Employees 12,000. ABCM member. 27 Blandford Street, London W1.

Barium Chemicals Ltd

Formed 1964 jointly by Laporte Chemicals Ltd (q.v.) and Imperial Smelting Corporation Ltd (q.v.), to merge their barium interests. It stems from Orr's Zinc White Ltd, incorporated 1898 to make zinc sulphide (Orr's Zinc White) and allied products, which in 1913 took over the Lugsdale Chemical Co., and merged with Imperial Smelting Corporation 1930. It makes a range of barium and zinc compounds. Employees 400. ABCM member. Widnes, Lancs.

Beecham Group Ltd

This stems from Thomas Beecham, born 1820, and the inventor of the well-known pills, which he started selling in 1842. His first factory was built in the early 1860's. Beecham Estates & Pills Ltd was registered 1924. The property and proprietary elements were separated 1928, when Beecham's Pills Ltd was formed to acquire Veno Drug Ltd, and the businesses of Sherley, Prichard and Constance and Lintox. In 1930, the five companies of Yeast Vite Ltd, Irving Yeast Vite Ltd, Holloways Pills Ltd, Iron Jelloid Co. Ltd and Dinneford & Co. Ltd were acquired. In 1933, Phensic Ltd was formed and followed by the acquisition of Phosferine (Ashton & Parsons) Ltd 1935, and Natural Chemicals Ltd (Phyllosan) 1936. Large scale expansion began in 1938 with the acquisition of Macleans Ltd (established 1917) and Eno Proprietaries Ltd (originating in the 1850's). Cephos Ltd was acquired 1959. In 1945 the name was changed to Beecham Group Ltd and Beecham Research Laboratories Ltd was formed. In 1954, their research was concentrated on antibiotics and resulted in the production of new synthetic penicillins. Associates in six other countries. Besides ethical and proprietary medicines, Beechams have interests in foods, health beverages, toiletries, confectionery and soft drinks, with which this summary is not concerned. Employed pharmaceutical capital £12·5m. Employees 2600. Great West Road, Brentford, Middlesex.

Berger, Jenson & Nicholson Ltd

Formed 1960 by merger of Lewis Berger & Sons Ltd (dating from 1760) and Jenson & Nicholson Group Ltd (originating 1821), manufacturers of paints and varnishes. In 1941, Lewis Bergers had acquired Keystone Paint & Varnish Co. (established from America 1911) and Sir W. A. Rose & Co. Ltd in 1957, while Jenson & Nicholson in 1948 acquired John Hall & Sons (Bristol & London) Ltd (founded 1788). There are interests in Cuprinol Ltd (wood preservatives), Spelthorne Metals Ltd (micronized metallic lead), Styrene Co-Polymers Ltd (synthetic resins) and Berger Traffic Markings Ltd. Associates in Australia, East Africa, Eire, France, India, New Zealand, Pakistan, Portugal, Rhodesia and West Indies. Capital £14m. Total employees 8000. Berger House, Berkeley Square, London W1.

F. W. Berk & Co. Ltd (since 1st Jan. 1966 Berk Ltd)

In 1870, the brothers Frederick William and Robert Berk formed a partnership to merchandise chemicals, chiefly German. Factory for superphosphate and sulphuric acid acquired 1873—still main manufacturing centre. F. W. Berk & Co. Ltd formed as private limited company 1891; became public company 1949. Group assumed name of Berk Ltd 1st January 1966. Present day main interests are heavy inorganics, sulphur, processed earths and clays, mercury, mercurials and agricultural products, filtration materials, organics and fine chemicals, drugs and botanicals, and general merchanting, still a large part of their business. Subsidiary companies include Powder Metallurgy Ltd, formed 1946 (metal powders), Berk Exothermic Ltd ("Thermofin"), Berk Leiner Ltd, formed 1963 jointly with P. Leiner & Sons (Wales) Ltd (manganese dioxide) and Berk Pharmaceuticals Ltd, formed 1961. There is also a subsidiary sand and gravel group. As associates there are Berk Spencer Acids Ltd, formed recently with Borax (Holdings) Ltd (q.v.) and Spencer, Chapman & Messel Ltd (acids), Abbey Chemicals Ltd, formed 1955 jointly with Hoyt Metal Co., a subsidiary of National Lead Co., U.S.A. (gelling agents, stabilizers and products for paint and plastics industries), and Detarex Ltd, in association with Rexolin Chemicals AB, Sweden (organic chelating agents). The company has substantial holdings in Bromine Compounds Ltd, of Israel (formed 1963), and interests in concrete, sand and gravel in the West Indies. There are also substantial chemical engineering and irrigation equipment activities. Employed chemical capital £5m. Employees 2000. ABCM member. Berk House, PO Box 500, 8 Baker Street, London W1.

Blundell-Permoglaze Ltd

A group of paint manufacturers, formed 1960, by the merger of Blundell, Spence & Co. Ltd (originated 1811) and Permoglaze Ltd (founded about 1900). Blundell, Spence was changed to Blundell-Permoglaze Ltd, Permoglaze Ltd was acquired and a new subsidiary founded in the name of Blundell, Spence & Co. Ltd. Associates in Australia, Chile, India and South Africa. Group capital £3m. Employees 1100. York House, 37 Queen Square, London W1.

William Blythe & Co. Ltd

Established 1845 to supply chemicals to the textile industry. Registered 1892. Public company 1928. Acquired 1922 John Riley & Sons Ltd (formed 1842 with similar object). Integrated into one 1958. Activities widely diversified. Supply extensive range of acids and metallic salts to practically all industries. Employed assets £2m. Employees 450. ABCM member. Holland Bank Works, Church, nr. Accrington, Lancs.

Boots Pure Drug Co. Ltd

Stems from 1863 when Jesse Boot at age 13 went into his parents' herbalist business in Nottingham. Jesse Boot & Co. Ltd founded 1883 as private company. Present company formed 1888, for business as retail chemists, supported by own pharmaceutical manufacture. Associates in Australia, Far

East, India, Pakistan and South Africa. In addition to retail chemist business (1280 branches), they are makers of pharmaceutical, cosmetic and household products, saccharin and weedkillers. Total employees 40,000, of which 750 are in chemical manufacture. ABCM member. 37 Station Road, Nottingham.

Borax (Holdings) Ltd

Incorporated as Borax Consolidated Ltd, 1899, to co-ordinate the activities of Pacific Borax and Redwoods Chemical Works Ltd (formed a few years before) and other organizations in England, France, South America and Turkey, dealing in borax and boron products. On reorganization in 1956, Borax (Holdings) Ltd was formed, with two principal operating subsidiaries in the U.K. and U.S.A. Borax Consolidated Ltd remains the U.K. operating company, making boron chemicals. Other U.K. subsidiaries are: Boroquimica Ltd, administering Argentine interests, Hardman & Holden Ltd (formed 1892 to unite Hardman & Co. and Hardman & Holden's), acquired 1960, makers of sulphuric acid, pigments and associated chemicals, Spencer, Chapman & Messel Ltd (incorporated 1899 to unite Spencer, Chapman & Co. and Chapman, Messel & Co.) (55% holding acquired 1960, with F. W. Berk & Co. Ltd (q.v.) (45%)), makers of inorganic acids (renamed Berk Spencer Acids in 1965), Theodore St. Just & Co. Ltd (incorporated 1944), makers of organic chemicals and wholly-owned subsidiary of Spencer, Chapman & Messel Ltd. There are associates in Argentina, Belgium, California (3), Chile, France, Germany, India, Netherlands, Peru, Spain and Turkey. The company is typical of British overseas initiative and enterprise in that it was the first to develop the new source of potash found in New Mexico in the late 20's, and is doing the same for the Saskatchewan potash deposits. U.K. capital £3·75m. Employees 1650. ABCM member. Borax House, Carlisle Place, London SW1.

J. C. Bottomley & Emerson Ltd

Founded 1851 as John Carr Bottomley, drysalter. Became private limited company 1919 with its present title by combination with Dawson & Emerson, founded 1912, as dye merchants. Original object was to make and market dyers' chemicals, including glauber salts, nitric acid and salts of iron and tin. The manufactures now include dyestuffs and paints. Capital £75,000. Employees 55. The present directors are the grandchildren of the founder. ABCM member. Brookfoot Works, Brighouse, Yorks.

Bowmans Chemicals Ltd

Established 1905 as private company called Bowmans (Warrington) Ltd to make lactic acid, sulphated oils and tanning auxiliaries, which it still manufactures *inter alia*. Only U.K. maker of lactic acid. Public company 1948. Capital £120,000. Employees 130. ABCM member. Moss Bank, Widnes, Lancs.

Bristol & West Tar Distillers Ltd

Formed 1952 to continue the tar distilling interests of Wm. Butler & Co. (Bristol) Ltd, which was established over 100 years ago to distil crude tar to produce creosote for the preservation of the sleepers on the London to Bristol railway then under construction by Brunel. Taken over 1962 by the South Western Gas Board, where it has a fellow subsidiary in Plymouth Tar Distillers Ltd, which was established 1952 as an associate of South Western Tar Distilleries, and transferred 1962 to the same gas board ownership. The company makes a range of tar distillation products. Merchanting activities handled by their trading company Bristol (Tar & Chemicals) Trading Ltd at same address. Capital £1m. Employees 210. ABCM member. St. Philips, Bristol 2.

The British Dyewood Co. Ltd

In 1898, four private companies, viz., Mucklow & Co. (founded 1842), Edward Milne & Co. (established 1810), John Dawson & Co. (started 1852) and W. R. Scott & Co. (formed 1883), combined to form The British Dyewood & Chemical Co. Ltd. This was reorganized in 1911 as The British Dyewood Co. Ltd. The original activity was the production of natural vegetable colouring and tanning materials. In 1939, the firm of J. L. Rose (founded 1924) was acquired and in 1959, Lambeth & Co. (Liverpool) Ltd (started 1930). The company was owned by the United Dyewood Corporation of New Jersey from 1911 to 1957; the share capital was then purchased by the British firm of Messrs Bullough Securities Ltd. Present manufacture includes tannic, gallic and pyrogallic acids, tanning extracts, colouring and medicinal products. Practically all raw materials are of vegetable origin. Capital £350,000. Employees 100. ABCM member. 19 St. Vincent Place, Glasgow C1.

British Glues & Chemicals Ltd

Incorporated 1920 to take over the businesses of The Grove Chemical Co. Ltd (formed 1856), Charles Massey & Son Ltd (founded 1815), Meggitts (1917) Ltd (originally established 1867), The Weaver Refinery Co. Ltd (formed 1900), J. & T. Walker Ltd (established 1795), and Williamson & Corder Ltd (founded 1893). The object was to present a united front against any German attempt to regain the control of the glue and gelatine industry, which had been exercised pre-war through S. Meggitt & Sons Ltd, which had been taken over in 1917 by the Controller of Enemy Businesses and sold to a group. The following joined or were acquired by British Glues & Chemicals Ltd as indicated: 1920 Lomas Gelatine Works Ltd (formed 1914) and The Improved Liquid Glue Co. (founded 1911) (now Croid Ltd); 1921 O. Murray & Co. Ltd (established 1907); 1926 B. Young & Co. (formed 1806); 1932 George Asprey & Son Ltd (formed 1918); 1933 The Standard Soap Co. Ltd (formed 1899); 1935 G. C. Russell Ltd (formed 1906) and many others such as The Tees Refinery Co. (originally the Tees Bone Mill Ltd, formed 1910), becoming 1960 International Protein Products Ltd, The Mitcham Poultry Food Co. (established 1923), Alfred Fairclough Ltd (formed 1923 and since 1962 called J. H. Fairclough Ltd), G. A. Shankland Ltd (over 50 years old); 1960 C. Simeon & Co. Ltd (formed 1874) and British Gelatine Co. Ltd (formed 1899); 1962 Wm. Oldroyd & Sons Ltd (started 1840 and half-owned since 1959), and Cleveland Product Co. (formed 1907); and 1964 R. Pintus

& Co. Ltd (founded 1907). There are U.K. associates as follows: International Protein Products Ltd (*vide supra*), makers of edible lipo-protein from peanuts; Cleveland Product Co. Ltd (*vide supra*), makers of gelatine; Wm. Oldroyd & Sons Ltd (*vide supra*), makers of gelatine; Calfos Ltd (formed 1939 to make calcium/phosphate food supplements); B. Young & Co. Ltd (*vide supra*), gelatine makers; C. Simeon & Co. Ltd and the British Gelatine Works Ltd (*vide supra*), makers of photographic gelatines; Croid Ltd (*vide supra*), glue makers; and B. Cannon & Co. Ltd (in association with Booth & Co. (International) Ltd). There are associates in Austria, Canada, Netherlands and U.S.A.

The firm's products include glues, gelatines, adhesives, soaps, fertilizers, fats and feeding stuffs. They also do merchanting through O. Murray & Co. Ltd (*vide supra*). Employed capital about £4m. Employees 2000. ABCM member. Berkshire House, 168/173 High Holborn, London WC1.

British Hydrocarbon Chemicals Ltd (BHC)

Formed 1947 under name of British Petroleum Chemicals Ltd by Anglo-Iranian Oil Co. Ltd (now BP) (q.v.) and The Distillers Co. Ltd (q.v.) (equal interest), to produce chemicals from petroleum. Present name adopted 1956. Products include butadiene, ethylene, ethylene dichloride, isobutylene, methanol, phenol, polyethylene, and propylene tetramer. In 1950 Forth Chemicals Ltd was incorporated; capital held two-thirds by BHC and one-third by Monsanto Chemicals Ltd (q.v.), for production of styrene. In 1955 Grange Chemicals Ltd was formed; two-thirds of capital held by BHC and one-third by Oronite Chemical Co. (now California Chemical Co.), a subsidiary of Standard Oil Co., California. Products: detergent alkylate and phthalic anhydride. Employed capital of above three firms, £60m. Employees 2800. ABCM member.

Associated with BHC is Border Chemicals Ltd, formed 1963 and shared equally by the British Petroleum Co. Ltd, The Distillers Co. Ltd and Imperial Chemical Industries Ltd (q.v.) to manufacture acrylonitrile. Plant operated by BHC. Capital £6m. Devonshire House, Mayfair Place, London W1.

The British Nicotine Co. Ltd

Formed 1906 as private company subsidiary of the Imperial Tobacco Co. (of Great Britain and Ireland) Ltd to carry on the business of chemists and to manufacture nicotine and any other products from the tobacco waste of the parent company. Nicotine and nicotine sulphate the main products. Employees 80. Lyster Road, Bootle 20, Lancs.

The British Oxygen Company Ltd (BOC)

Incorporated 1886 as Brins Oxygen Co. Ltd to produce oxygen and promote its uses in chemical and metallurgical processes. Present name 1906. Acquired Allen Liversidge Group of companies 1930, and interest in dissolved acetylene. British Oxygen Chemicals Ltd formed as subsidiary 1955. Associates in major Commonwealth countries, also in South Africa, Italy and Norway. British Oxygen Group manufactures include oxygen, nitrogen, hydrogen, helium, the rare gases, nitrous oxide, ethylene, acetylene, synthetic resin emulsions, dicyandiamide and melamine. Extensive activities in the engineering field include manufacture of gas producing plant and of equipment for the use of their products. U.K. chemical and allied capital over £60m. Employees in chemical and chemical engineering 7500 out of total of 14,500. ABCM member. Hammersmith House, London W6.

British Paints (Holdings) Ltd

Holding company controlling subsidiaries manufacturing in Great Britain, Australia, Barbados, British Guiana, Canada, India, Ireland, Japan, New Zealand, Netherlands, Nigeria, South Africa, Southern Rhodesia, Trinidad and U.S.A. Manufacturers of paint and surface coatings, synthetic resins, moulding powders, dry colours, elastomers, jointing compounds and sealants and, outside the chemical field, laminated plastics (North British Plastics Ltd). Registered as private company 1930, converted to public company 1931, and name changed in 1941 from J. Dampney & Co. (which started at the turn of the century). Associated with Resinous Chemicals Ltd (formed 1949), makers of synthetic resins, pigments, dyestuffs and moulding materials. Overseas associates in Barbados, British Guiana, Far East, Japan, Middle East, Netherlands and U.S.A. British Paints (Holdings) Ltd acquired in 1965 by the Celanese Corporation of America. U.K. employees 2000. Portland Road, Newcastle-upon-Tyne 2.

British Tar Products Ltd

Registered 1920. Public company 1925. Object to refine crude tar and benzole. As the result of nationalization of the coal mines, ceased to distil crude tar in 1958 and crude benzole in 1961 but continued using semi-refined products as starting materials. Also hydrogenates coal tar products and produces a range of esters and industrial methylated spirits. Also provides for the storage and handling of petroleum products and liquid chemicals. Febrail Ltd formed in 1965 in conjunction with Briggs & Townsend Ltd for production of basic plastics materials. Capital, exclusive of storage, £250,000. Employees 130. ABCM member. 418A Glossop Road, Sheffield 10.

British Titan Products Co. Ltd (BTP)

Formed 1930 but did not attempt production till 1933 when it was reorganized as a consortium of British companies and one American company with a controlling interest. The latter withdrew 1948 and the British members took control. The present shareholders are Goodlass Wall & Lead Industries Ltd, Imperial Chemical Industries Ltd, Imperial Smelting Corporation Ltd (a subsidiary of Rio Tinto-Zinc Corporation Ltd), and R. W. Greef & Co. Ltd. The original objective, which still operates, was the manufacture of titanium pigments (titanium oxide). Apart from the shareholders, one U.K. manufacturing associate, Titanium Intermediates Ltd, a wholly-owned subsidiary formed 1954 jointly with Peter Spence & Sons Ltd and acquired 1961, which makes titanium tetrachloride and organic titanates. Overseas associates in Australia, Canada, France, India, Sierra Leone and South Africa. Employed chemical capital £5·8m. Employees 2500. ABCM member. 10 Stratton Street, London W1.

Brock's Fireworks Ltd

Founded by John Brock about 1700. Since then the firm has been associated with most firework displays, both official and private, at home and abroad, and contributed to the production of service pyrotechnics in the two world wars. In addition to display fireworks, the firm makes Government stores, Board of Trade signals, engine starters, furnace igniters, and pyrotechnics for agriculture, including insecticidal smokes, gas cartridges and bird scarers. It has four factories. Hemel Hempstead, Herts.

Burt Boulton & Haywood Ltd (BB & H)

Formed 1848 to deal with railway engineering, mainly sleepers and their preservation, hence their association with timber and tar. Became public company 1924. Its activities fall into three broad groups, viz., timber (a major business but outside present scope), chemicals, and tar and allied industries. Burts & Harvey Ltd (incorporating Alchemy Ltd) is a subsidiary covering the main chemical manufactures and the crop protection chemicals of the Agricultural Division. U.K. associates are PR Chemicals Ltd (refiners of tar acids) and Baywood Chemicals Ltd (distributors of crop protection products in co-operation with the Bayer Co.). Another associate is Printar Industries Ltd (tar distillers and manufacturers of disinfectants). There are also interests, *inter alia*, in tarmacadam and road contracting. There are overseas associates in Belgium, Italy and Portugal. Apart from tar products, the chemical manufactures cover a wide range of plasticisers, driers and alkyl phenols. U.K. employees in the chemical group 1200. ABCM member. Brettenham House, Lancaster Place, Strand, London WC2.

Butler Chemicals Ltd

Formed 1964 as successor to Wm. Butler & Co. (Bristol) Ltd, which was founded as tar distillers in 1843, became limited 1863 and public 1947. Consequent on the nationalization of the gas industry, the company had, in 1961, to relinquish its interests in tar distillation and its distillery went to the South Western Gas Board. Manufactures include paper makers' size, disinfectants, antiseptics, wood and cordage preservatives and solvents. Concerned also with the refining of lubricating oils, etc.; also importers and distributors of fuel oils. Associates in Belgium and Sweden. Chemical capital £750,000. Employees 250. ABCM member. Rockingham Works, Avonmouth, Bristol.

Cabot Carbon Ltd

Founded 1948 as subsidiary of Cabot Corporation, Boston, for production of carbon black chiefly for tyres, to obviate imports from U.S.A., with encouragement of Board of Trade. Assets over £4m. Employees 390. Has also a plant supplying carbon black-in-plastics dispersion service. Associated companies in Argentina, Canada, France and Italy. Stanlow, Ellesmere Port, Cheshire.

Carless, Capel & Leonard Ltd

Dates back to 1859 when Eugene Carless set up in Hackney Wick as distiller and refiner of mineral oils. From that time, fractional distillation has been the keynote of its activities. The partnership of Carless, Capel & Leonard was established 1872, and was converted to a private liability company in 1949. The firm introduced the name "petrol" for the high volatility spirit it pioneered for the new internal combustion engine in the early 90's. Its products fall into three main classes, fuels, solvents and chemicals, the last-mentioned being highly refined hydrocarbons as starting materials for chemical synthesis. Employees 200. ABCM member. Hope Chemical Works, Hackney Wick, London E9.

Carson–Paripan Ltd

Formed 1962 by merger of Walter Carson & Sons Ltd (founded 1795) and Paripan Ltd (founded 1855), both paint makers and merchants. Associates are Runnymede Dispersions Ltd (incorporated 1958 to make pigment dispersions) and Penetone Co. Ltd (registered 1960 as Penetone-Paripan Ltd, name changed 1964), owned jointly by Carson-Paripan Ltd, Penetone International Corporation of America and C. Jungdahls & Fabriks, of Sweden, and makers of chemical degreasers, strippers, etc. Manufactures include all kinds of paints and resins therefor. Subsidiary of Bell's Asbestos & Engineering (Holdings) Ltd. Capital £550,000. Employees 450. Holman Road, Battersea, London SW11.

Catalin Ltd

Formed 1936 for the manufacture of plastics (originally the phenol-formaldehyde type), with a share-holding by the Catalin Corporation of America. In 1947, by agreement, it became an independent, wholly British-owned public company. Capital £200,000. Employees 200. Waltham Abbey, Essex.

J. W. Chafer Ltd

Formed 1901 for manufacture of potato fungicides. Later developed specialized machinery for hire for application of agricultural chemicals. Now specialize in the production of micronized dust fungicides. Associates are J. W. Chafer (Scotland) Ltd, Nutritional Consultants Ltd, Winnipeg, and in the U.S.A., Amchem, Dow Chemicals and Rohm and Haas. Employed capital £500,000. Employees 350. Milethorne Lane, Doncaster, Yorks.

Chemical Compounds Ltd

Formed 1946 for the manufacture of fine organic chemicals, not commercially made in the U.K. Products mainly for pharmaceutical and agricultural industries; also intermediates, photographic chemicals and materials for plastics. Employees 60. ABCM member. Groat Avenue, Aycliffe Trading Estate, nr. Darlington, Co. Durham.

The Chemical Supply Co. Ltd

Formed 1910 to make fine chemicals; became limited company 1922. It is the principal holding company for the group: BNR Ltd (formaldehyde makers, formed 1932), Rex Campbell & Co. Ltd (merchants, formed 1928), The Research and Manufacturing Co. Ltd (makers of aromatic chemicals, formed 1912), Century Chemical Co. Ltd (phosphoric acid makers established 1911), and Chemical Utilities Ltd (makers of sundry chemicals, formed 1922). Its products include solvents, plasticizers, pharmaceuticals, aromatics, pigments, and molybdenum compounds. Capital £1m. Employees 200. ABCM member. 7 & 8 Idol Lane, Eastcheap, London EC3.

Chemstrand Ltd

Established 1955 as private company wholly-owned by Monsanto Co., U.S.A., to manufacture and market man-made fibres, with nominal capital of £800,000. Capacity in Northern Ireland for 27m lb of " acrilan " acrylic fibre in 1963. Production of Nylon 66 in Ayrshire due 1966 at 18m lb p.a. U.K. associates are Lansil Ltd (formed 1921), makers of cellulose acetate, and Polythane Fibres Ltd and Stretchables International Ltd, both subsidiaries of Monsanto Co., U.S.A., and makers of synthetic elastomeric and latex yarns. Associates through Monsanto Europe S.A. in Israel and Luxembourg. 10–18 Victoria Street, London SW1.

CIBA United Kingdom Ltd

Formed 1961 as the holding company for the CIBA group of firms in Great Britain. The interests of CIBA (from Basle) with a history back to 1859, started in the U.K. over 50 years ago. The associated companies include the Clayton Aniline Co. Ltd (q.v.), manufacturers of dyestuffs and allied products, CIBA Clayton Ltd, formed 1953 for dyestuffs and chemical sales, CIBA Laboratories Ltd, founded 1934 for pharmaceutical production and sales, CIBA Chemicals Ltd, formed 1962 for pharmaceutical chemical manufacture, and CIBA (ARL) Ltd for synthetic resins and adhesives. This last-mentioned company stemmed from a small research company founded at Duxford 1934; it became a member of the CIBA group 1947 and received its present name 1958. Associated with the group are the CIBA Foundation for the Promotion of International Co-operation in Medical and Chemical Research, opened in 1949, and the CIBA Fellowship Trust, set up in 1958. Employed capital of group £16·25m. Employees 2850. ABCM member. 96 Piccadilly, London W1.

The Clayton Aniline Co. Ltd

Formed 1876 to produce basic raw materials for the coal-tar colour industry. In 1911, The Society of Chemical Industry in Basle (now CIBA Ltd) purchased whole of shareholding. In 1919, J. R. Geigy, SA, and Sandoz Ltd, of Switzerland, admitted to minority shareholding. Manufactures cover dyestuffs, dyestuffs intermediates, pigment intermediates and textile application products. Employed capital £13m. Employees 1130. ABCM member. Clipstone Street, Ashton New Road, Clayton, Manchester 11.

Coalite and Chemical Products Ltd

The fore-runner of this company was Low Temperature Carbonisation Ltd, formed in 1917, absorbing two earlier companies, to develop a process of low temperature carbonization based on Thomas Parker's 1906 patent. The present name dates from 1948. The main product is smokeless fuel. The chemical interest lies in the by-product tar, gas spirit and ammoniacal liquor and the chemicals extracted and further processed therefrom. These include phenols and other tar acids, weedkillers, and phosphate plasticizers. There are four operating subsidiaries: Doncaster Coalite Ltd, The Derbyshire Coalite Co. Ltd, Coalite Oils & Chemicals Ltd and the recently acquired Duramis Fuels, which supplies oil additives. Total capital £6m. Employees 1600. ABCM member. PO Box 21, Chesterfield, Derbyshire.

Coates Brothers & Company Ltd

Started 1877 as a partnership to supply printing inks. Became a private company 1887. Reconstructed as public company 1948, as the parent manufacturing organization with two selling subsidiaries for printing inks and paints respectively and a third, Cray Valley Products Ltd, formed 1938 to make synthetic resins. The production range covers printing inks, synthetic resins, paints and varnishes, and printing equipment. There are associates in Australia, Central Africa, Denmark, East Africa, Far East, France, India, New Zealand, Pakistan and South Africa. Capital £3–4m. Employees 1200. 1–7 Easton Street, London WC1.

Colgate-Palmolive Ltd

Formed 1922 as subsidiary of the U.S. company of same name. Originally concerned with soap and dental creams. Now also large makers of fatty acid chlorides and N-acyl sarcosinates, and grinders of silica. Close liaison with U.S. parent and its sister companies in Europe. Employees 2000. 371 Ordsall Lane, Manchester 5.

Commercial Plastics Industries Ltd

Commercial Plastics Ltd, founded 1946, to make PVC sheeting as first company of the group. Second company, Anglo-American Plastics Ltd, formed 1951 to make polythene film. 1957 Iridon Ltd formed to make polystyrene sheeting and Greenwich Plastics Ltd, previously Greenwich Leathercloth Co., acquired. Other smaller companies set up or acquired between 1947 and 1953. Group reorganized 1953 with main trading companies as follows: Commercial Plastics Ltd, consolidating PVC sheeting interests, Anglo-American Plastics Ltd—polythene sheeting, Iridon Ltd—polystyrene sheeting, Rilite Ltd—glass fibre products, and Plastic Containers Ltd—vacuum formed containers. There are two other companies outside the main field, *viz.*, Thermalon Ltd—building materials and Mondart Ltd—aerosol products. Group acquired jointly by Unilever and Hambros Bank and Associates, 1964. Associates in Belgium, France, Netherlands and U.S.A. Employees 3000. Berkeley Square House, Berkeley Square, London, W1.

The Co-operative Wholesale Society Ltd

This is a large group with some 190 factories producing food, clothing, footwear, furniture, etc. Manufacture of paints commenced 1913 to cater for its other manufactures and the co-operative retail trade. Range covers cellulose, polyurethane, and melamine finishes for furniture and stoving finishes; Halcyon colour-blend dispensing units throughout the country producing 1200 colours from nine colorants and two base paints. Capital employed at paint works £250,000. Employees 150. CWS Paint & Varnish Works, Stockbrook Street, Derby.

Horace Cory & Co. Ltd

Formed 1828 to make chemical pigment colours. Associates in U.S.A. Present products are organic pigments and food colours. Nathan Way, London SE16.

Courtaulds Ltd

The original business of Samuel Courtauld & Co. was started in 1816 first as silk-throwers then as weavers specializing in mourning crape. Public company formed 1904; reconstructed 1912 as Courtaulds Ltd to exploit viscose, while maintaining weaving business. Viscose manufacture started in U.K. 1905, and in 1910 by subsidiary American Viscose Co. (later American Viscose Corporation), which had to be sold in 1941. In 1920's viscose interests established or acquired in France, Germany, Italy and Canada. Acetate production commenced in 1928, followed in 1930's by patent litigation with British Celanese Ltd (successor of British Cellulose and Chemical Manufacturing Co. Ltd established 1916), which was acquired in 1957 with several smaller U.K. cellulose fibre makers in ensuing years. In 1934, British Cellophane Ltd established with 75% interest. In 1940, British Nylon Spinners Ltd was formed jointly with ICI. Courtauld's share was transferred to the latter in 1964 in exchange for ICI's holdings in Courtaulds. In the decade following World War II, the firm re-entered viscose production in U.S.A. and set up interests in Australia and South Africa. Their fibre range in the U.K. was also greatly extended. From 1958, interests were acquired in paints (especially Pinchin Johnson & Associates Ltd, registered 1899 as Pinchin Johnson & Co. Ltd), packaging, and plastics moulding, while large outside engineering contracts were undertaken at home and abroad. More recently, extensive interests were acquired in the textile industry including Lancashire Cotton Corporation and Fine Spinners & Doublers.

In the chemical field, the principal U.K. subsidiaries are now British Celanese Ltd, British Cellophane Ltd, Pinchin Johnson & Associates Ltd, James Nelson Ltd (originally registered as private company 1914 and owns Nelson Silks Ltd, makers of acetate fibres, and 50% of Nelson Acetate Ltd, making acetate flake), and British Enka Ltd, making viscose fibres. There is a 50% interest in EF Co. Ltd, manufacturers of polyurethane elastomeric fibres, and 25% in the United Sulphuric Acid Corporation Ltd (q.v.), making sulphuric acid from anhydrite. Joint company Hedon Monomers Ltd formed 1965 through its subsidiary, British Celanese, with the Distillers Co. Ltd (q.v.), to make vinyl acetate.

There are overseas subsidiaries in Australia, Canada, France, South Africa and the U.S.A., in addition to those of British Cellophane Ltd and Pinchin Johnson & Associates Ltd. Also associates in Italy, Swaziland and West Germany.

The group manufacture a wide range of man-made fibres, paints and allied products, as well as carbon disulphide, carbon tetrachloride and sulphuric acid. There are many other activities outside the chemical field such as textiles, plastic moulding, engineering and toys. U.K. employed chemical capital £105m. Employees 32,000. British Celanese is ABCM member. 18 Hanover Square, London W1.

The Croda Organization Ltd

Formed 1925 as private company called Croda Ltd. Present name adopted 1960. Converted to public company 1964. Originally manufactured lanolin from wool-grease from textile industry. Range expanded since 1945 to include comprehensive series of fatty acids. This group consists of 17 wholly-owned subsidiaries, of which four are abroad, with two in U.S.A., one in Germany and one in Italy. U.K. production is divided as follows: chemicals (49%), rust preventive oils and coatings (29%), paints (11%), packaging (7%) and edible compounds (pure fatty acids) (4%). Employed assets £600,000. U.K. employees 325. ABCM member. Cowick Hall, Snaith, Goole, Yorks.

The Crookes Laboratories Ltd

Formed originally as Crookes Collosols Ltd 1912. Acquired 1918 by British Colloids Ltd, formed 1918. Present name adopted 1946. Public company 1951. Acquired 1961 by Arthur Guinness, Son & Co. Ltd (60%) and Philips-Duphar, of Holland (40%). The original objective was the production of colloids for use in medicine. This is still a considerable part of the activities which now cover a substantial part of the present pharmaceutical field, including vitamins, hormones, vaccines and synthetic drugs. Capital £1m. Employees 420. Park Royal, London NW10.

Cussons Group Ltd

Business commenced 1897, public company 1947. Reorganized 1963 and the group formed with operating activities transferred to newly-created subsidiaries. Cussons Sons & Co. Ltd, manufactures soap and toilet preparations; Gerard Brothers Ltd (formed 1897, acquired 1955) makes soap milling base, glycerine, soap products and detergents; PC Products (1001) Ltd (formed 1947, acquired 1958) makes detergent products; Richmond Aerosols Ltd (incorporated 1962, acquired 1963) makes aerosols for household and toiletry use; Sinclair, Owen & Co. Ltd makes cardboard boxes and luggage. Associates in Australia, New Zealand and South Africa. Overseas business handled by Cussons (International) Ltd. Capital £1·75m. Employees 1500. Kersal Vale, Manchester 7.

Cyanamid of Great Britain Ltd

Incorporated 1923 as Cyanamid Products Ltd, to market chemicals made by its parent American Cyanamid Co., and to purchase chemical raw materials for its U.S. associates. Name changed 1957. Entered pharmaceutical field 1945, marketing ethical products of Lederle Laboratories. Manufacture started 1950. Now makes antibiotics, other ethical pharmaceuticals, agricultural and veterinary chemicals and melamine. Capital about £700,000. Employees 750. Bush House, Aldwych, London WC2.

L. Dennis & Co. Ltd

Founded 1922. Manufacturers of magenta dye, dyestuffs intermediates, bleaching liquor and sodium hypochlorite. Capital £400,000. Employees 60. ABCM member. Thorpe Chemical Works, Walkden, Manchester.

The Distillers Co. Ltd (DCL)

Formed 1877 by the amalgamation of six grain distillers, some of whom were also producing fermentation alcohol for industrial use; from this stem DCL's activities as a major producer of chemicals. Only the chemical and plastics interests of DCL are relevant to the present summary of its activities.

With the acquisition of a small chemical works at Carshalton and of a distillery at Hull in 1925–26 and the subsequent erection of the Hull works of its subsidiary, British Industrial Solvents Ltd (incorporated 1929, liquidated 1953), DCL began the large-scale synthesis of alcohol-based industrial chemicals—particularly of solvents such as acetates and acetone for use in paints, lacquers and printing inks; plasticizers such as phthalates, for use in thermoplastic materials; and intermediates such as acetic acid, acetic anhydride and higher alcohols for synthetic fibres, dyes, pharmaceuticals and plastics. DCL also took an active part in establishing the U.K. production of acetone and butanol by the deep-fermentation process, first in a factory at King's Lynn and subsequently at the Bromborough works of Commercial Solvents (GB) Ltd, which continued in operation until 1957. Commercial Solvents (GB) Ltd was formed 1935 as U.K. subsidiary of Commercial Solvents Corporation (US). It was bought in 1939 by DCL (75%) and Barter Trading Corporation (25%); it ceased production in 1958 and ceased to exist in 1964.

During the 1939–45 war, DCL's chemical production came almost entirely under Government direction. In 1939, with the technical assistance of Shawinigan Chemicals of Canada, DCL built and subsequently operated for the Government a factory at Kenfig, South Wales, to make calcium carbide; on this production DCL based its post-war development of vinyl plastics—polyvinyl chloride and polyvinyl acetate. These carbide-based routes are now technologically outmoded, and the Kenfig factory will close in 1966. DCL's experience of deep-fermentation techniques led to its building the Government penicillin factory at Speke, Liverpool, towards the end of the war; here, from 1946, DCL made penicillin and other antibiotics until 1962, when these interests were sold to the American organization Eli Lilly & Company.

Until 1946, DCL's chemical production had been based on the fermentation initially of grain and later of molasses. The production of acetone from imported petroleum-based isopropanol was then commenced. The establishment shortly afterwards of BP's new oil refinery at Grangemouth, Stirlingshire, gave DCL the opportunity to begin the steady changeover to petroleum as the main basis of its chemical (and plastics) production. This development centres essentially around British Hydrocarbon Chemicals Ltd (q.v.), the joint DCL/BP company whose first Grangemouth plants were commissioned in 1951.

DCL's direct interests in the plastics industry (which dated from 1937) were confined to its ownership of British Resin Products Ltd (BRP) and its holdings in BX Plastics Ltd and F. A. Hughes & Co. Ltd, until the end of World War II, when a large new works was established at Barry, South Wales, to which the BRP and F. A. Hughes plastics manufacture was transferred, and where in 1946 the first PVC plant of British Geon Ltd (formed 1945) was commissioned. To ensure adequate supplies of chlorine, DCL (jointly with Fisons) acquired in 1954 Murgatroyd's Salt & Chemical Co. Ltd (q.v.). In 1938 British Resin Products had been the first U.K. company to produce polystyrene, starting from alcohol; in 1953, when petroleum-based styrene monomer became available from Forth Chemicals Ltd (a BHC subsidiary), modern production of polystyrene began at Barry, by Distrene Ltd, a joint DCL/Dow International company formed in 1954.

DCL's earlier chemical and plastics interests had been mainly in the production of basic materials; but its holding in BX Plastics Ltd led to the acquisition of The British Xylonite Co. Ltd, and to the merging in 1963 of this company and its subsidiaries with the U.K. plastics interests of the Union Carbide Corporation in a new group Bakelite Xylonite Ltd (q.v.).

As DCL's chemical and plastics activities had greatly expanded in size and scope since 1946, the DCL Chemicals and Plastics Group was established in 1963 to co-ordinate them. The Group is organized in four sub-groups dealing respectively with production and sales in the U.K. of (a) chemicals and (b) plastics, by wholly-owned divisions and subsidiaries, and by associates managed by DCL, (c) DCL's interests in other U.K. associates, and (d) DCL's interests in overseas associates, together with the necessary research, engineering, technical and general services departments. The main U.K. units are as follows: (W = wholly owned by DCL)

Chemicals

(i) *Industrial Solvents Division* (W). Makes solvents, plasticizers and intermediates and operates the plants of three associates: Hedon Chemicals Ltd (50% Shawinigan Chemicals Ltd, Canada—vinyl acetate), Orobis Ltd (50% California Chemical Co., U.S.A., which changed its name to Chevron Chemical Co. in 1965—lubricating oil additives) and Grange Chemicals (a BHC subsidiary—phthalic anhydride). The Methylating Co. Ltd (W) is part of this division.

(ii) *Carshalton Division* (W). Produces fine and special-purpose chemicals, including surfactants and industrial cleaning materials for two associates, Honeywill-Atlas Ltd (50% Atlas Chemical Industries, U.S.A.) and Honeywill-DuBois Ltd (50% Du-Bois Chemicals, U.S.A.). Honeywill & Stein Ltd (W), chemical merchants, are part of the division.

(iii) *Carbon Dioxide Division* (W). Handles the production and distribution of solid and liquid carbon dioxide (particularly bulk liquid).

(iv) *Carbide Division* (W). Operates on lease the Government calcium carbide works at Kenfig, which will close in 1966. Synthite Ltd (q.v.), producers of formaldehyde, are associated with this sub-group, and with the Industrial Solvents Division.

Plastics

(i) *British Resin Products Ltd* (W). Produces synthetic resins for plastics moulding, adhesives, impregnation and surface coatings. Markets " Rigidex " polyethylene for British Hydrocarbon Chemicals, polystyrene for Distrene Ltd, and (in the U.K.) polychloroprene for Distugil (French associate).

(ii) *British Geon Ltd* (45% B. F. Goodrich, U.S.A.). Producers of vinyl plastics; and Hedon Monomers Ltd jointly with Courtaulds (q.v.) through its subsidiary, British Celanese.

(iii) *Distrene Ltd* (50% Dow International). Formed 1954. Makers of polystyrene. Accounting, purchasing and commercial information services for this sub-group, and common factory services at Barry, are provided by Distillers Plastics Services Ltd (W). F. A. Hughes & Co. Ltd (W), plastics merchants, are also members.

U.K. Associate Companies include British Hydrocarbon Chemicals Ltd (q.v.), with its two subsidiaries, Forth Chemicals Ltd and Grange Chemicals Ltd, Border Chemicals Ltd, Bakelite Xylonite Ltd (q.v.), Murgatroyd's Salt & Chemical Co. Ltd (q.v.), and Orobis Ltd (*vide supra*).

The manufacturing and trading interests of the Distillers Co. (Biochemicals) Ltd, formed in 1947, were sold to Eli Lilly (q.v.) in 1962.

There are overseas associates in Australia, France, India and South Africa.

DCL Chemicals and Plastics Group has some 8000 employees. ABCM member. 21 St. James's Square, London SW1.

Diversey (U.K.) Ltd

Formed 1958 as subsidiary of the American Diversey Corporation and became sole owner of Deosan Ltd. Makers of industrial detergents, bactericides, metal-treatment compounds and the like, also associated equipment. Employed capital £500,000. Employees 500. Parent has many overseas associates. 42–46 Weymouth Street, London W1.

Dunlop Chemical Products Division

Formed 1957 by Dunlop Rubber Co. Ltd to produce adhesives, latex compounds, emulsions, reclaim dispersions and sealing compounds. Manufactures now include building adhesives and synthetic gutta-percha (transpolyisoprene). Employees 450. Chester Road Factory, Erdington, Birmingham 24.

Du Pont Company (United Kingdom) Ltd

A subsidiary company of E. I. du Pont de Nemours & Co., U.S.A., formed 1956 to manufacture, import and distribute in the U.K. the products of its parent. Production of neoprene synthetic rubber commenced 1960 and of toluene diisocyanate 1963 in Northern Ireland. The company contracts out the manufacture of du Pont herbicides and distributes these and imported products in the textile fibres, elastomers, plastics, industrial film and other fields. Associates with all its parent's subsidiaries. Employed capital about £12m. Du Pont House, 18 Breams Building, Fetter Lane, London EC4.

Durham Chemical Group

This group started 1918 as The Ouseburn Trading Co. Ltd. Its main manufacturing company, Durham Chemicals Ltd, began simultaneously, but from 1928 to 1947, it was called the Newcastle-upon-Tyne Zinc Oxide Co. Ltd. It absorbed the subsidiary, Typke & King, 1956. It manufactures zinc oxide, zinc salts, metal powders, inorganic silicates, and metal soaps. Durham Raw Materials Ltd, with world-wide agents, is its marketing organization. Two principal members of the Group with American associates are Nuodex Ltd, makers of paint driers and additives and fungicides, and Durham & Bonny Ltd, makers of a specialized plastic product for electrical insulation. Associates in the U.S.A. and Germany. A subsidiary fabricates PTFE parts for electronics and engineering. Employed capital £1·5m. Employees 450. ABCM member. Birtley, Co. Durham.

Esso Petroleum Company Ltd (ESSO)

Registered 1888 as the Anglo-American Oil Co. Present name adopted 1951. As the first foreign affiliate of the Standard Oil Co. of New Jersey, the original objective was to extend the market for American produced kerosene and paraffin wax. With total assets of over £250m, the company now refines and markets petroleum products throughout the U.K. and is searching for indigenous crude oil.

Entered the petrochemical industry 1957 to supply close cut range of heptenes—for ICI (q.v.)—extended to butadiene and ethylene. Three companies built alongside the plant to take products by pipeline: The International Synthetic Rubber Co. Ltd (q.v.) for butadiene to make SBR synthetic rubber, Monsanto Chemicals Ltd (q.v.) to polymerize ethylene, and Union Carbide Ltd (q.v.) to convert ethylene to ethylene oxide and to ethylene glycol. Ethylene also sent by pipeline to ICI works at Severnside. Manufacture of butyl rubber started 1963. Styrene plant to be commissioned 1966. Sulphur recovered as refining by-product. Lubricating oil additives plant started 1960. Makes also a number of hydrocarbon solvents, e.g. white spirit. Petrochemical operations account for 30% of capital invested in the £87m refinery. Chemical employees about 1000. Esso Chemical Ltd formed late 1965 to control its U.K. chemical industries. Parent has world-wide interests. ABCM member. Victoria Street, London SW1.

Federated Paints Ltd

Formed 1951 by the amalgamation of Williamson Morton & Co. (founded 1856), Strathclyde Paint Co. Ltd (established 1889) and Boyd Stewart & Co. (formed 1919). Subsequently Brown & Sons (founded 1783) was acquired. The objectives which still operate were to manufacture paints, particularly for decorative and industrial use. Associates in South Africa. Capital £199,000. Employees 120. 309–361 Dobbies Loan, Glasgow C4.

Fisons Ltd

Fisons Ltd had its roots in the manufacture of fertilizers (mainly superphosphate) by Edward Packard, the Prentice Brothers, and James and Joseph Fison in the middle of the nineteenth century. This led in 1929 to the formation of Fison, Packard and Prentice Ltd by amalgamation. Since then the company has broadened its interests till it is now a major group of 65 companies (31 in the U.K. and 34 overseas). The name was changed to Fisons Ltd in 1942. Its interests span agricultural fertilizers and pesticides with Fisons Fertilizers Ltd (which absorbed West Norfolk Fertilizers in 1965) and Fisons Pest Control Ltd (formed in 1939 and acquired in 1954), horticultural products with Fisons Horticulture Ltd, which took over the horticultural products side of International Toxin Products in 1965, industrial chemicals with Whiffen & Sons Ltd (established in 1854 and acquired in 1947), pharmaceuticals with Fisons Pharmaceuticals Ltd consolidated in 1964, and comprising Benger Laboratories Ltd (formed 1891 and acquired in 1947), Genatosan Ltd (formed in 1916 and acquired in 1937), and Fulford-Vitapointe, acquired in 1964. Other activities include scientific apparatus and laboratory chemicals, foods, man-made fibres and a 50% interest in Murgatroyd's Salt & Chemical Co. Ltd (q.v.) and in a U.K. factory Nypro (U.K.) Ltd, for making the nylon material caprolactam with the Dutch State Mines. An International Division is responsible for subsidiary and associated companies abroad, and controls Fisons Overseas Ltd, which handles the overseas sales of the products manufactured by the U.K. companies. The most important overseas interests are in Canada, U.S.A., Belgium, India, South Africa, Sudan, Zambia, Rhodesia, and central and east Africa. Employed capital is £40m. U.K. employees 8600. ABCM member. Harvest House, Felixstowe, Suffolk.

Joint U.K. company (50/50)—Aquitaine-Fisons Ltd—formed 1966 with Societé Nationale des Pétroles d'Aquitaine (S.N.P.A.) to develop plastics interests.

Forestal Industries (U.K.) Ltd

From its incorporation in 1902, it was for many years a manufacturer of vegetable and synthetic tanning extracts. In recent years, it has extended its chemical interests by the acquisition of a number of companies, now operating as Divisions, the principal being Farnell Carbons Ltd (incorporated 1937, acquired 1960)—activated charcoal, Kaylene Chemicals Ltd (incorporated 1953, acquired 1960)—chemicals and pharmaceuticals, V. W. Eves & Co. Ltd (formed 1949, acquired 1960)—vitamins, antibiotics and food supplements for animals, R. Cruickshank Ltd (formed 1902, acquired 1961)—industrial chemicals and metal finishing processes, plants and ancillaries. Tanning materials are made by the old established Calder & Mersey Extract Co. Ltd, which is now named FLT Ltd. There is also a water treatment Division. There are associates in Argentina, Germany, Kenya, Natal, Rhodesia and South Africa. The company's capital of £285,000 is all held by the Forestal Land, Timber & Railways Co. Ltd. ABCM member. The Adelphi, John Adam Street, London WC2.

Formica International Ltd

Formed 1957 as a subsidiary of the De La Rue Group. Sixty per cent of share capital owned by De La Rue Co. Ltd and 40% by the American Cyanamid Co. De La Rue started plastics manufacture 1915. The range now includes phenolic and melamine resins and industrial and decorative laminates; also Formica wood chipboard and plywood by the J. & H. Rosenberg Group of Companies acquired 1963. There are subsidiaries in Australia, Belgium, France, Germany, India, Netherlands, New Zealand, Spain and Switzerland. U.K. employees 2250. De La Rue House, 84/86 Regent Street, London W1.

Geigy (Holdings) Ltd

Registered in the name of The Geigy Colour Co. Ltd 1922 to take over the existing U.K. selling agencies for dyestuffs made by its parent company in Basle. U.K. manufacture began in the early 30's. The name was changed to The Geigy Co. Ltd 1945 and to the present title 1953. The scope has been extended to include dyestuffs, intermediates, pigments, textile auxiliaries, plasticizers, pesticides, pharmaceuticals and a wide range of other industrial chemicals. It has as subsidiaries Ashburton Chemical Works Ltd, James Anderson & Co. (Colours) Ltd, The Geigy Co. Ltd and The Geigy Pharmaceutical Co. Ltd, and is associated with The Clayton Aniline Co. Ltd (q.v.). In 1965 they were merged into one company, called Geigy (U.K.) Ltd with four divisions for dyestuffs and textile chemicals, industrial chemicals, pigments and pharmaceuticals. Its parent in Switzerland has subsidiaries throughout the world. Employed U.K. capital £12·5m. Employees 2350. Simonsway, Manchester 22.

Gerhardt–Penick Ltd

Began 1864 for drug and chemical merchanting as C. F. Gerhardt Ltd. Manufacture commenced 1952. Acquired small fine chemical company, Huffer & Smith Ltd, 1957, and 125-year-old Benbows Dog Mixture Co. 1959. To indicate the link with their close American associate, S. B. Penick & Co., the companies were consolidated under the above name 1961. In addition to merchanting activities on behalf of American, Swiss and Turkish associates, it makes a range of fine chemicals for proprietary and ethical pharmaceuticals and of products in the insecticide, rodenticide and veterinary fields. ABCM member. Thornton Laboratories, Purley Way, Croydon.

Givaudan & Co. Ltd

Incorporated 1950 as a wholly-owned private company subsidiary of L. Givaudan et Cie, SA, Switzerland, for the manufacture of synthetic aromatic chemicals and perfumery specialities. Activities extending into field of flavouring materials. Associates in Argentina, Brazil, France, Spain and U.S.A. Employed capital £1·25m. Employees 130. Whyteleafe, Surrey.

Glaxo Group Ltd

The group was formed 1962 to facilitate the control of the some sixty subsidiary and sub-subsidiary companies already operating under the aegis of Glaxo Laboratories Ltd, which was the largest unit of the group. The origin goes back to Joseph Nathan & Co., founded in New Zealand 1873, to export local produce. A London office was opened 1876, and Joseph Nathan & Co. Ltd formed 1899. In 1908, Glaxo Baby Food was first advertised in the national press. In 1935, Glaxo Laboratories, up to then a branch of Joseph Nathan & Co. Ltd, became a private limited company. In 1947, it became a public company and absorbed its parent. Then in 1962, with the group formation, it became a subsidiary company to carry on business as before.

Over the years, the Glaxo interests have extended widely in the field of pharmaceuticals, including immunological products, veterinary medicines and agricultural and horticultural products. Other activities include infant and invalid foods, animal feed supplements, surgical instruments and hospital equipment.

The following are the main subsidiary companies in the U.K.:

1. Allen & Hanburys Ltd. Founded originally 1715. After various changes of name over the years, it assumed its present title 1893. It merged with Glaxo 1958. Their field was pharmaceuticals, surgical instruments and hospital equipment.

2. Edinburgh Pharmaceutical Industries Ltd (EPI Group)—joined Glaxo 1963. It was established in 1961 and included the following: (a) Macfarlan Smith Ltd, formed 1960 by merger of J. F. Macfarlan & Co. Ltd and T. & H. Smith Ltd, two businesses over 150 years old, specializing in the manufacture of opium alkaloids and fine chemicals; (b) Duncan Flockhart & Co. Ltd, started as druggists 1806, later makers, *inter alia*, of anaesthetics, joined EPI Group 1952; (c) Pinkerton, Gibson & Co. Ltd, started as Clark & Pinkerton nearly 100 years ago, merged with T. & H. Smith 1964 as wholesalers; (d) William Paterson & Sons (Aberdeen) Ltd, started 1838 as wholesalers and retail chemists, joined EPI Group 1956; (e) W. & R. Hatrick Ltd, started about 1830 as wholesale and retail chemists, joined EPI Group 1935.

3. Evans Medical Ltd. Formed 1925 as Evans, Sons, Lescher & Webb, Ltd; name changed to Evans Medical Supplies 1945, assumed present title 1959. Joined Glaxo 1961.

4. Glaxo-Allenburys (Export) Ltd. Formed 1960.

5. Glaxo Research Ltd. Formed 1962.

6. The Murphy Chemical Co. Ltd—formed 1931, makers of agricultural and horticultural insecticides and allied products; joined Glaxo 1955.

7. Glaxo Laboratories Ltd—as described above. (see Addenda p. 329).

Joint company formed with BDH Group (q.v.) to combine their wholesale pharmaceutical interests in 1965.

There are overseas subsidiaries and associates in Argentina, Australia, Belgium, Brazil, Canada, Ceylon, Colombia, Cuba, Eire, France, Ghana, India, Italy, Kenya, Malaysia, New Zealand, Nigeria, Pakistan, South Africa, Switzerland, Thailand, Turkey, Uruguay, Venezuela. There are representatives in practically every country. U.K. chemical employees 10,000. ABCM member. 47 Park Street, London W1.

Glovers (Chemicals) Ltd

Incorporated 1931 to make chemical auxiliaries for the textile industry. Public company 1965. Main manufactures are surface active agents of all kinds. Capital £350,000. Employees 100. ABCM member. Wortley Low Mills, Whitehall Road, Leeds 12.

Golden Valley Colours Ltd

Formed 1894 under a slightly different title to exploit a nearby deposit of ochre and to make other mineral colours. Present name adopted 1938. Its subsidiaries are Golden Valley Ochre & Oxide Company (Wick) Ltd (making mineral pigments as the successor of the original company), Golden Valley Ochre and Oxide Co. (Yate) Ltd (makers of synthetic colours, established 1938), and Micronised Pigments Ltd (formed 1946 to make pigments of great fineness). There are two companies making oxides of iron in South Africa and Spain. Capital £500,000. Employees 400. Wick, Bristol.

Goodlass Wall & Lead Industries Ltd

A holding company formed1930 to cover a wide range of activities, including chemicals, paints, die-castings and non-ferrous alloys. Its chemical subsidiary is Associated Lead Manufacturers Ltd. This was formerly Walkers, Parker & Co. Ltd, started 1778. The name was changed in 1948, when there was an amalgamation with seven subsidiaries of Goodlass Wall & Lead Industries Ltd, when some went into liquidation, viz., The Cookson Lead & Antimony Co. Ltd (founded 1704), Locke, Lancaster and W. W. & R. Johnson & Sons Ltd (both established in 1790 and united 1894), Foster, Blackett & James Ltd (lead departments only) (originating 1774), The Librex Lead Company Ltd, previously Rowe Bros. Ltd (reformed 1949), The London Lead Oxide Co. Ltd (reformed 1949), A. T. Becks & Co. Ltd, and the Oidas Metals Co. Ltd. Champion Druce Ltd (dating back to about 1850 and assuming its present name in 1890, becoming limited 1930) was acquired 1954. Other subsidiaries are Goodlass Wall & Co. Ltd, makers of paints, etc., Frys Metal Foundries Ltd, Alexander Fergusson & Co. Ltd, makers of paint and lead compounds, Frys Diecasting Ltd, Durastic Ltd (floorings, etc.), and Harrison & Son (Hanley) Ltd, makers of colours for paints and plastics. Associates in Argentina, Australia, Germany, India, Ireland, Italy and South Africa. Subsidiaries and associates number 50. The main chemical manufactures include antimony, lead, lithium and zirconium compounds, paints, pigments and plastics. Among other products are building and constructional materials, largely based on lead, and printers' metals. Total capital £24m. U.K. employees 4000. Associated Lead Manufacturers is ABCM member. 14 Gresham Street, London EC2.

W. R. Grace Ltd

Formed 1930 as Dewey & Almy Ltd, acquired by W. R. Grace & Co., U.S.A., 1954. Took present name 1962. Object to make sealing compounds, soldering fluxes and adhesives for can making and food packaging; additional products are plastic packaging materials and battery separators. U.K.

associate is Silica Gel Ltd, formed 1924, to make silica gel, and equipment for gas and air drying, and acquired 1964. American parent has associates in 10 overseas countries. Capital £850,000. Employees 450. Elveden Road, Park Royal, London NW10.

R. Graesser Ltd

Robert F. Graesser, a German, came to England in 1863, established works at Ruabon in 1867 and traded as Robert Graesser, after brief partnership as Crowther & Graesser. He distilled tar to make phenol, cresols and xylenols. R. Graesser Ltd was formed as a limited company in 1916. In 1920, it amalgamated with the European interests of Monsanto Chemicals, Inc. of St. Louis, as Graesser-Monsanto Ltd. Monsanto bought out the Graesser interest in 1928. In 1935, tar acid refining was recommenced by N. H. Graesser, the son of the founder, and in 1945 the business was converted to a limited company under the old name of R. Graesser Ltd. In the same year, it became a subsidiary of Lancashire Tar Distillers Ltd and responsible for refining all the tar acids of the group. Capital £750,000. Employees 250. ABCM member. The Chemical Works, Sandycroft, Chester.

Graesser Salicylates Ltd

Originally established as F. R. Graesser-Thomas & Co., 1932, to make salicylic acid and derivatives. Converted into limited company with present name 1934. In 1960, became member of Aspro-Nicholas Group, which makes and markets proprietary and ethical pharmaceuticals. Products include salicylates, aspirin and a wide range of pharmaceutical chemicals for the Group. Capital £230,000. Employees 176. ABCM member. Sandycroft, Chester.

Walter Gregory & Co. Ltd

Formed 1910, to make and sell veterinary products for livestock, in continuation of the business done by Walter Gregory for the previous 25 years. It has as subsidiaries Somerset Pharmaceuticals Ltd (founded 1953 for retail and professional business), Swallowfield Aerosols Ltd (formed 1957 to expand, as formulators and packers, the aerosol trade started 1947), and Holyoak Products Ltd (founded 1962 as formulators and packers for non-aerosol products). Little chemical manufacture, as main efforts are devoted to formulation and packaging of materials required by the farmer. Capital £300,000. Employees 250. Wellington, Somerset.

Griffiths Bros. & Co. London Ltd

Established 1869 by Thomas and William Griffiths to make a patented non-toxic pigment as substitute for white lead, since developed as lithopone. Thomas credited as "inventor and patentee of enamel paint". In 1889, invented Griffith Bros. "Balloon Brand anti-sulphuric enamel" to protect storage batteries for new electricity requirements. In 1902 absorbed Daniel Judson & Son. Became private limited company 1916. Products cover paints, varnishes and finishes. Associate in Australia and world-wide agents. Employees 120. Well Lane, Wednesfield, Wolverhampton.

Edward Gurr, Ltd

Formed 1946 to produce biological stains and reagents for use in microscopy. Now supplies over 2000, with substantial export trade. Employed capital £25,000. ABCM member. 42 Upper Richmond Road West, London SW14.

Hadleigh Crowther Ltd

Formed 1952 under the name of Gascoigne-Crowther Ltd to pioneer the introduction of quaternary ammonium sterilizers. It specializes in the field of surface disinfection for the food industries. Private company. Capital £50,000. Employees 100. Caversham Laboratories, Reading, Berks.

Robert Haldane Ltd

Formed 1940 to produce sodium acetate for the Ministry of Supply. Now make chlorinated and other coal tar products, antiseptics and disinfectants. Capital £55,000. Employees 20. ABCM member. Underwood Chemical Works, Murray Street, Paisley, Renfrewshire.

Harshaw Chemicals Ltd

Formed 1936 as subsidiary of Harshaw Chemical Co., U.S.A., to exploit their patented processes in electroplating. Associates in France and Netherlands. Present products are addition agents for electroplating and industrial cleaners, with plating engineering and medical specialities outside the chemical field. Employed capital £75,000. Employees 50. ABCM member. Daventry, Northants.

B. Hepworth & Co. Ltd

Established 1868, incorporated 1899, to make chemicals for the carpet industry. Now produces chemicals for textile processing, and for the photographic, pharmaceutical and detergent industries. Two subsidiaries: Healey Meters Ltd (electrical instruments) and Dudleys (Redditch) Ltd (small motors and wipers). Capital £150,000. Employees 40. ABCM member. PO Box 10, Chemical Works, Kidderminster, Worcs.

Hercules Powder Co. Ltd

Registered 1944 with capital of £300,000 as subsidiary of Hercules Powder Co., U.S.A. Into it was incorporated Paper Makers Chemicals Ltd (formed 1924), a manufacturing organization owned by Hercules. U.K. manufactures include paper makers chemicals, synthetic resins, plasticizers, synthetic rubber emulsifiers and industrial defoamers. Also distributes U.S. parent's products. Owns half interest in Nelson Acetate Ltd (formed 1948), cellulose acetate makers. Fred H. Wrigley Ltd (formed 1948) is subsidiary for crushing and pulverising industrial raw materials. Associate in Portugal. Employees 250. 1 Great Cumberland Place, London W1.

Hickson & Welch Ltd

The Hickson group of companies owes its foundation in 1893 to Ernest Hickson, who had previously been a director of Brook, Simpson & Spiller. Originally he merchanted the new aniline dyes, then made inorganic chemicals for the textile trade and sulphur black dye. In 1915, Hickson & Partners Ltd was established to make TNT and later dyestuffs and intermediates. This was succeeded by Hickson & Welch Ltd in 1931. In 1932, the U.K. and Commonwealth rights in Wolman wood preservatives were secured and developed. During World War II, they made anti-gas chemicals and DDT. In 1946, the range of dyestuffs intermediates was expanded to cope with shortages. The production of optical whitening agents was begun in 1949. These are now sold in over 60 countries.

In 1954, John W. Leitch & Co., with its two subsidiaries, was acquired. This firm dates from 1890 and made dyestuffs and intermediates. In 1934, in conjunction with H. Th. Bohme, Germany, it had formed the Gardinol Chemical Co. Ltd to make sulphated fatty alcohols and related products. The firm is now defunct, and its production has been absorbed into Hickson & Welch Ltd, for whom its old selling company of Ronsheim & Moore Ltd still operates, as a subsidiary. Timber preservatives are sold through Hickson's Timber Impregnation Co. (GB) Ltd, formed 1946, a sister company of the Hickson & Welch (Holdings) Group, formed 1951. There is a subsidiary in South Africa and an associate in India. U.K. chemical capital £2·5m. Employees 750. ABCM member. Ings Lane, Castleford, Yorks.

Thos. Hinshelwood & Co. Ltd

Began 1878 as oil refiners and drysalters. Commenced paint production 1887. Now manufacturers of paints, pigments and lubricants. Capital £105,000. Employees 120. Glenpark Street, Glasgow E1.

L. B. Holliday & Co. Ltd

Formed 1918 by Major L. B. Holliday, also a well-known racehorse owner, when recalled from active service to manufacture sorely needed picric acid. He was the grandson of Read Holliday, who founded Read, Holliday & Sons Ltd in 1830 to distil ammonia from gas works liquor; this led to tar distillation so that in 1860, following Perkin's discovery of mauveine, activities were naturally extended into the dyestuffs field; the firm was absorbed when British Dyes Ltd was formed in 1915. In 1919, British Dyes Ltd and Levinstein Ltd were amalgamated to become The British Dyestuffs Corporation Ltd (see ICI).

L. B. Holliday & Co. also turned to the manufacture of dyestuffs; they now produce a wide range and have agencies in over 40 countries. Employees 1000. ABCM member. Huddersfield.

Hopkin & Williams Ltd

Founded 1850 to make pharmaceutical products and chemicals for research and analysis. In 1880 ownership passed to Howard & Sons. Incorporated as limited company 1903. Laid the technical foundations of the U.K. "rare-earth" industry. In 1911, issued the first British publication on "Analytical Reagents, Standards and Tests". In 1929, Baird & Tatlock (London) Ltd, makers of scientific instruments and apparatus, acquired controlling interest. In 1934, in collaboration with The British Drug Houses Ltd, compiled the first edition of "AnalaR Standards for Laboratory Chemicals," which now comprises over 300 specifications. The Baird & Tatlock group, with which is also associated W. B. Nicolson (Scientific Instruments) Ltd, and Optica United Kingdom Ltd, is since 1959 a subsidiary of the Derbyshire Stone group of companies, and supplies a wide range of reagents and laboratory chemicals. There are branches in Australia, East Africa, Rhodesia and South Africa, and agents in every major country. ABCM member. Freshwater Road, Chadwell Heath, Essex.

Hull Chemicals Ltd

After trading many years, became incorporated 1930 as Hull Chemical Works Ltd. Present name adopted January 1965. Manufactures soaps, cleansing materials, disinfectants, insecticides, antiseptics and agricultural preparations. Kirkby Street, Hull.

Imperial Chemical Industries Ltd (ICI)

ICI was formed 1926 to acquire the interests of Brunner, Mond & Co. Ltd, United Alkali Co. Ltd, British Dyestuffs Corporation Ltd and Nobel Industries Ltd, which then had a total share capital of £38m. A major objective was to form an organization with financial, technological and commercial resources which would enable it to match those of the largest industrial units which had been established since World War I, such as Du Ponts, Union Carbide & Carbon Corporation and Allied Chemical & Dye Corporation in the U.S.A., and the German IG.

Each of the ICI merging companies had an interesting history:

1. *Brunner, Mond & Co. Ltd.* In 1872/3, Ludwig Mond and John Brunner formed a partnership to operate the Solvay ammonia–soda process. This became Brunner, Mond & Co. Ltd 1881. In 1916, they exchanged shares with the Castner-Kellner Alkali Co. Ltd, formed 1895 by the Aluminium Company Ltd to make electrolytic caustic soda. The Aluminium Company sold Castner's sodium cyanide process for a holding in Cassel Gold Extracting Co. Ltd (formed 1884 and later renamed Cassel Cyanide Co.). Castners absorbed the Aluminium Company 1901. In 1917, a controlling interest was acquired in Chance & Hunt, formed 1898, by the amalgamation of Oldbury Alkali Co. Ltd (of Chance Bros. & Co.) and William Hunt & Sons, of Wednesbury. In 1919, Brunners acquired the Ammonia Soda Co., and a controlling interest in Buxton Lime Firms Ltd (formed 1891). In 1920, Castners became a subsidiary of Brunners, Electro-Bleach & By-Products Co. (formed 1899) was acquired, and Synthetic Ammonia & Nitrates Ltd was formed to make synthetic ammonia. Control of Magadi Soda Co., Kenya, was obtained 1924–25. In 1927, Cassel Cyanide Ltd became a subsidiary of Brunners.

2. *United Alkali Co. Ltd.* Formed 1890 by amalgamation of over 40 companies, including James Muspratt & Sons Ltd (formed 1851 succeeding James Muspratt who established the LeBlanc process in 1823), Gaskell, Deacon & Co. Ltd (formed 1855), John Hutchinson & Co. (formed 1881 succeeding John Hutchinson, who started manufacture 1847), Wm. Pilkington & Sons (formed 1865), and most of Charles Tennant & Partners Ltd (formed 1885).

3. *British Dyestuffs Corporation Ltd (BDC)*. Formed 1919 by merger of British Dyes Ltd (formed 1915 with Government backing by Read, Holliday & Sons, active since 1860) and Levinstein Ltd (formed 1895, succeeding I. Levinstein established 1864), which in 1917 had acquired Claus & Co. Ltd, whose predecessors, Claus & Rée, had taken over Brook, Simpson & Spiller, which had purchased the works of Perkin & Sons in 1874.

In 1925, BDC acquired Scottish Dyes Ltd, formed 1919 with Government help as successor to Solway Dyes Ltd, formed 1917 from the dye manufacturing activities of Morton Sundour Fabrics Ltd.

After the formation of ICI, BDC in 1928 acquired Oliver Wilkins (successor 1907 to Derby Chemical Colour Co., formed 1904 to make pigments). In 1931, it merged with British Alizarine Co. Ltd (founded 1882 by various dyers to buy works of Burt, Boulton & Haywood, who in 1876 had bought Perkin's alizarine process and plant). Leech, Neal & Co. Ltd (successor 1888 to Leech & Neal, founded 1864 to produce pigments) was acquired in 1936.

4. *Nobel Industries Ltd*. Formed 1920 as successor to Explosives Trades Ltd, formed 1918 as a merger of many companies, including Nobel's Explosives Co. Ltd (formed 1877 as successor to British Dynamite Co.), Curtis's & Harvey Ltd, Bickford, Smith & Co. Ltd, Kynoch Ltd and Eley Brothers Ltd. Subsequently set up manufacture of nitrocellulose lacquers (Nobel Chemical Finishes Ltd, established 1926), acquired interests in leathercloth manufacture, and metal fabricating firms. By the time of the ICI merger had reorganized into the following groups: explosives, metal and ammunition, artificial leather, collodions and varnish, and sundries.

Organizational Changes Since 1926

1929: activities of merging companies reorganized into eight groups; Alkali, Explosives, Metals, General Chemicals (including United Alkali Co., Castner-Kellner Alkali Co., Cassel Cyanide Co., and Chance & Hunt), Dyestuffs, Fertilizer and Synthetic Products, Lime, and Leathercloth. Each group became ICI (X) Ltd in 1931, X being the name of the group. 1933/36: Croydon Mouldrite Ltd acquired, ICI (Plastics) Ltd formed 1938. 1935: Nobel Chemical Finishes became wholly-owned subsidiary; renamed ICI (Paints) Ltd in 1940. 1942: Imperial Chemical (Pharmaceuticals) Ltd formed. 1944: Groups were renamed Divisions; Explosives Division renamed Nobel Division in 1948. 1956: "Terylene" Council (formed 1951) became Fibres Division. 1957: Heavy Organic Chemicals Division formed. 1958: Leathercloth Division became ICI (Hyde) Ltd and made part of Paints Division. 1959–61: Lime Division and Salt Division absorbed into Alkali Division. 1962: Metals interests segregated with formation of Imperial Metal Industries Ltd as a holding company for a number of wholly- and partly-owned subsidiaries and associates. 1964: Alkali and General Chemicals Divisions merged to become Mond Division. 1965: ICI Fibres Ltd formed as a wholly-owned subsidiary to combine the activities of Fibres Division and those of British Nylon Spinners Ltd (formed jointly with Courtaulds Ltd 1939 and wholly acquired by ICI 1964).

ICI is now organized under the following manufacturing divisions: Mond,

Nobel, Agricultural, Heavy Organic Chemicals, Dyestuffs, Plastics, Pharmaceuticals, and Paints (including the former ICI (Hyde) Ltd) in addition to the wholly-owned ICI Fibres Ltd and Imperial Metal Industries Ltd. Other wholly-owned subsidiaries include Plant Protection Ltd (formed 1937 with Cooper, McDougall & Robertson Ltd and acquired 1958) and Withins Paper Staining Co. Ltd acquired 1960. Among the partly-owned and associated chemical companies are British Titan Products Co. Ltd, formed 1933, with Imperial Smelting, Goodlass Wall & Lead Industries, Titan Co. and R. W. Greef; The United Sulphuric Acid Corporation, formed 1950 with other acid makers and users, to manufacture from anhydrite; Border Chemicals Ltd, formed 1963 with Distillers Co. Ltd and British Petroleum Ltd to make acrylonitrile, etc.; Scottish Agricultural Industries Ltd, formed 1928 with Scottish producers of agricultural products; Richardsons Fertilisers Ltd, formed 1961, formerly Richardsons Chemical & Manure Co.; Ulster Fertilisers Ltd, formed 1960, formerly Ulster Manure Co.; the Weardale Lead Co. Ltd, majority shareholding acquired 1962; British Visqueen Ltd, formed 1953 with Visking Corporation, wholly-owned from 1959 to 1961 since when E. S. & A. Robinson have a minority shareholding; Robinson Plastic Films Ltd, formed 1961 with E. S. & A. Robinson; and Ilford Ltd, with shareholding acquired 1958. There are also several associates in the non-chemical fields, such as oil and textiles.

ICI has a large number of manufacturing subsidiaries and associates overseas; these operate in Argentina, Australasia, Belgium, Canada, Ceylon, Denmark, Eire, Finland, France, Germany, India, Italy, Kenya, Malaysia, Mexico, Netherlands, Pakistan, Portugal, South Africa, Spain, Sweden and the U.S.A.

ICI's manufactures range over the whole field of chemical industry, as is indicated by their divisions, with the exception of coal carbonization products. They have also substantial interests in the fields of metals, textiles, wallpaper and oil refining. In the U.K. parent company of ICI, employed capital is about £760m and employees, excluding the metal group, some 94,000. ABCM member. Imperial Chemical House, Millbank, London SW1.

Imperial Smelting Corporation Ltd (ISC)

Formed 1929 to develop the U.K. zinc producing industry. Its forerunner, the National Smelting Co., formed 1917, took over a zinc and sulphuric acid works dating from 1876. ISC obtained an interest in Orr's Zinc White Ltd (founded 1898) 1931, and acquired it 1954. In 1934, Aluminium Sulphate Ltd was founded, jointly with British Aluminium Ltd, and in 1959 its activities were extended jointly with The Alumina Co. Ltd (a British Aluminium subsidiary) and Peter Spence & Sons Ltd. In 1932, an interest was acquired in Fricker's Metal & Chemical Co. Ltd (founded 1910 and now wholly-owned), as makers of zinc oxide. Pure Chemicals Ltd (founded 1944 and makers of fine chemicals) was acquired 1957. There is a 50% interest with Dow Chemicals Ltd in Thorium Ltd (q.v.), and with Laporte Chemicals Ltd in Barium Chemicals Ltd (q.v.). Since 1962, ISC has been a wholly-owned subsidiary of the Rio Tinto-Zinc Corporation Ltd.

Among its non-zinc products are hydrofluoric acid and highly fluorinated organic compounds. Employees (including smelting) 3400. ABCM member. 6 St. James's Square, London SW1.

International Nickel Ltd

Originated in the Mond Nickel Co. Ltd, established 1900, as refiners of nickel, cobalt and silver ores. It became the International Nickel Company (Mond) Ltd 1961. On 1st January 1965, the name was changed to International Nickel Ltd, to emphasize its identification with its parent company, The International Nickel Co. of Canada, Ltd, which was formed 1902 from the remains of the Canadian Copper Co. and that it is a part of the world-wide international Nickel organization, providing service to nickel-using industries throughout the world. Its chemical business originated with its subsidiary, Henry Wiggin & Co. Ltd, founded 1835, as Evans & Askin, to make and refine cobalt oxides for the pottery industry; present name adopted about 1870; limited company 1892; taken over by Mond Nickel Co. 1919. Present manufactures include nickel and cobalt chemicals, selenium compounds and tellurium; the main activity is metal production. Chemical employees 220. ABCM member. Thames House, Millbank, London SW1.

The International Synthetic Rubber Co. Ltd

Formed 1956 by a consortium of the major tyre companies—Dunlop, Firestone, Goodyear, Avon, Michelin, Pirelli, North British Rubber and BTR Industries—to make the U.K. independent of imported materials, costing $25m a year. Production started 1958 of styrene butadiene rubbers. Capacity 1964 120,000 tons p.a. Extensions in hand for extra polybutadiene and butadiene copolymers. Capital about £10m. Employees 1000. ABCM member. Brunswick House, Brunswick Place, Southampton, Hants.

Jeyes Group Ltd

This group was based on Jeyes Sanitary Compounds Co. Ltd, founded by John Jeyes 1877 to manufacture and deal in disinfectant and preservative compounds; incorporated 1885; became public company 1955. In 1935, independent companies were established in Ireland and South Africa. The following have been acquired as indicated; 1955 Ibbetson & Co. (incorporated 1947) and its associates, making disinfectants, toilet cleansers, etc. (wound-up 1960); 1963 The Parazone Co. Ltd (established 1891); Wimsol Ltd (incorporated 1934), makers of household bleaches, and Three Hands Products Ltd, makers of disinfectants and dishwashing liquids (founded 1932 as the British Disinfectant Co., incorporated 1943, present name adopted 1955); and 1964 Serta Ltd (incorporated 1963), manufacturers of aerosols. Other subsidiaries have been incorporated or acquired but no longer trade. Jeyes Group Ltd was registered 1964. There are associates in Australia and New Zealand. The group makes a wide range of disinfectants, insecticides, toilet bowl cleansers, household bleaches, dishwashing liquids, window cleaners, aerosols and air purifiers. It has an important interest in toilet tissue products. One of its subsidiaries represents an American maker of drain and window cleaning compounds. U.K. employed capital £3m. Employees 1350. 31 River Road, Barking, Essex.

Johnsons of Hendon Ltd

The main operating company of the group Johnsons of Hendon (Holdings) Ltd (founded 1953). Originated 1743 for the assaying of precious metals. By 1829 chemical manufacture had started and firm became Johnson & Sons. In 1882 became Johnson & Sons, Manufacturing Chemists Ltd, and began pharmaceutical production. In 1948, took its present name. The other chemical company in the group is Pictograph Ltd (acquired 1964), (originally the Chemical Division of Pictorial Machinery Ltd), compounding preparations for the photo-engraving and allied trades. In 1965, merged with Hunter-Penrose Littlejohn, makers of printing machines, to form a holding company, Johnsons-Hunter-Penrose Littlejohn Ltd. Johnsons of Hendon will remain as the fine chemical makers. Makers of organic, inorganic and photographic chemicals, pharmaceutical intermediates, formulations for photographic use and test papers; dealers in cameras and projectors. Associates in Belgium, India and Pakistan. Group capital £1·5m. Employees 750. ABCM member. Hendon Way, London NW4.

Johnson, Matthey & Co. Ltd

Formed in 1817, it stems from a business started by the Johnson family in 1743 for the assaying and refining of precious metals of the gold, silver and platinum group, and the preparation of vitreous colours from these metals. The Mattheys joined the firm about 1840. The company's chemical activities now include the refining of the precious metals, the preparation of their salts, the manufacture of glazes, inorganic colours and pigments, and the production of catalysts and substances in a very high state of purity. It also fabricates apparatus, and has considerable business in the precision engineering of precious metals and alloys and in chemical engineering. The company has in recent years acquired the following: in 1926, Johnson & Son's Smelting Works Ltd and Johnson & Sons (Assayers) Ltd, which were parts of the old Johnson business; in 1961, Cowan Bros. (Stratford) Ltd (founded 1910), makers of pigments and colours; in 1963, Blythe Colour Works Ltd (established 1870), producers of glazes and inorganic colours and pigments; and at the same time F. J. Dean Ltd (founded 1925), specializing in oxides for vitreous enamels and pottery colours. In 1965, the Blythe Colour Works and the ceramic division were merged and reorganized as the Transfers Division and the Blythe Colour Division. There are overseas associates in Australia, Austria, Belgium, Canada, France, Germany, India, Italy, Netherlands, New Zealand, South Africa, Sweden and U.S.A. Total employees 6500. ABCM member. 73–83 Hatton Garden, London EC1.

Koch-Light Laboratories Ltd

Original company was L. Light & Co. Ltd, formed 1936 to make unusual organic compounds for research, analytical reagents, the rarer synthetic drugs and flavouring materials. In 1964, joined with Koch Laboratories Ltd (established 1961), to form the present company. Organic Research Chemicals Ltd (formed 1950) is a manufacturing subsidiary; another, International Technical Development Ltd (founded 1936) makes apparatus. Manufactures include synthetic organic chemicals, biochemicals, deuterated compounds, ultra-pure elements, single crystals and pure inorganic compounds. Capital £300,000. Employees 150. ABCM member. Poyle Trading Estate, Colnbrook, Bucks.

Lake & Cruickshank Ltd

Formed 1955, as manufacturing chemists. It makes fine chemicals, primarily alkaloids from natural materials and synthetically. Capital £150,000. Employees 31. ABCM member. North Bridge Road, Berkhamsted, Herts.

The Lancashire Chemical Works Ltd

Founded 1936, to make chrome products for the tanning and textile industries and activated carbons. Swan Chemical Co. Ltd, formed 1940, is an associate and makes special products from its intermediates. Capital £120,000. Employees 20. ABCM member. Glossop, Derbyshire.

Lancashire Tar Distillers Ltd (LTD)

Formed 1929 as a co-operative scheme to merge the interests of family tar distilling concerns in the Lancashire area, viz., North Western Co-operative Tar Distillers Ltd (formed 1926), J. E. C. Lord (Manchester) Ltd (founded 1896), C. Lord & S. A. Lord (formed 1914), Thos. Horrocks & Sons Ltd (established 1910), and Hardman & Holden Ltd (formed 1892). Since then, the following have been acquired: 1931 Manchester Creosote & Storage Co. Ltd (formed 1910), creosote storage; 1935 James E. Dawson (formed 1921), tar distillers; 1937 Metcalfe (Tar Distillers) Ltd (established 1900); 1945 R. Graesser Ltd (q.v.), tar acid refiners; 1950 Diggory & Co. Ltd (ounded 1908), tar distillers; 1951 Brotherton & Co. Ltd (formed 1878), tar distilling and tar acid refining works; 1952 W. C. Smithie & Co. Ltd (established 1910), tar distillers; and 1952 Nemlin Chemicals Ltd (formed 1935), tar distillers. The company's objective is still the distillation and refining of tar and its products. In 1963, it acquired a controlling interest in the Manchester Computor Centre Ltd. Capital £2·5m. Employees 600–650. ABCM member. Chronicle Buildings, 74 Corporation Street, Manchester 4.

Lankro Chemicals Ltd

Founded 1937 as private company by Dr. F. H. Kroch, a German refugee, to make speciality chemicals for the leather, fur and textile industries. In conjunction with Petrochemicals Ltd, established Oxirane Ltd in 1949, as first British manufacturer of glycol ethers and ethanolamines. These two companies were taken over by the Shell Chemical Co. Ltd in 1955 but Lankro maintains its interest. Production now includes non-ionic detergents and emulsifiers, stabilizers, plasticizers and systemic weed killers. Since 1962, Shell Chemical Co. Ltd has minority interest. Associates in Australia and New Zealand. Capital £3m. Employees 700. ABCM member. Bentcliffe Works, Salters Lane, Eccles, Manchester.

Laporte Industries Ltd

The business of B. Laporte was established in the early 80's as importers of continental chemicals for textiles. Manufacture of hydrogen peroxide commenced 1888 in Yorkshire and in 1898 at Luton for straw hat industry. Other chemical manufacture followed. In 1908, the business became a public company—B. Laporte Ltd. In 1948, the name was changed to Laporte Chemicals Ltd, which functioned both as a holding and operating company till 1953, when Laporte Industries Ltd was formed as the holding company. The following companies had been merged in Laporte Chemicals Ltd and liquidated: Associated Phosphate Manufacturers Ltd (incorporated 1927 to take over the production of food phosphates from The Monarch Chemical Co., and acquired 1930), Genoxide Ltd (formed 1926 to bottle hydrogen peroxide), Laporte Minerals Ltd (incorporated 1925 as Malehurst Barytes Co. Ltd to mine barytes, acquired 1932 and name changed 1951; mines now sold), and Wm. Burton & Sons (Bethnal Green) Ltd (incorporated 1927 to take over the old-established firm of Wm. Burton & Sons, makers of hydrogen peroxide and related by-products, and acquired 1929).

There are also the following subsidiaries in the group:

1. *Laporte Titanium Ltd.* Incorporated 1927 as National Titanium Pigments Ltd; Laporte acquired interest 1932; became wholly-owned 1952 and present name adopted 1953.

2. *Laporte Acids Ltd.* Incorporated 1894 as John Nicholson & Sons Ltd to acquire the old family business of John Nicholson & Sons, makers of sulphuric acid; acquired 1946 and present name adopted 1951. The businesses of the following were merged in this company and liquidated: Hunt Brothers (Castleford) Ltd (incorporated 1932 to acquire the business of Hunt Brothers, established about the middle of the nineteenth century, makers of acids; acquired by Laporte 1949 and merged 1953), Cleckheaton Chemical Co. Ltd (incorporated 1904 to take over the acid business of Saunders & Saunders Ltd; acquired 1954 and liquidated 1959), Bierley Chemical Co. Ltd (incorporated 1919 as merchanting subsidiary of Cleckheaton Chemical Co. and liquidated 1959), James Wilkinson & Sons Ltd (incorporated 1924 to take over the business of James Wilkinson & Sons, makers of acids and fluorine compounds, acquired 1959 and merged 1962), The Sheffield Chemical Co. Ltd (incorporated 1955 as SC Co. Ltd and name changed the same year, makers of sulphuric acid; acquired 1959 and merged 1962).

3. *The Fullers' Earth Union Ltd.* Incorporated 1890 to purchase and operate fullers' earth mines; acquired 1954.

4. *Peter Spence & Sons Ltd.* Incorporated 1900 to take over the manufacture of alum and other chemicals by Peter Spence & Sons, founded 1846; acquired 1960; has financial interest in Thames Alum Ltd, formed 1960, and Don Alum Ltd, established 1961, producers of liquid aluminium sulphate.

5. *Glebe Mines Ltd.* Incorporated 1945 as subsidiary of James Wilkinson & Sons Ltd to take over its lead and fluorspar mining interests; acquired with James Wilkinson & Sons Ltd (q.v.); in 1962 its business was merged with that of the Cupola Mining & Milling Co. Ltd (formed 1918 and acquired by

Laporte 1960), under the name of Glebe Mines Ltd. In 1964, Fluorspar Ltd (incorporated 1941) was acquired by Laporte and merged with Glebe Mines Ltd.

6. *Dales Chemicals Ltd.* Incorporated 1955, acquired 1964, with mineral leases for fluorspar and barytes.

7. *Howard & Sons Ltd and Howards of Ilford Ltd.* Business founded 1797 as chemical manufacturers; limited company formed 1903. As a result of technical reorganization, Howards & Sons Ltd, now a holding company, was formed 1920 and Howards of Ilford Ltd as their operating company 1953; acquired by Laporte 1961. Howards has interest in Bowmans Chemicals Ltd (q.v.), makers of lactic acid.

8. *A. W. Brook Ltd.* Incorporated 1918; acquired 1934; marketing company.

9. *Barium Chemicals Ltd* (q.v.). Incorporated 1964 jointly by Laporte and Imperial Smelting Corporation (q.v.) to take over the production of barium compounds and associated products carried on by the two companies.

Laporte Industries Ltd has subsidiaries in Australia (making hydrogen peroxide, titanium oxide, etc.) and in Germany (making hydrogen peroxide and organic peroxy compounds). There are also associates in Algeria, France, India, Japan and Spain. In 1965 the company was reorganized into two U.K. and one overseas division. The General Chemicals Division will include Laporte Chemicals, Laporte Acids, Peter Spence & Sons and Glebe Mines. The Organics and Pigments Division will comprise Laporte Titanium, Howards of Ilford, The Fullers' Earth Union and G. D. Holmes Ltd. The overseas division will cover all interests abroad.

The Laporte group makes a wide range of products including hydrogen peroxide, titanium oxide, fullers' earth, acids, etc. In addition to the mining interests already noted, it has subsidiaries concerned with water-way carriage and fresh water supply. U.K. chemical capital £22m. Employees 5600. ABCM member. Hanover House, 14 Hanover Square, London W1.

Lawes Chemical Co. Ltd

Founded by John Bennett Lawes of Rothamsted fame, who took out the first patent for the manufacture of superphosphate in 1841. The present company was incorporated 1872 to acquire the business of J. B. Lawes under title of Lawes Chemical Manufacture Co. Ltd, which adopted present name 1935. Gwalia Fertilisers (Briton Ferry) Ltd is subsidiary, incorporated 1934 as agricultural merchants for lime, slag and compound fertilizers. In addition to superphosphate, Lawes makes full range of compound fertilizers; it also warehouses and distributes. Capital over £1m. Employees 350. Creeksmouth, Barking, Essex.

Leicester, Lovell & Co. Ltd

Formed 1936, the two principal shareholders being the Casein Co. of America, Inc. (now the Borden Chemical Co.) and Lovell & Christmas Ltd, London, to manufacture casein glues for the wood-working industries. In

1953 became a subsidiary of the Borden Chemical Co., U.S.A. U.K. subsidiaries are M. Sondheimer (Sondal Works) Ltd, London (glues and adhesives), Crystal-brite Products Ltd, Feltham (detergents, seals, polishes and adhesives), The Arabol Manufacturing Co. Ltd, London (adhesives and adhesive tapes), and E. D. Edwardson & Co. (Ware) Ltd (adhesives for bookbinding and packaging). There is a subsidiary in France. The range of products includes resins of the urea/, phenol/ and resorcinol/formaldehyde, furane and epoxide types and adhesives based on casein, polyvinyl acetate and synthetic rubber. U.K. employed capital about £2m. Employees 500. North Baddesley, Southampton.

Lennig Chemicals Ltd

A subsidiary of Rohm & Haas Co., U.S.A., started in the U.K. 1936 as importers of its parent's chemicals. Manufacture commenced 1955. Products include acrylate monomers and emulsions, plastics stabilizers, ion exchange resins, oil additives and pigment colours. Employees 500. 26/28 Bedford Row, London WC1. (See Addenda p. 329.)

Eli Lilly International Corporation

Eli Lilly & Co., founded in the U.S.A. in 1876, established the above corporation 1943 to direct its external operations. A branch headquarters was set up in London 1963 to deal with the U.K., Europe, N. Africa and the Middle East. In addition to the corporation, the following associates operate in the U.K.: Lilly Industries Ltd, Dista Products Ltd, Elanco Products Ltd, Eli Lilly & Co. Ltd, and Lilly Research Laboratories Ltd. In 1962 the Corporation acquired the manufacturing and trading activities of the Distillers Company (Biochemicals) Ltd. The original objectives were to provide medicinal products of the highest quality and to contribute to the progress of medicine through research. It covers a wide range of chemical manufactures in the medicinal field from analgesics to vitamins. It employs about 2000 in the U.K. Henrietta Place, London W1.

Howard Lloyd & Co. Ltd

Originated 1880 as family concern to make pills and emulsions. Private limited company in the 1930's. Member of the Lloyd group which includes Lloyd-Hamol, Lloyd Anphar and Lloyd's Research. Products include pharmaceuticals, synthetic organic chemicals based on furoic acid, clinical reagents, and high protein foods. Over 100 employees. ABCM member. Clerk Green, Batley, Yorkshire.

Donald Macpherson & Co. Ltd

This is a group of companies specializing in paint products, with its origin in 1885. Public company 1948. Formation of group started in 1954 with acquisition of R. Cruickshank (Cellulose) Ltd (founded in the 1860's); Sherwoods Paints Ltd, founded 1777, purchased 1957; Mody & Co. Ltd (polishes and varnishes), founded 1925, acquired 1958; the Wilkinson Group of companies (formed 1921 and acquired 1960), consisting of L. G. Wilkinson

Ltd, Gerald Carter Ltd and Wilkinson Paints Ltd, all making wood finishes; and Thos. Parsons & Sons Ltd (established 1802, acquired 1964). Associates in Eire and U.S.A. In addition to paints, the group makes synthetic resin emulsions, powders and solutions. Capital £2·2m. Employees 1800. Warth Mills, Radcliffe Road, Bury, Lancs.

Marfleet Refining Company Ltd

In 1934, a co-operative company of the trawler owners of Hull was formed as British Cod Liver Oil Producers (Hull) Ltd, to produce high quality cod liver oil, superseding the technical grade oil previously made by the Hull Fish Meal & Oil Co. Ltd. In 1939, a similar company was formed in Grimsby and came into a co-operative scheme under a controlling company known as Portaccord Ltd. The name of the latter was changed in 1946 to British Cod Liver Oils (Hull & Grimsby) Ltd, and the Hull company became the Hull Cod Liver Oils Ltd. The parent took the name of Marfleet Refining Co. Ltd 1956. Refiners of fats and oils and makers of technical oils and medicinal and nutritional products based on fish and marine mammals. Employees 370. ABCM member. PO Box 3, Hull, Yorks.

May & Baker Ltd (M. & B.)

This firm, of sulphonamide fame, was founded 1834 by J. May and W. G. Baker, to manufacture fine chemicals such as mercurials, bismuth salts and cyanides. Incorporated 1890. Pharmaceutical production stimulated by World War I. Sulphonamide work commenced 1937. Manufactures now include pharmaceuticals, pesticides, veterinary products and photographic, laboratory and industrial chemicals. Two U.K. associated companies for plastics and pharmaceutical specialities (marketing). Overseas, in addition to the close connections with the French Rhone-Poulenc, there are associates in Australia, India, New Zealand, Pakistan, South Africa and West Africa. ABCM member. Dagenham, Essex.

Mechema Ltd

Formed 1927 as Metallurgical Chemists Ltd to make chemicals. These include metallic inorganics such as salts of copper, lead, manganese and zinc and various crop protection materials. In 1952, Agricola Plant Protecting Chemicals Ltd was established to sell the agricultural products. In 1964, the names were changed to Mechema Ltd and Agricola Chemicals Ltd, respectively. Associates in Norway. Employed capital £125,000. Employees 95. ABCM member. Talbot Wharf Chemical Works, Port Talbot, Glam.

The Midland Tar Distillers Ltd (MTD)

Formed 1923 by the amalgamation of Robinson Bros. Ltd (founded 1869), Lewis Demuth & Co. (formed 1865), and Major & Co. (established 1856), to create a co-operative scheme with the gas industry, to distil gas works' tar and produce primary coal tar chemicals. A number of smaller companies were subsequently absorbed. After nationalization, adjacent Gas Boards acquired a shareholding. In 1960, Schenectady Midland Ltd was created in association with Schenectady Chemical Inc., New York, to manufacture in

U.K. the latter's range of resins and develop other products. Subsidiaries in U.S.A. In addition to a full range of primary coal tar products, the company makes a number of derivatives, particularly in the pyridine field, and synthetic pyridine bases. Total assets £6m. Employees 1000. ABCM member. Oldbury, Birmingham.

Mirvale Chemical Company Ltd

Incorporated 1925, to distil and refine coal tar by-products. The manufactures now include phenols, cresols, xylenols, farm and general purpose disinfectants and weedkillers. The firm also handles agricultural and horticultural control chemicals for U.S. company. Joseph Weil & Son Ltd act in certain fields as their selling organization and distributors to spraying contractors. The company finances a scholarship for investigation of weed problems. Employed capital £400,000. Employees 150. ABCM member. PO Box 1, Mirfield, Yorks.

Mobil Chemicals Ltd

The Kleemann group was founded 1905 as O. & M. Kleemann Ltd, a merchanting business. Plastic production began 1938 with cellulose acetate. Kleestron Ltd formed 1951 to make polystyrene. Erinoid Ltd (founded in the early 1900's) acquired 1957. O. & M. Kleemann Ltd taken over by Socony Mobil 1961, and name changed to Mobil Chemicals Ltd. Socony Mobil is international group of oil companies with associates in most countries. The U.K. plastics interests were acquired in 1965 by BP; these included the above-mentioned firms with Wokingham Plastics Ltd and Mobil Chemical Sales Ltd and the name was changed to BP Plastics Ltd. Plastic products include polystyrene, polyethylene, cellulose acetate and casein. Capital over £7m. Employees 800. West Halkin House, West Halkin Street, London SW1.

Monsanto Chemicals Ltd

The American parent company was founded in 1901. In 1920, it purchased a half-share in the British company, Robert Graesser (q.v.) (dating back to 1867—makers of fine chemicals), which then changed its name to Graesser-Monsanto Chemical Works. The other half was bought in 1928 and in 1934 it became a public company with its present name of Monsanto Chemicals Ltd, which has, over the years, expanded its activities.

There are three wholly-owned subsidiaries, viz., Jablo Plastics Industries Ltd (formed 1936, acquired 1963), Flamingo Foam Ltd (formed 1957, acquired 1960, and making polystyrene board), and Polyglaze Ltd (formed 1960, acquired 1963, marketing polystyrene sheet). It has interests in Forth Chemicals Ltd (q.v.) and a half share, acquired 1958, in R. H. Cole & Co. Ltd (formed 1935, which with its subsidiaries makes plastic raw materials). There are interests overseas in Australia and India.

Its manufactures include basic organic chemicals and intermediates, fine chemicals and pharmaceuticals, lubricating oil additives, plastics and polymers, rubber chemicals, and specialized technical chemicals. U.K. capital about £25m. Employees 3750. ABCM member. Monsanto House, 10–18 Victoria Street, London SW1.

Thomas Morson & Son Ltd

Founded 1821 as makers of alkaloids and fine chemicals. Incorporated 1915. Bought 1957 by Merck & Co. Inc., America, and is their U.K. chemical manufacturing subsidiary. Their U.K. parent is Merck Sharp & Dohme Ltd, which started 1928 as H. K. Mulford & Co. Ltd, became Sharp & Dohme Ltd 1930 and took its present name 1956. The products include synthetic organic chemicals for pharmaceuticals, glycero-phosphates and fine chemicals for pharmaceutical, photographic and other purposes. Capital £850,000. Employees 250. ABCM member. Summerfield Chemical Works, Wharf Road, Ponders End, Enfield, Middlesex.

Murgatroyd's Salt & Chemical Co. Ltd

Established 1947 as private company (originally as Murgatroyd's Vacuum Salt Co. Ltd), absorbing Murgatroyd's Mid-Cheshire Salt Works Co., formed 1890. The objective was to utilize the local brine wells to make salt, chlorine and caustic soda. Became a jointly owned subsidiary of Distillers Co. Ltd (q.v.) and Fisons Ltd (q.v.) 1954. Its production is confined to salt and the inorganic chemicals derived from it. Employed capital £4m. Employees 700. ABCM member. Elworth, Sandbach, Cheshire.

National Coal Board (Coal Products Division) (NCB–CPD)

On vesting day—1st January 1947—the NCB became the owner of 55 coking plants. In due course, on 1st January 1963, this led to the formation of the Coal Products Division responsible for carbonization, briquetting and chemicals with production organized on the basis of three divisions—Northern, Midlands and South Western. In addition to the normal coal carbonization products, the NCB has a number of proprietary materials derived therefrom. In 1957, the firm of Thomas Ness Ltd, formed about a hundred years ago, tar distillers and chemical manufacturers, was acquired. A company, Pitch Plastic Polymers Ltd, was formed jointly with the Ruberoid Co. Ltd in 1964, to market a new damp-proof course. Value of fixed assets of the CPD about £35m. Employees 6000. 26–28 Dorset Square, London NW1.

Newton, Chambers & Co. Ltd

Originally established 1793 primarily as iron-founders. Later, collieries were established which were lost on nationalization in 1947. The development of coal by-products in the middle 80's from the associated carbonization plants led to their original Chemical Division. The company has a wide range of other interests, including chemical engineering, covered by subsidiaries. Today, the chemical business is concentrated in the subsidiary, Izal Ltd, formed 1963. The name Izal was registered as a trade mark in 1893 and there is a range of Izal products, including disinfectants and insecticides. The Ronuk polishes, waxes, etc., were added when Ronuk Ltd, established 1896, was acquired 1960. Izal (Overseas) Ltd is an export organization. There are associates in Australia, Ireland and New Zealand. Total group capital £13m. Izal Ltd employees 1200; group 6000. ABCM member. Thorncliffe, Sheffield.

North Thames Gas Board

The commencement of the commercial production of gas created the Gas Light & Coke Co. in 1812. With the nationalization of the gas industry in 1949, it became the North Thames Gas Board, with certain modifications to its operating area. The organization has always treated the large tonnages of its own by-product tar, benzole and ammoniacal liquor and has manufactured sulphuric acid for that purpose. At its Beckton Works, which is unique and probably the largest of its kind, it makes a full range of tar products. Employed chemical capital is estimated at £10m. There are 1000 employees out of a total of 22,000 for the Board as a whole. ABCM member. Tar and Ammonia Products Works, Beckton, London E6.

Novadel Ltd

Formed 1929, for sale of flour improvers and materials for paint and allied industries made by its Dutch parent, Koninklijke Industrieele Maatschappij Noury & Van der Lande NV. Manufacture started 1937. Products include organic peroxides, white lead, lead and barium/cadmium stabilizers, fatty acids, synthetic organic acids, metallic driers and flour improvers, in addition to powder feeders and flourmilling machinery components. Selling agents in many countries. Capital £450,000. Employees 275. St. Ann's Crescent, Wandsworth, London SW18.

Nutfield Manufacturing Co. Ltd

Business started 1924 to manufacture hydrofluoric acid and was taken over by a private limited company in 1944. It now makes sodium fluoride, hand cleaners and wax removing fluids, and distributes hydrofluoric acid, fluorides, battery acids and distilled water. Capital £70,000. Employees 30. ABCM member. King's Mill Works, South Nutfield, Redhill, Surrey.

Organon Laboratories Ltd

Formed 1939 as subsidiary of NV Organon, Holland, makers of hormones and related products and an offshoot of the Zwanenberg Meat Co., which supplied the raw materials (glands, etc.). Became public company after the war. British shareholding bought out 1957, so the company is now part of Organon International Group with associates in many countries. Hormones and other steroids and cosmetic hormone cream are main products. Markets quaternary ammonium compounds for Onyx Chemical Corporation, U.S.A. Employed capital about £600,000. Employees 500. Crown House, London Road, Morden, Surrey.

Ortho Pharmaceuticals Ltd

Formed 1947 as part of the world-wide Johnson & Johnson group, to develop and manufacture in the pharmaceutical field, particularly for obstetrics and gynaecology. U.K. associates are Johnson & Johnson (Great Britain) Ltd, Ethicon Ltd, Johnson's Ethical Plastics Ltd and Decro Ltd. Associates in 26 other countries. Employed capital £500,000. Employees 150. Saunderton, Bucks.

Osmond & Sons Ltd

Formed as private company 1854. Became limited company 1909. Makers of veterinary medicines, sheep and cattle dips, vitamin and mineral supplements, disinfectants, insecticides and aerosols. Also stockbreeders and farmers through Osmond Estates Ltd. Associates are Osmonds Aerosols Ltd (formed 1964 by changing the name of Osmonds Pharmaceuticals Ltd, formed 1949) and Barnoldby Crop Driers Ltd (formed 1949). Associates in Australia, Eire, Netherlands and South Africa. Chemical capital £100,000. Employees 200. Grimsby, Lincs.

Ozalid Company Ltd

Business started 1926 as branch of Agfa. Ozalid formed 1928 as private company, subsidiary of Kalle, part of the German IG. Reconstituted as wholly British public company 1942. Its object was the manufacture of photo-sensitized reproduction materials. Prior to 1939, sensitizing preparations were imported from Germany—all major constituents now made from domestic raw materials. Main fine chemical manufactures include stabilized diazo compounds and coupling components. Other products cover sensitized materials, carbon papers, typewriter ribbons, electrophotographic substances and machines for reprographic processes. U.K. associates Block & Anderson Ltd and Kolok Manufacturing Co. Ltd are not chemical makers. Capital between £50,000 and £100,000. ABCM member. Debden Industrial Estate, Langston Road, Loughton, Essex.

Pains-Wessex Ltd

Formed 1965 by merging James Pain & Sons Ltd, Mitcham, with Waeco Ltd, Salisbury, as a subsidiary of the British Match Corporation, which includes a range of non-match interests and as Bryant & May has associates throughout the world. James Pain & Sons have been making " Pains Fireworks " for over a century. Waeco Ltd was founded 1933 as the Wessex Aircraft Engineering Co. to service civil light aircraft, but as a result of the then depression, it took up experimental pyrotechnic work with the encouragement of a nearby Government Experimental Establishment At the end of World War II, Government work ceased. It turned to commercial markets and abbreviated its name to its initials Waeco. The products of the new company include fireworks, marine and railway signals, flares, fuses for stage and film effects and smoke pellets and generators for the dispersion of insecticides, acaricides and fungicides. 250 employees. High Post, Salisbury, Wilts.

Parke, Davis & Co.

Formed in America 1866 to supply reliable medical products. London Branch opened 1891. It makes synthetic medicinal agents. U.K. employees 1300. Company has branches in 30 countries. Staines Road, Hounslow, Middlesex.

William Pearson Ltd

The Pearson companies were started in Hamburg in 1880 by William Edward Pearson, an Englishman, for the manufacture of improved types of disinfectants and antiseptics and the exploitation of the non-poisonous disinfectant "Creolin". William Pearson Ltd was registered 1901. Pearson's Antiseptic Co. Ltd, a subsidiary, was formed 1908. In 1965, these companies were dissolved and William Pearson (Holding) Co. Ltd and William Pearson (Hull) Co. Ltd formed. There are subsidiaries or associates in Brazil, Canada, Italy and Venezuela. Present production covers germicidal and anti-oxidant compounds for the disinfectant, petroleum and plastics industries, disinfectants, and germicides. Employees 150. ABCM member. Clough Road, Hull, Yorks.

Peboc Ltd

Founded 1947 by C. T. Bowring & Co. (Fish Oils) Ltd, which was part of an organization of varied commercial interests with roots back in the early nineteenth century, in conjunction with the Premier Yeast Co. and Crookes Laboratories, for the production of vitamin D3. In 1956, it became a wholly-owned subsidiary of Bowrings, who in 1959 arranged for Duphar Ltd, a subsidiary of the well-known NV Philips Gloilampfabriken to become equal partners. Peboc makes, in addition to vitamins, fine chemicals such as muscle-relaxing drugs, analgesics and synthetic organic compounds. Capital £60,000. Employees 45. ABCM member. Belvue Road, Northolt, Middlesex.

The Permutit Company Ltd

Established 1899, registered 1913, to specialize in the design and construction of water treatment plant of all types. For this purpose it manufactures a range of ion exchange materials, used also in the chemical, metallurgical and allied industries. U.K. subsidiary Zerolit Ltd, registered 1963. Subsidiaries in Australia, Canada and South Africa. ABCM member. Permutit House, Gunnersbury Avenue, London W4.

The Pfizer Group

The Pfizer organization dates back to 1848, when two Germans, Charles Pfizer and Charles Erhart, migrated to America and established a partnership as manufacturing chemists with the title Charles Pfizer & Co. The business developed and in 1951, the first companies of Pfizer International were set up in Belgium, Brazil, Canada, Mexico, Panama, Puerto Rico and the U.K. The U.K. group now includes Pfizer Ltd, Coty (England) Ltd, Leas Cliff Products Ltd, Universal Laboratories Ltd, Agricare Products, Harvey Pharmaceuticals, the House of Romney, and British Alkaloids Ltd. In 1958, Pfizers acquired the old-established firm of Kemball Bishop & Co., fine chemical manufacturers. It has a diversified range including antibiotics, vaccines, steroids, drugs, animal health products, petrochemicals, plastics, industrial chemicals, cosmetics, over-the-counter medicines, food products, and a range of consumer goods. Employed chemical capital £6·5m. Employees 1900. ABCM member. Sandwich, Kent. (See Addenda, p. 329.)

Philblack Ltd

Incorporated 1936 as Palatine Gas Corporation Ltd. Present name adopted 1948, when decision taken to make oil-furnace carbon black, previously imported for the rubber industry. The group includes Jones Gas Process Co.

Ltd, makers of plant for production of medium thermal carbon black and domestic gas. Philblack Ltd makes and sells under licence from Phillips Petroleum Co., U.S.A., but is entirely British-owned. Agents in some 30 countries. Employed capital £5·5m. Employees 400. PO Box 1, 1 Henbury Road, Westbury-on-Trym, Bristol.

L. J. Pointing & Son Ltd

Started 1916 as a merchanting concern. In 1925, commenced manufacture of foodstuff colours and acid dyestuffs. Partnership of L. J. Pointing & Son formed 1931. Became limited company 1949. Activities now much expanded in field of food colours and oil soluble and leather dyestuffs. Employees 50. ABCM member. 45 Hallstile Bank, Hexham, Northumberland.

Proctor & Gamble Ltd

In 1837, Thomas Hedley founded a company in Newcastle-upon-Tyne to make soap and candles. In the same year, two British emigrants, William Proctor and James Gamble formed a partnership in Cincinnati, also to make soap and candles. The American company purchased the share capital of the British firm 1930, which changed its name to Proctor & Gamble Ltd 1962. The products now include soaps and detergents, glycerine and fats for the catering and bakery trades. There are member companies of the organization in over 10 countries. Employed capital about £14m. Employees 4000. Newgate House, Newcastle-upon-Tyne.

Prodorite Ltd

Formed 1925 to make corrosion resisting products based on a new substance, "Prodorite Acid Resisting Pitch Concrete", and used in most industries. Chemical resisting cements also made from basic materials. W. Chattaway Ltd is U.K. subsidiary concerned with fabricated steel. Associates in 14 overseas countries. Employed capital £600,000. Employees 700. ABCM member. Eagle Works, Wednesbury, Staffs.

The Radiochemical Centre

In the early days of World War II, it was necessary to increase the output of radioactive luminous compounds. Thorium Limited (q.v.) established in 1940 a small refinery to treat radium concentrates at Amersham. By 1943, the products included mesothorium, radiothorium, radium D and polonium. In 1944, the scheme was broadened to include radioactive isotopes, resulting from developments in atomic energy. In 1946, the Radiochemical Centre was established as a trading organization to process and distribute radioactive materials. The assets of Thorium Limited were purchased by the Ministry of Supply, and the company remained as managers until 1950. Then the establishment came under direct management of the Department of Atomic Energy (M.O.S.), and was transferred to the United Kingdom Atomic Energy Authority on its formation in 1954. The Centre was closely associated with AERE Harwell at this time, and by a reorganization in 1960 it became solely responsible for production and distribution of all radioactive products offered by the Authority for civil purposes. The Centre has steadily increased in size and activity. The staff is now about 450. The number of products is now many hundreds. In 1963/4, sales were valued at over £1·5m; there were about 50,000 despatches, nearly half for export. There are three main groups: pure radioisotopes, extracted from materials irradiated in a nuclear reactor or cyclotron; radioactively "labelled" compounds containing radioactive atoms; and appliances for medical or industrial use containing radioactive materials. Amersham, Bucks.

Reckitt & Colman Holdings Ltd

This group's interest in the chemical industry was built up by Reckitt & Sons Ltd, one of its members, under two main heads, *viz.*, pigments and antiseptics. In 1852, Isaac Reckitt & Sons started the manufacture of the well-known Reckitt's Blue, based on ultramarine, then imported. Production of ultramarine started 1882. In 1926, Reckitts absorbed SA des Usines Destree, one of the major Continental makers, with factories in Belgium and France. Reckitts now have factories also in Brazil, India and Spain. In 1962, the fount of the ultramarine industry, Usines Guimet, started in 1829, was acquired from the great-grandson of the founder, Jean-Baptiste Guimet. Reckitt's (Colours) Ltd was formed 1950 to run this side of the business and now provides a range of pigments covering ultramarine, yellow oxide, cadmium yellow and reds, and phthalocyanine blues and greens.

In 1932, Reckitt & Sons Ltd commenced the marketing of Dettol antiseptic and later extended the Dettol range. To ensure a supply of the principal raw material, para-chloro-meta-xylenol, a joint company was formed during World War II by amalgamation with Dickinson & Son, Oswaldtwistle, under the name of Cocker Chemical Co. Ltd, after Sir William Cocker, the founder of Dickinsons. Dickinsons were then making DDT and other chlorinated compounds for the services. The firm now covers a range of antiseptic and disinfectant materials and of flavouring and perfumery agents. In 1965 Reckitt & Sons acquired the firm of F. W. Hampshire.

Another member of the group, Sissons Brothers & Co. Ltd, have been in the paint industry since 1803. They were acquired 1956. Capital employed in above activities is some £4m. Employees 1500. The group, which also manufactures a wide range of food and household articles, has a capital of about £40m, and 20,000 employees. ABCM member. Hull.

Reichhold Chemicals Ltd

The company, formed 1952, controls the following subsidiaries: Beck, Koller & Co. (England) Ltd, founded 1933 (makers of synthetic resins and resin compounds); James Beadel & Co. Ltd, formed 1923 (sole selling agent of Beck Koller); Vinyl Products Ltd, formed 1939 (polyvinyl acetate, polystyrene and acrylics); and Vinatex Ltd, formed 1947 (polyvinyl chloride compounds and pastes). There are many associate companies overseas. Capital employed £2m. Employees over 900. Hillgate House, 26 Old Bailey, Londo EC4.

Rentokil Group Ltd

Incorporated 1927. Developed from a Danish company making the Ratin preparation as a biological rodent control. The following companies have been acquired as indicated: Chelsea Insecticides Ltd (1941) (name changed to Disinfestation Ltd 1953), Rentokil Ltd and Woodworm and Dry Rot Control Ltd (1957), Fumigation Services Ltd (1958), Insecta Laboratories Ltd (1959), Rid-Pest Ltd (1960), Wood Preservation Ltd (1960), Associated Fumigators Ltd (1961), Pestcure Ltd (1961), Thomas Harley & Co. Ltd (1961), Scientex Ltd (1960), and Yorkshire Fumigation Services Ltd (1964). In 1962, all the U.K. trading companies were given the title of Rentokil Laboratories Ltd. Harry Hardy Ltd and Hilton Hirst & Co. Ltd, specializing in chemicals and wool waste recovery, joined the group in 1965. Starting with rats and mice, the company now covers all forms of industrial and domestic pest control and wood preservation. For this it manufactures certain chemicals and formulates others. The group is international, with associates in 12 countries. Capital over £900,000. U.K. employees 1600. Felcourt, East Grinstead, Sussex.

Richardsons Fertilisers Ltd

Formed as Richardson Bros. Ltd about middle of nineteenth century to make sulphuric acid, chemical fertilizers and feeding stuffs. Purchased 1897 by W. & H. M. Goulding to become Richardsons Chemical Manure Co. Ltd. Present name adopted 1961. ICI purchased 51% interest 1960. Ulster Fertilisers Ltd, Londonderry, is under same control. Products are sulphuric and nitric acids and concentrated compound fertilizers. Capital employed in the two companies £7·4m. Employees 500. 1 Short Strand, Belfast 5.

Robinson Brothers Ltd

Formed 1869. Limited company 1894. Original purpose was the distillation of coal tar and ammonia. In 1921, tar interests were merged into the formation of the Midland Tar Distillers Ltd. Manufacture of organic chemicals continued, e.g. rubber accelerators, esters and derivatives of piperidine, and thioglycollic acid. Employees 310. ABCM member. Oldbury, Birmingham.

James Robinson & Co. Ltd

Formed 1840 for the extraction of natural dyes. Manufacture of sulphur black commenced 1913. Products include sulphur and chrome dyes and intermediates for dyestuffs and for pharmaceutical and veterinary products. Capital £400,000. Employees 150. ABCM member. Hillhouse Lane, Huddersfield, Yorks.

Roche Products Ltd

Founded 1908. Member of world-wide group of F. Hoffmann–La Roche & Co., Basle. Makers of pharmaceuticals such as analgesics, sulphonamides, vitamins and psychotropic drugs. Employees 850. ABCM member. Welwyn Garden City, Herts.

Sandoz Products Ltd

Incorporated as the Sandoz Chemical Co. Ltd 1911, as subsidiary of Sandoz, Basle, itself established 1886, to make and deal in dyestuffs, pigments, pharmaceuticals and textile auxiliaries. There are now several subsidiaries including Cotopa Ltd (chemical modification of textile fibres), The Albion Winding Co. Ltd (yarn winders), Sandoz Products (Ireland) Ltd (established 1947) (sales), and The Sandoz Trading & Shipping Co. Ltd (importers and exporters). The parent Sandoz International has affiliates in over 20 countries. Capital about £4m. Employees 600. Calverley Lane, Horsforth, Leeds.

Schermuly Limited

Founded 1897 by William Schermuly, British seaman and inventor of the first practical marine line-throwing apparatus. The Pistol Rocket Line-throwing Apparatus was invented 1920. The original company, the Schermuly Pistol Rocket Apparatus Ltd, was formed 1926. Present name adopted 1964. Activities, which are both chemical and engineering, also include the production of a range of pyrotechnic devices and light flares and signals, many of them used for marine purposes, and a Blakodizing service. Member of the Charterhouse Group of Companies. Employees 300. SPRA Works, Newdigate, Dorking, Surrey.

Scott Bader & Co. Ltd

Founded as E. Scott Bader 1920 as merchants of chemicals and plastic materials. Incorporated under present title 1923. Manufacture started 1932. Now makers of polyester resins, polymer emulsions, PVC pastes, plasticizers, alkyd resins and resin-based textile pigment printing pastes. Products produced under licence in 16 overseas countries. Employees 310. It is a co-ownership and community organization. ABCM member. Wollaston, Wellingborough, Northants.

Scottish Agricultural Industries Ltd

Formed 1928 as a public company, by a merger of six leading Scottish fertilizer manufacturers, in which ICI took a controlling financial interest. The object was to rationalize in the interests of efficiency and economy and to create a large retail trading organization primarily for fertilizers. It has as subsidiaries: SAI Horticulture Ltd, Leith; W. & A. Geddes Ltd, Wick; Scottish (Pentland) Fertilisers Ltd, Edinburgh; and J. & J. Cunningham Ltd, London. It manufactures compound and other fertilizers. Its non-chemical activities are processing of basic slag, manufacture of animal feeding stuffs, importers, blenders and merchants of farm seeds, factors for farm produce, and Scottish agents for ICI's agricultural products. Employed chemical capital approximately £13m. Employees 1725. ABCM member. 39 Palmerston Place, Edinburgh 12.

Scottish Tar Distillers Ltd (STD)

Formed 1929 to co-ordinate the larger tar distilling interests in Scotland, particularly those of James Ross & Co. (Lime Wharf) Ltd (founded in 1845) and Henry Ellison (Glasgow) Ltd. Became public company 1949. The company now handles under a co-operative scheme all the tar from gas works and coke ovens in Scotland and produces a full range of coal-tar products. Associated with Richard Smith Ltd, chemical manufacturers and merchants since 1841. Employed capital £1·25m. Employees 300. ABCM member. Falkirk, Scotland.

G. D. Searle & Co. Ltd

Formed 1953 as subsidiary of G. D. Searle & Co., Chicago (founded 1888) for the provision to U.K. and overseas markets of pharmaceutical products, viz., steroids, anticholinergics, anti-histamines and tranquillizers. Chemical manufacturing capital £150,000. Total U.K. employees over 300. Lane End Road, High Wycombe, Bucks.

Shell Chemical Co. Ltd

Shell is part of the Royal Dutch Shell group of companies, with a world-wide business in the production, refining and distribution of petroleum products. Its U.K. chemical operations began with the formation in 1928 of Technical Products Ltd, which initially marketed petroleum-based products for domestic use and packaged agricultural and horticultural chemicals. Other chemicals from Shell plants abroad were added, including the detergents based on secondary alkyl sodium sulphates, to which was given the name of "Teepol", derived from Technical Products Ltd. Plant to make Teepol came into operation at the start of World War II. Technical Products Ltd kept its identity till 1945, when two new companies were formed, viz., Shell Chemical Manufacturing Ltd to organize production, and Shell Chemicals Ltd to sell. Rapid development ensued and in 1955 Shell Chemical Co. Ltd was formed to combine the manufacturing and marketing organizations. At the same time, Petrochemicals Ltd (formed 1946) was acquired with the Weizman and Ziegler patent rights which it was exploiting.

Petrochemicals Ltd had two principal subsidiaries, viz., Styrene Products Ltd (owned 60–40 with Erinoid Ltd) and Styrene Co-Polymers Ltd, in which Petrochemicals, Pinchin Johnson and Lewis Berger had each a one-third interest. Shell soon obtained complete ownership of the former and has maintained its holding in the latter. Shell Chemical Co. has also extended its interests into agriculture. A joint company called Shellstar Ltd was formed in 1965 with Armour & Co. of America to make fertilizers and agricultural chemicals. It has now major activities in the petrochemical, plastics, fertilizer and insecticides field. Employees about 5000. ABCM member. Shell Centre, Downstream Building, London SE1.

Smith & Nephew Associated Companies Ltd

Originated in Hull 1856, when T. J. Smith set up as "an analytical and pharmaceutical chemist and sole agent for Lipscombe's patent filters". Refined cod-liver oil was an early venture. In 1896, it became a partnership, T. J. Smith & Nephew. Other medicinal oils, glycerine, petroleum jelly and flowers of sulphur were added and surgical dressings developed. Limited company 1907 as Smith & Nephew. Expansion continued and other companies were formed or acquired. The Group of Smith & Nephew Associated Companies, formed 1937, has now 22 subsidiaries and associated companies in the U.K., and 31 overseas in Australia, Bahamas, Belgium, Canada, Denmark, Eire, Ethiopia, France, India, Italy, Malaysia, New Zealand, Nigeria, Pakistan, South Africa and Spain. Its main chemical manufactures are in the field of drugs for tuberculosis. There is a wide range of pharmaceutical, medical, hospital, household, toiletry and sanitary products. The chemical production company is T. J. Smith & Nephew, Hull, an ABCM member. Chemical capital £150,000. Employees 80. 2 Temple Place, Victoria Embankment, London WC2.

Smith Kline & French Laboratories Ltd

In 1897, A. J. White, an American medicine vendor, established a company, A. J. White Ltd, in England. In 1907, Menley & James was formed as a subsidiary. These companies were purchased 1956 by Smith Kline & French Laboratories, Philadelphia, which had been long established in the pharmaceutical field and in 1957, the above name was adopted instead of A. J. White Ltd. In 1962, Bridge Chemicals Ltd was formed as a subsidiary for the manufacture and sales of fine chemicals. In the same year, Menley & James Laboratories was formed as a division of Smith Kline & French Laboratories for the marketing of proprietary products in the pharmaceutical field. There are associates in Australia, Canada, India and South Africa. Total U.K. employees 840. Bridge Chemicals is ABCM member. Welwyn Garden City, Herts.

The South Eastern Gas Board (SEGB)

This Board, formed 1948, embodies the former South Metropolitan Gas Company, established 1824 as the South Metropolitan Gas Light & Coke Co., but changed its name on amalgamating three or four other London companies in 1880. In 1897, it commenced the production of sulphuric acid and ammonium sulphate and in 1903, the distillation of tar and benzole and the preparation of the various products therefrom. The SEGB has a fully-owned subsidiary, South Eastern Tar Distillers Ltd, formed 1927, which specializes in road tar and special pitches for fibre pipes and enamels, marketed through a subsidiary Metrotect Ltd, formed 1957. In 1961, the London Tar & Chemical Co. Ltd was constituted to market the tar, benzole, ammonia and sulphur products of the North Thames (q.v.) and South Eastern Gas Boards and to plan their production and development programmes, which include the South Eastern Tar Distillers, who carry out their own sales in conformity with the programmes. In World War I, the South Met. was the first U.K. producer of synthetic phenol. Present chemical production confined to products of tar and benzole, sulphuric acid and ammonium sulphate. Chemical capital £1·7m. Employees 480. ABCM member. Katherine Street, Croydon, Surrey.

South Western Tar Distilleries Ltd (SWTD)

Formed 1918 as a co-operative tar scheme between gas companies in south and south west England and tar distillers with works at Totton, Portsmouth and Plymouth. In 1952, the Southern and South Western Gas Boards took half share and in 1962 purchased the rest of the shares, thus winding up the co-operative scheme. For some years, Burts & Harvey Ltd, the tar distillers at Totton and Portsmouth, was half-owned by Burt, Boulton & Haywood Ltd (q.v.), who later made it a wholly-owned subsidiary. In 1952, Burts & Harvey hived off their tar distillation activities and are now concerned only with manufacture of chemicals. At Plymouth, tar distillation has been taken over by the South Western Gas Board. The SWTD distils crude tar and crude benzole. Capital £250,000. Employees 180. ABCM member. Eling House, Totton, Southampton.

E. R. Squibb & Sons Ltd

Formed 1949 as part of the world-wide American organization created in 1858 and merged 1952 in the Olin Mathieson Chemical Corporation, to service the Commonwealth market with pharmaceuticals from a U.K. source and to conduct medical research. Associated with it in U.K. is Olin Mathieson Ltd, makers of compressed air mining equipment. Squibb's medical research is conducted in six countries.

Their range of pharmaceuticals includes antibiotics, anti-rheumatic agents, antiseptics, tranquillizers, diuretics, vitamins and veterinary products. Capital £200,000. Employees 400. Regal House, Twickenham, Middx.

Standard Fireworks Ltd

Formed 1910, as Standard Fireworks Co., became limited 1910 and later public. Largely family concern. Engaged solely in firework manufacture, with associated production of paper tubes and cardboard boxes and printing. Capital £350,000. Employees 600. Standard House, Half Moon Street, Huddersfield, Yorks.

The Staveley Iron & Chemical Co. Ltd

Originally incorporated 1863 as The Staveley Coal & Iron Co. Ltd to carry on business in coal, iron and ironstone and as a manufacturing company. Chemical production commenced just prior to World War I. With nationalization, the company was split into two. The manufacturing side at Staveley was incorporated 1948 as a private company called The Staveley Iron & Chemical Co. Ltd, which included the manufacture of chemicals among its activities, as well as iron ore and iron production and iron founding. Manufacturing business not at Staveley remained under the original company. The chemical production includes sulphuric acid, caustic soda, chlorine, chlorates, aniline, nitrobenzene, sulphate of ammonia and coal tar products. Employees 400. ABCM member. Near Chesterfield, Derbyshire.

Stevenson & Howell Ltd

Founded 1882, with wide business objectives including manufacturing chemists, importers and refiners of essential oils and makers of flavouring and perfumery essences. Their main activity now is the manufacture of flavouring essences for the food and beverage trade, but they also make some fine chemicals for flavours and process essential oils; there is some trade in ancillary equipment. The Standard Essence Company Ltd, London (formed 1919) and The Standard Essence Co. (Ireland) Ltd (formed 1935) sell the main company's products. Associates in Australasia. Total capital £1m. Employees 200. Standard Works, Southwark Street, London SE1.

Joseph Storey & Co. Ltd

Founded 1860 to make coal tar products, picric acid, sizes, pigments and driers. Of these only pigments and driers remain, but other items, such as stabilizers, have been added. Subsidiary of the Dental Manufacturing Co. Ltd. ABCM member. Heron Chemical Works, Lancaster.

John & E. Sturge Ltd

John Sturge commenced pharmaceutical chemical manufacture in 1822. Partnership with brother formed 1830 as John & Edmund Sturge. Private limited company 1910. Public company 1960. Production concentrated on citric acid and precipitated calcium carbonate. Sturge (Citric) Ltd formed 1930 with Rowntree organization for fermentation manufacture of citric acid. Rowntrees interest bought out 1949. Associates in France, India and Italy. Capital £2m. Employees 530. ABCM member. Wheeley's Road, Birmingham 15.

Sutcliffe, Speakman & Co. Ltd

Formed 1902 to develop ideas of Mr E. R. Sutcliffe in the engineering field. Their chemical interests are in active carbon (with plants for solvent recovery and water treatment) and soda-lime. Other activities concern briquetting, brickmaking, dust collection and general engineering. Associates in Canada. Employed chemical capital £450,000. Employees 330. Leigh, Lancs.

Synthite Ltd

Formed 1920 to make synthetic materials and chemicals. It was derived from D. R. Syndicate Ltd, which during World War I was conducting research into the manufacture of phenol-formaldehyde type resins. The formaldehyde was then entirely of foreign origin and costly, so manufacture was started in 1920 with large scale production in 1928, when Synthite became a member of the Tennant Group of Companies (q.v.). It has since expanded to meet the increased demands of the plastics industry. It produces formaldehyde both in concentrated form and in solution in alcohols for a variety of uses. Its U.K. associates are Tennant's Textile Colours Ltd, Charles Tennant & Co. Ltd, Tennants (Lancs.) Ltd, James M. Brown Ltd and Pan Britannica Industries Ltd. It has an American associate. Issued capital £600,000. Employees 224. ABCM member. Ryders Green, West Bromwich, Staffs.

Tennants Consolidated Ltd

This group had its origin in Charles Tennant & Co., a partnership of chemical and soap manufacturers formed 1797, which 1885 became limited as Charles Tennant & Co. of St. Rollox Ltd. It has had an interesting but somewhat complex history. The United Alkali Co., when formed 1890, took over much of its chemical production. It was reorganized in 1891 as Charles Tennant & Co., a private company, which became limited 1913. The present company was formed 1930 as a holding company to control the various interests which extend far beyond the chemical field. There were early offshoots, one of which, Charles Tennant & Co., of Carnoustie Ltd, was merged 1928 into Scottish Agricultural Industries Ltd, in which ICI holds a controlling interest.

The chemical manufacturing companies associated with the group are: James M. Brown Ltd (established 1926, limited 1943), makers of cadmium oxide, zinc oxide and sundry chemicals; Pan Britannica Industries Ltd (incorporated 1932), makers of fertilizers, insecticides and garden products; Synthite Ltd (q.v.) (established 1920), makers of formaldehyde; and Tennant's Textile Colours Ltd (formed 1949), makers of pigments etc. for textiles. The capital is over £2m. James M. Brown Ltd, Synthite Ltd and Charles Tennant & Co. Ltd, are ABCM members. 69 Grosvenor Street, London W1.

Thorium Ltd

On the outbreak of World War I, the U.K. had no supply of thorium nitrate required for the manufacture of gas lighting mantles, so in 1914 Thorium Ltd was established by Hopkin & Williams, the Volker Lighting Corporation Ltd and Howard & Sons Ltd to remedy this deficit. In 1928, ICI became a major shareholder because of its lighting industry interests. In 1959, ICI and Howard & Sons Ltd, the then shareholders, sold their shares to Rio Tinto Corporation Ltd and Dow Chemie International AG, Zurich, who held equal shareholdings. Subsequently, Rio Tinto merged with Consolidated Zinc Corporation to form the new Rio Tinto-Zinc Corporation Ltd, which has passed their shareholding to their subsidiary, Imperial Smelting Corporation. Associated with Rio Tinto Dow of Canada, for supply of thorium concentrates. Manufactures now cover a range of rare earth products such as compounds of thorium, cerium, lanthanum and neodymium. Employed capital £160,000. Employees 100. ABCM member. 6 St. James's Square, London SW1.

Turner & Newall Ltd

In addition to its extensive interests in the field of asbestos, Turner & Newall operates in the chemical field through the following companies:

1. Newalls Insulation & Chemical Co. Ltd, which arose 1964 from a merger of Newalls Insulation Co. Ltd, insulation contractors and suppliers, and the Washington Chemical Co. Ltd, makers of insulating materials and magnesia chemicals. The old names disappear. The Washington Chemical Co. Ltd was founded 1842 to exploit patents for the manufacture of oxychloride and basic carbonate of lead and to make the same products (soda ash, sulphuric acid, bleaching powder and alum) as the associated Felling Chemical Works, which started 1833, but was closed 1886. Later it concentrated on the production of magnesium carbonate and oxide from local dolomite. Subsequently it was found that the magnesium carbonate, when mixed with asbestos, produced a revolutionary thermal insulation material, so a company was

formed 1903 to exploit this, and was registered 1908 as Newalls Insulation Co. Ltd. In 1920, these two companies joined forces with Turner Brothers Asbestos Co. Ltd and this amalgamation was the commencement of Turner & Newall Ltd. The present merger in effect brings them back to where they were in 1903 but with a much extended scope of activity as regards insulating materials.

2. British Industrial Plastics Ltd, which originated in 1894 in the British Cyanide Co., formed to make sodium cyanide to extract gold from low-grade ores. By 1936, it was so deeply involved in plastics that it changed its name to British Industrial Plastics Ltd, having already acquired two subsidiaries, the Beetle Products Co. Ltd (formed 1925) and the Streetly Manufacturing Co. Ltd (acquired 1929). In the same year, it founded BIP Tools Ltd. In 1949, BIP Engineering Ltd was born. In 1954, there was a reorganization. BIP Chemicals Ltd was formed to take over the functions of the Beetle Products Co. Ltd and Beetle Bond Ltd (a sales company created 1943), while British Industrial Plastics Ltd became the holding company for the BIP group, which since then has formed BIP Reinforced Products Ltd 1957, BIP Gaydon Ltd 1961, and acquired Hornflowa Ltd 1960. In the same period, companies were established in Germany, Mexico, South Africa and U.S.A., all in the field of plastics or machinery therefor. In 1961, the BIP group merged with Turner & Newall Ltd. Employees 1450.

BIP Chemicals and Newalls Insulation & Chemical Co. are ABCM members. Turner & Newall Ltd, Asbestos House, Fountain Street, Manchester 2.

3. Ferodo Ltd. Founded 1897 to provide brake blocks to meet the growing needs of mechanical transport. Joined the Turner & Newall organization 1926. Phenol-formaldehyde compounds are now utilized in the production of the linings and Ferodo has a sizeable section of its activities devoted to making these resins.

Uclaf Ltd

Formed 1956 as wholly-owned subsidiary of Uclaf SA (now Roussel-Uclaf SA), Paris. Original objective was manufacture of cortico-steroids. In addition, hormones, hypotensive products and semi-conductor silicon now made. Its associate, Roussel Laboratories Ltd, is its sole selling agent for chemicals. Upsil Ltd, jointly owned with Pechiney SA, sells silicon and other products of the group. Capital £675,000. Employees 130. ABCM member. Marshgate Lane, Stratford, London E15. (See Addenda, p. 329.)

Unilever Limited

The Unilever Group of Companies covers a wide range of manufacturing interests, many of them based on oils and fats. A great deal of their business lies in consumer goods. Unilever was brought into existence in 1929 by a merger between Lever Brothers Ltd, soapmakers, which dates from 1885, and the Anglo-Dutch Margarine Union, formed 1927, but with origins dating back to 1897. In 1960, a U.K. Chemical Group was formed, most of the products of which are raw materials for industry and not consumer goods. It includes the following: Joseph Crosfield & Sons Ltd (formed 1815 and acquired 1896, with silicates as its main interest), Price's (Bromborough) Ltd (formed 1854 to make candles, taken over 1937, main interests fatty acids and oleo-chemicals), John Haigh & Co. Ltd (a small oleo-chemical firm, acquired 1962), Proprietary Perfumes Ltd (took over the firm's Central Perfumery Department 1960), Associated Adhesives Ltd (formed 1964), selling agents for John Knight Ltd (began soap making 1810, acquired 1920, now producers of glues and other adhesives), Gloy & Empire Adhesives Ltd (acquired 1963), and Fixol & Stickphast Ltd (acquired 1961), J. L. Thomas & Co. Ltd and Tellam Products Ltd (acquired 1960), Glycerine Ltd (formed 1929 to centralize glycerine interests), Charles Lowe & Co. (Manchester) Ltd (founded 1864, incorporated 1929, acquired 1964, makers of synthetic resins and refined tar acids), Walker Chemical Co. Ltd (established 1953, taken over 1963, manufacturers of formaldehyde and associated products), Birmingham Chemical Co. Ltd (formed 1924, acquired 1964, makers of brominated vegetable oils, food flavourings and essences), and Advita Ltd (formed 1946, with interests in vitamins and emulsifiers for margarine). In 1965, Food Industries Ltd formed to supply ingredients to makers of food and drink; it absorbed Advita Ltd and Birmingham Chemical Co. Ltd with others. Outside their chemical group, but within the scope of this survey, Unilever have extensive interests in detergents of all kinds, representing about 20% of the firm's turnover. The main manufacturers in the U.K. are Lever Brothers & Associates Ltd (the largest), Gibbs Proprietary (formerly Pepsodent) Ltd, Industrial Soaps Ltd, and Domestos Ltd.

Unilever have chemical as well as detergent interests in Australia, Canada, Chile, France, Germany, Mexico, Netherlands, South Africa and Spain, and world-wide sales arrangements. Employed capital in above activities in the U.K. about £30m. Employees 11,000. ABCM member. Unilever House, London EC4.

Union Carbide Ltd

A subsidiary, formed 1954, of the American Union Carbide Company, till 1957 the Union Carbide and Carbon Corporation, formed 1917 by the amalgamation of the Union Carbide Company, established 1898 to make calcium carbide, and the National Carbon Company, founded 1886 to make carbon electrodes, to which the Bakelite Company was added 1939. The British company amalgamated Gemec Ltd (importers and makers of chemicals), British Electro Metallurgical Co. Ltd (selling ferro-alloys), and Kemet Products Ltd (originally established to make barium products). It operates in three divisions, viz., chemicals, alloys, and engineering products. Only the first comes within the present purview. U.K. production includes ethylene oxide and derivatives, glycol ether solvents, ethanolamines, polyethylene glycols, surface active agents, propylene oxide derivatives, and polyalkylene glycols. Silicones, plastic monomers and polymers and acrylic fibre are marketed for its U.S. parent.

The plastics interests of Union Carbide Ltd, with those of its associated Bakelite Ltd were merged in 1963 in a 50–50 deal with The Distillers Co. Ltd (q.v.) to form Bakelite Xylonite Ltd (q.v.). ABCM member. 8 Grafton Street, London W1.

United Coke & Chemical Co. Ltd

Formed 1923 as wholly-owned subsidiary of the United Steel Companies Ltd, with a share capital of £150,000, to distil the tar from its parent's coke ovens. Took over 1945 the selling organization of Coke Oven Products Ltd (founded 1935), which was then wound up. Makes a wide range of benzole and coal tar derivatives, coumarone resins, polyester resins, phthalic anhydride, ammonium sulphate, a resin-impregnated fabric bearing (Orkot), special pitches for electrodes, impregnation purposes, coal briquetting, and extenders for epoxy resins, and paints. Many overseas selling agencies. Capital assets over £15m. Employees 1150. ABCM member. PO Box 25, The Grange, Treeton, Rotherham.

The United Sulphuric Acid Corporation Ltd

Formed 1951 as a co-operative venture between eleven participants which include Courtaulds Ltd, Imperial Chemical Industries Ltd, Fisons Fertilizers Ltd, The Clayton Aniline Co. Ltd, Thomas Bolton & Sons Ltd, The Alumina Co. Ltd, McKechnie Brothers Ltd and Transparent Paper Ltd, to make sulphuric acid from indigenous raw materials, particularly anhydrite (calcium sulphate) to avoid the shortage of acid due to the then-limited supplies of American sulphur because of the Korean War. Portland Cement Clinker is a by-product of the manufacture. Capital £4·75m. Employees 250. Green Oak Works, Tan House Lane, Widnes, Lancs.

Upjohn Ltd

The American Upjohn Co. was founded 1886. In 1952, it embarked on a programme of international development, since when Upjohn products are now made in Argentina, Australia, Belgium, Canada, France, Great Britain, Japan and Mexico, and marketed in 104 countries through 22 subsidiaries and branches. The U.K. firm was established 1952 as a private company to make and sell ethical pharmaceuticals, which include corticosteroids, antibiotics, medical specialities and veterinary products. Nominal capital £210,000. Over 200 employees. Fleming Way, Crawley, Sussex.

Vantorex Ltd

Incorporated 1961, with two trading divisions, viz., The Rexall Drug Co. Ltd and Riker Laboratories Ltd. In 1964, it became the name of what was formerly The Rexall Drug and Chemical Co. (U.K.) Ltd, within which were established five divisions, viz., Riker Laboratories Ltd (incorporated 1951), The Rexall Drug Co. Ltd (marketing), an established U.K. subsidiary of the Rexall organization in the U.S., Imco Container Co. Ltd (makers of plastic containers), Tupperware Manufacturing Co. Ltd (makers of plastic food containers), and Tupperware Sales.

Carnegie Brothers, a firm started 1911 to make fine chemicals was incorporated 1931. In 1936, Carnegie Chemicals (Welwyn) Ltd was established as a new business by one of the Carnegie brothers. In 1941, Carnegie Quinine Works Ltd was established at Welwyn to meet a government request for larger quinine manufacturing facilities. A further associated company, Carnegie

Organics Ltd, was formed 1945. In 1953, these four companies were grouped into one entity, Carnegies of Welwyn Ltd, which was acquired 1958 and absorbed 1964 by Riker Laboratories Ltd, which is the pharmaceutical and chemical manufacturing division of Vantorex. Riker Laboratories employees 460. ABCM member. Morley Street, Loughborough, Leics.

Vitamins Group

The first company of the group was formed 1921 as Agricultural Food Products Ltd, to manufacture animal feed supplements for farm livestock. An essential constituent was Vitamin B. In 1927, the success in animal nutrition was applied in the human field. Vitamins Ltd, formed to market the products, developed a range of pharmaceutical vitamin preparations. Nutritional Information Centre set up during World War II to assist in the dissemination of relevant nutritional knowledge. The present manufactures include bulk pharmaceutical fine chemicals. Associates in Belgium, Ireland and Norway. Employees 1100. Upper Mall, London W6.

Vitax Ltd

Established 1939 on modest lines. In 1959, became a member of the Reckitt & Colman Holdings Group. Manufacturers of organic manures and agricultural and horticultural pesticides for direct sale to users. Employees 118. Burscough Bridge, Ormskirk, Lancs.

Ward, Blenkinsop & Co. Ltd

Formed 1939 as wholly British unit to manufacture fine chemicals. Its present production covers some 300 organic chemicals for pharmaceutical and industrial use. Outside the chemical field, it makes medical and veterinary specialities. It has two U.K. subsidiaries:

1. Sorex (London) Ltd, formed 1948 with 50% shareholding, wholly-owned 1960, for rodenticides and insecticides and through its associate Mark Smith Ltd (founded 1893) for horticultural chemical preparations.

2. Harker Stagg Ltd, formed 1820 and acquired 1956, formulating, compounding and packing for the pharmaceutical, cosmetic and allied industries.

There are associates in Australia, India and Netherlands. U.K. chemical capital £500,000. Employees 600. ABCM member. Fulton House, Empire Way, Wembley, Middx.

The Wellcome Foundation Ltd (trading as Burroughs, Wellcome & Co.)

This is an organization without parallel in the pharmaceutical industry. Founded 1880 by Silas M. Burroughs and Henry S. Wellcome as a partnership, named Burroughs Wellcome & Co., it was registered under its present title by Wellcome in 1924. Sir Henry Wellcome died in 1936 and bequeathed his sole ownership of the company's shares to The Wellcome Trust to which all distributed profits are paid. The Trust is a recognized public charity, with, as its main objects, the world-wide advancement of research in human and veterinary medicine and allied sciences and the establishment or endowment of research museums and libraries. To date over £6m has been allocated as benefactions. The Wellcome Foundation Ltd trades as Burroughs, Wellcome & Co., which has built its reputation on pharmaceutical products of high quality, many the result of original research in the Foundation's laboratories, founded in 1894. Its overseas organization comprises subsidiaries and associates, many of them actual manufacturers, in Argentina, Australia, Belgium, Brazil, Canada, Eire, France, India, Italy, Kenya, New Zealand, Nigeria, Pakistan, South Africa, Rhodesia and the U.S.A. U.K. employees about 4000. Outside the chemical field, the Foundation has interests in biological and laboratory diagnostic and culture media.

In 1959, Cooper, McDougall & Robertson Ltd (formed 1925) joined the group as a subsidiary. McDougall & Robertson had combined in 1921 and had amalgamated with Cooper & Nephews in 1925; those companies were of long standing. McDougall entered the field of sheep dips in 1852, Robertson well before 1900 and Coopers in 1843. The last mentioned had established their Cooper Technical Bureau for advisory and research functions, which now has world-wide recognition. Minsal Ltd, makers of animal feed supplements, was acquired 1964. The products of Cooper, McDougall & Robertson Ltd include animal dips, veterinary medicines, fly sprays, insecticides, vitamins and food supplements, poultry remedies, domestic insecticides and aerosols. Capital £1·75m. U.K. employees 1400. There are subsidiaries and associates in Argentina, Australia, Belgium, Brazil, Eire, Italy, Kenya, New Zealand, South Africa, Rhodesia and the U.S.A. Burroughs, Wellcome & Co. and Cooper, McDougall & Robertson are ABCM members. The Wellcome Building, Euston Road, London NW1.

Whey Products Ltd

Formed 1926 as a private company to make lactose (sugar of milk) from the otherwise waste whey from the cheese and casein making industries. After it was a going concern and had established a company in New Zealand, it was taken over by United Dairies. Unigate is now its parent. Besides lactose, it makes animal foodstuffs. About 400 employees. ABCM member. 35 Crutched Friars, London EC3.

Whitmoyer Reed Ltd

Incorporated 1922, as R. F. Reed Ltd, to make citrates, iodides and bismuth salts. Present name adopted 1960 when majority interest acquired by Whitmoyer Laboratories Inc., U.S.A. Became wholly-owned subsidiary 1963 and in 1964 the latter became subsidiary of Rohm & Haas, of Philadelphia. Now mainly concerned with veterinary and animal health products of parent. Capital £150,000. Employees 62. ABCM member. Riverside Works, Hertford Road, Barking, Essex. (See Addenda, p. 329.)

Williams (Hounslow) Ltd

Charles Hanson Greville Williams, a pupil of Hofmann and a collaborator of Sir William Perkin, discovered the first quinoline dye (quinoline blue) 1857 and safranine (a basic colour) 1859. In 1877 he established a company which continues today as Williams (Hounslow) Ltd, with family control. It has as U.K. associates Williams (Hounslow) Yorkshire Ltd, Williams Ansbacher Ltd (owned jointly with Sun Chemical Corporation, U.S.A.) and Morton-Williams Ltd (jointly with the Morton Chemical Co. (Chicago) for polymer-based packaging materials). Associates also in Canada, Netherlands and Switzerland, and agents in over 60 countries. Specialists in dyestuffs and products for the food, printing ink, plastic, leather, varnish, petroleum, cosmetic and pharmaceutical industries. Employees 450. ABCM member. Hounslow, Middlesex.

Witco Chemical Co. Ltd

Formed 1935 as subsidiary of Witco Chemical Co., Inc., U.S.A., to merchant industrial chemicals. Commenced manufacture 1954. Products include chemicals for rubber, paint and plastics, synthetic resins, metallic stearates, and adhesives. Associates in Belgium, Canada, France, Italy, Netherlands, and U.S.A. Capital about £500,000. Employees 130. North West Wing, Bush House, Aldwych, London WC2.

John Wyeth & Brother Ltd

Originated in America 1860. U.K. company appeared as such 1937. Now subsidiary of American Home Products Corporation. It manufactures ethical pharmaceuticals, infant foods and nutritionals. U.K. associates are International Chemical Co. Ltd (counter-line medicinals and household products) and E. R. Howard Ltd (household goods). Overseas associates through its parent in some 25 countries. U.K. chemical capital £2m. Employees 850. Huntercombe Lane South, Taplow, Maidenhead, Berks.

The Yorkshire Dyeware & Chemical Co. Ltd

Formed 1900 by the amalgamation of several old-established concerns supplying natural dyes, soaps and chemicals to textile, leather and other industries. Its manufactures now include dyestuffs (synthetic and natural), intermediates, synthetic tannins, leather chemicals, adhesives, flocculating and dispersing agents, synthetic resins and coated fabrics. It has a U.K. associate, Brifex Ltd, making coated fabrics. It has an Australian branch. It has a German associate. Employed capital £4m. Employees 800. ABCM member. Black Bull Street, Leeds 10.

Yorkshire Tar Corporation Ltd

Incorporated as private company 1955, by acquiring The Yorkshire Tar Distillers Ltd (YTD) (established 1926), and the Vulcan Chemical Co. Ltd (founded 1908). Became public company 1956. In 1954, YTD had acquired Henry Ellison Ltd (formed 1928). The two operating companies are YTD and Synthetic Chemicals Ltd (90% owned), formed 1946. Other associates are YTD Plant Hire Ltd and Central Oil Refining Co. Ltd (50% owned). In addition to YTD's wide range of tar products and hydrogenated derivatives, Synthetic Chemicals Ltd manufacture, *inter alia*, sulphuric acid, sodium sulphite and various sulphonated hydrocarbons. Capital about £3m. Employees 740. YTD is ABCM member. Cleckheaton, Yorks.

Robert Young & Co. Ltd

Formed 1895. Became limited 1912. Original object was production of soaps and disinfectant fluids. Manufacture of animal dips commenced 1918. Activities now include sheep and cattle dips, veterinary preparations, disinfectant fluids and mineral feed supplements. Youngs (Pharmaceuticals) Ltd is subsidiary formed 1959 for the sale of vaccines, sera and veterinary preparations. Associates in Argentina and New Zealand. U.K. capital £250,000. Employees 200. Cranstonhill Chemical Works, Elliot Street, Glasgow C3.

ADDENDA

1. Bakelite Xylonite Ltd (see p. 275)

   From 1st January 1966, organized into three divisions, viz., Cascelloid for bottle making and like operations; consumer products; and components. British Xylonite Co. Ltd ceased to trade 31st December 1965.

2. Glaxo Group (see p. 294)

   On 1st January 1966, formed new subsidiary—Glaxo International Ltd—to co-ordinate overseas pharmaceutical operations. Glaxo-Allenburys (Export) Ltd became its subsidiary.

3. Lennig Chemicals Ltd (see p. 307)

   From 1st January 1966, Whitmoyer-Reed Ltd (q.v.) amalgamated as its Animal Health Division.

4. The Pfizer Group (see p. 313)

   Acquired 1965 the Bridge Colour Co. and the Hull & Liverpool Red Oxide Co.

5. Uclaf Ltd (see p. 323)

   New U.K. Co.—Roussel-Uclaf Ltd—formed 1966 to co-ordinate the activities of Roussel Laboratories Ltd and Uclaf Ltd.

6. Whitmoyer-Reed Ltd (see p. 327)

   From 1st January 1966, became Animal Health Division of Lennig Chemicals Ltd (q.v.).

PART C

Trade Associations

THIS section is confined to those organizations, closely connected with the chemical industry, of which the membership is primarily by firms or companies, as distinct from scientific, technical and professional societies, where membership is constituted by individuals. A list of such societies is included as an Appendix; they play a very important and in many cases a vital part in the activities of the industry. There have also been excluded those trade organizations which operate for the benefit of industry as a whole as, for example, The British Employers' Confederation, The British Institute of Management, The British Productivity Council, The British Standards Institution, The Federation of British Industries, The Institute of Export, The National Association of British Manufacturers, and The Royal Society for the Prevention of Accidents; also the new Confederation of British Industries (CBI), recently formed (1965) by the amalgamation of the Federation of British Industries, the British Employers' Confederation and the National Association of British Manufacturers.

After consultation with the relevant Government Departments, it was found that there were nearly 60 trade associations connected with the chemical industry. Of these, the two most comprehensive in their scope are The Association of British Chemical Manufacturers, dealing with general, economic and technical questions, and The Association of Chemical and Allied Employers, handling labour problems, recently (1st Jan. 1966) combined to form the Chemical Industries Association Ltd (CIA). Many of the others relate to specialized interests. The information set out in the following pages has been derived from the organizations themselves. Some of those approached failed to reply to repeated reminders, while a few asked to be excused because they felt they were on the periphery of the industry, or were becoming of diminished importance because of organizational changes such as amalgamations. None of these was of major importance.

Trade associations are often the target of ill-informed critics who allege they are engaged in price control or other forms of market regulation. This has always been far from the truth as regards the chemical industry. Any such activities as may have existed have been definitely terminated by The Restrictive Trade Practices Act, 1956. It will be seen from what follows that the normal functions of most trade associations fall into two categories, internal and external. internal, to provide for co-operation among the members and to promote efficiency, and external, to act as the mouth-piece of the membership when dealing with other organizations, and in particular with Government Departments in regard to legislation and regulations.

Another point which is often not appreciated is that trade associations are voluntary organizations. There is no compulsion on any firm to become a member. While members have to undertake to comply with certain membership rules, e.g., in regard to subscription or notice of withdrawal, decisions of the association or of its governing body are, in general, not mandatory; they are in effect recommendations only and their acceptance depends entirely on moral suasion. Furthermore, trade associations as a general rule are non-political.

The associations are listed alphabetically.

The information was collected in the latter half of 1964, to which the membership figures apply. Since then, the Department of Scientific & Industrial Research (DSIR) has been abolished (1965).

Adhesives Manufacturers' Association

Formed 1940 as the Association of Manufacturers of Vegetable Adhesives, to further the interests of the industry and maintain close liaison with Government Departments. With the introduction of synthetic raw materials, the present name was adopted in 1953. Over 20 members making adhesives for industrial purposes. Affiliated to National Association of British Manufacturers (NABM). 6 Holborn Viaduct, London EC1.

Associated Manufacturers of Veterinary and Agricultural Products (AMVAP)

Formed 1927 as the Animal Medicine Makers' Association. Changed its name 1956. Has the usual objectives of a trade association. The emphasis is on products for the treatment of animals. Membership 17. Affiliated to ABCM. Cecil Chambers, 86 Strand, London WC2.

Association of British Chemical Manufacturers (ABCM)

The main association for the chemical industry. Founded 1916, to promote co-operation between chemical manufacturers, to act as co-ordinating body for the industry in dealing with Government Departments, and to promote efficiency in the widest sense. The Association does not normally concern itself with wages or labour questions, which come within the purview of the Association of Chemical & Allied Employers (q.v.). It is strictly non-political, is not a trading organization and does not concern itself with price or quota conventions. Membership is restricted to established bona-fide makers of chemicals in the U.K. The subscription is based on the proportion of the total capital employed in the chemical business of the member.

A Standing Joint Committee with the AC & AE maintains consultation and joint action on matters of major policy affecting the industry as a whole. The British Chemical Industry Safety Council (q.v.) similarly co-ordinates safety activities. A productivity organization on a geographical area basis is maintained jointly with the AC & AE and the trade unions connected with the industry.

ABCM maintains a close watch on all new legislation and Governmen regulations, overseas trade, trade agreements, and European Free Trade Association matters. Other subjects on which the Association has taken a leading part include patents and trade marks, poisons legislation, safety in chemical works, standardization, industrial information, fuel efficiency, packaging, productivity, trade effluents and chemical engineering research. Publications (members only: *Proceedings of Council* (monthly), *Bulletin of Information* (weekly), *Productivity News* (bi-monthly). A Directory, listing over 10,000 chemicals and their manufacturers, is published every two years, with a supplement in the intervening year. It is available, free of charge, to bona-fide purchasers of chemicals anywhere.

Publications (purchasable) include *Safety Guide, Quarterly Safety Summary, Labels for Gassing Casualties* and booklets on Marking Containers, Marking Techniques, Instrumentation, Packaging, Work Study, and Productivity Techniques. Membership 235 and 13 affiliated associations. Cecil Chambers, 86 Strand, London WC2.

This has been combined (1st January 1966) with the Association of Chemical & Allied Employers to form the Chemical Industries Association Ltd (CIA).

Association of British Manufacturers of Agricultural Chemicals (ABMAC)

Formed 1928 as the Association of British Insecticide Manufacturers when the Insecticide Section of the London Chamber of Horticulture was dissolved. Changed its name 1958. A normal trade association in the field of agricultural and horticultural insecticides, fungicides, weed-killers, growth regulators and other chemicals for crop protection. Membership open to merchants of British and foreign products as well as manufacturers. Membership 31. Co-operated with the Ministry of Agriculture in the formulation in 1943 of the scheme for the approval of proprietary products for the control of plant pests and diseases. Initiators of the periodic Weed Control and Insecticide and Fungicide Conferences. Member of the European Association of National Pesticide Associations, formed 1959. Affiliated to ABCM. Cecil Chambers, 86 Strand, London WC2.

Association of British Organic and Compound Fertilisers Ltd

Formed 1939 and incorporated 1940 as the Association of British Organic Fertilisers, to assist the Government in a scheme for the voluntary control of organic fertilizers. Present name adopted 1962. Its objects are broadly to promote the trade and protect the interests of members. Firms in the U.K. engaged in the manufacture, importation and wholesale distribution of organic fertilizers are eligible. Membership 71. Audrey House, 5–7 Houndsditch, London EC3.

The Association of the British Pharmaceutical Industry (ABPI)

The Association had its origin in the Drug Club, a trade association formed 1891; this was succeeded by the Wholesale Drug Trade Association 1930. The passage of the National Health Service Act, 1946, foreshadowed new problems for the industry, so a review of the Constitution and Rules took place, and in 1948 the present name was chosen as conveying more accurately the scope and functions of an association in which manufacturing interests predominated. In 1949 it merged with the Pharmaceutical Export Group set up in 1940. The main objects are to promote efficiency, develop exports, and represent the industry *vis-à-vis* Government Departments and other organizations. Membership is open to manufacturers or wholesale distributors of drugs and pharmaceutical chemicals and preparations. It does not represent manufacturers of proprietary medicines advertised to the public, which is the function of the Proprietary Association of Great Britain (q.v.). It has four divisions, dealing with standard drugs, medical specialities, veterinary products, and pharmaceutical wholesalers. Membership over 150. It is an associate member of the International Pharmaceutical Federation. Affiliated to ABCM. Mercury House, 195 Knightsbridge, London SW7.

In 1965, the Wholesale Druggists Association, the Scottish Wholesale Druggists Association, the Ulster Wholesale Druggists Association and the Wholesale Sundries and Proprietaries Association decided to establish a new organization to represent the interests of pharmaceutical wholesalers. It will be affiliated to the ABPI.

Association of Chemical & Allied Employers (AC & AE)

Originally formed 1917 as the Wages Committee of Chemical Manufacturers to provide a national approach to the war-time authorities controlling wages. In the same year, it changed its name to Chemical Employers' Federation and later to the Chemical & Allied Employers' Federation. Its present name dates from 1936. In 1945, there was established a Chemical & Allied Industries Joint Industrial Council (JIC), with subordinate chemical, fertilizer and plastics JIC's. In 1946, the Association of Drug & Fine Chemical Manufacturers merged with the AC & AE, and a corresponding JIC was set up. In 1954, the Scottish Association of Chemical Manufacturers, which had been in existence since World War I, dissolved and its members joined the AC & AE.

The main objects of the AC & AE are the negotiation of wage rates and conditions with the trade unions, and labour problems in general, such as training, safety and industrial health. It is a constituent member of the British Employers' Confederation (BEC) for national and international (ILO) industrial policy and legislation. A full member must agree to operate and be bound by the terms and conditions of the relevant JIC. There is provision for affiliation on a limited basis. Under 300 members. Booklets on *A Career in Chemicals* (for school leavers) and on the training of engineering apprentices are generally available. Imperial House, 15 Kingsway, London WC2.

This has been combined (1st January 1966) with the Association of British Chemical Manufacturers to form the Chemical Industries Association Ltd (CIA).

Association of Tar Distillers (ATD)

Formed 1885—claims to be the oldest U.K. trade association—to promote co-operation between distillers of coal-tar; to be a clearing house of information on technique and development, and to represent the views of the industry. Membership 26. Handbook on *Coal Tar Fuels* publicly available. Affiliated to ABCM. 9 Harley Street, London W1.

The British Acetylene Association

Formed 1901 to promote acetylene engineering and manufacture and to facilitate the interchange of information among members. Membership open to those with active interest in the acetylene and welding industries. Issues technical handbooks with world-wide distribution. Carbide House, 55 Gordon Square, London WC1.

British Aerosol Manufacturers Association (BAMA)

Formed 1961 to enable manufacturers to discuss common problems and to form a link with Government Departments. Objectives include the stimulation of the use of aerosols and the formulation of desirable standards. Membership limited to firms in the U.K. who are fillers of aerosols or makers of the necessary raw materials, component parts or machinery. Membership 36. Organized the Fourth International Aerosol Congress and the Second International Aerosol Exhibition at Brighton 1963. Member of the Federation of European Aerosol Associations. Affiliated to ABCM. Cecil Chambers, 86 Strand, London WC2.

British Aromatic Compound Manufacturers' Association

Formed 1942 to provide collective action to promote and protect the interests of manufacturers of aromatic compounds in the U.K. and of the industry generally. Membership is open to persons, firms and companies engaged in the blending of materials to produce aromatic compounds in the U.K. for wholesale as such for manufacturing purposes. 69 Cannon Street, London EC4.

The British Chemical & Dyestuffs Traders' Association Ltd (BCDTA)

Formed 1923 by the amalgamation of two associations set up about 1919 with similar objects but representing different sections of the trade. Its objects are to watch and protect the interests of merchants and distributors in the chemical, dyestuffs and allied industries; to represent these interests *vis-à-vis* Government Departments and other organizations; to provide an information service; and to handle, in strict confidence, the individual problems of members. Importers, exporters, merchants or distributors of commodities of the above-mentioned industries, if domiciled or carrying on business in the U.K., are eligible for membership; this can include manufacturers here or overseas. Some 80 members. 12B Westminster Palace Gardens, Victoria Street, London SW1.

British Chemical Industry Safety Council (BCISC)

Formed 1956 by the Association of British Chemical Manufacturers (q.v.) and the Association of Chemical and Allied Employers (q.v.). Its objectives are to maintain the industry's high safety standards and to improve and extend safe working practices. It co-ordinates the safety and medical activities of the two organizations, collects and analyses accident statistics, gives advice on medical and technical matters affecting the health and safety of employees, and organizes appropriate courses. Imperial House, 15 Kingsway, London WC2.

British Colour Makers' Association (BCMA)

A Colour Makers' Association was formed 1919 as a section of the National Paint Federation. Became incorporated as the British Colour Makers' Association 1932. In 1939 it withdrew from the National Paint Federation and became affiliated to the ABCM. It is the normal type of trade association for U.K. firms engaged in the manufacture of earth colours and/or coloured pigments, both inorganic and organic. Cecil Chambers, 86 Strand, London WC2.

British Disinfectant Manufacturers' Association (BDMA)

Originally formed 1919. Dissolved 1929 but soon reformed on modified basis. Main objects are to encourage the production of efficient disinfectants and antiseptics, to stimulate research, to promote the standardization of tests and to express the views of the industry *vis-à-vis* Government Departments and other organizations. Standard tests promulgated through the British Standards Institution. Open to firms in the U.K. Membership 43. Affiliated to ABCM. Cecil Chambers, 86 Strand, London WC2.

British Essence Manufacturers' Association

Formed 1917, the main objects being to deal with such matters as affect the trade, and to promote and protect the interests of its members. Full membership is open to persons, firms or companies actively engaged in carrying on in the U.K. the manufacture of flavouring essences, the majority of whose capital is owned by British nationals. Associate membership is open to others similarly engaged. 69 Cannon Street, London EC4.

The British Industrial Biological Research Association (BIBRA)

Formed 1960, with DSIR grant, to provide authoritative information on possible toxicity hazards from chemicals used in connection with foodstuffs and cosmetics. Membership, 250, embraces U.K. food, chemical, plastic, cosmetic and packaging material industries. Associate membership open to educational and sponsored research organizations and to foreign firms. Annual income £130,000. Bi-monthly international journal, *Food & Cosmetics Technology*, generally available. Monthly bulletin for members only. Woodmansterne Road, Carshalton, Surrey.

British Man Made Fibres Federation

Formed 1943 as the British Rayon Federation, becoming 1950 the British Rayon and Synthetic Fibres Federation. Present name dates from 1954. It is a federation of interested textile trade organizations to provide a voice for the industry as a whole and to promote its interests. Membership confined to appropriate trade organizations, which now number eight. Publication, *Facts about Man-Made Fibres*, freely available. 58 Whitworth Street, Manchester 1, and 41 Dover Street, London W1.

British National Committee on Surface Active Agents

Formed 1962, to enable U.K. interests to join the Comité International de la Detergence, formed 1957 to standardize terminology, methods of analysis and performance tests for surface active agents. This body organizes international congresses on the subject, the fourth being held in 1964. Membership of the U.K. committee confined to U.K. manufacturers of surface active agents or their raw materials. Membership 27. Cecil Chambers, 86 Strand, London WC2.

The British Plastics Federation (BPF)

Formed 1933. Previously there had been a plastics' moulders association till it became apparent that there was a need for the collaboration of all sections of this new and rapidly expanding industry through a single organization with the usual trade association functions. Membership covers British manufacturers of primary materials, semi-finished and finished products and machinery required for the various processes. Membership about 370, organized in 12 groups, each operating in a specific sector of the plastics industry. Range of publications, including monthly technical abstracts, generally available on payment. 47–48 Piccadilly, London W1.

British Road Tar Association (BRTA)

Formed 1927 by the gas, coke oven and tar distilling industries to further the use of road tar by research (in co-operation with the Department of Scientific and Industrial Research), technical service and publicity. Commonwealth firms can be associates. Membership about 40. Quarterly journal, *Road Tar*, and *Road Tar Manual*, with free wide distribution. Founder member of International Road Tar Conference. 9 Harley Street, London W1.

British Sulphate of Ammonia Federation Ltd

Formed 1920 to succeed the Sulphate of Ammonia Association, which had existed since 1914, and which was the outcome of the work of the Sulphate of Ammonia Committee, formed by the main U.K. producers in 1897, to handle home and export sales. The Federation ceased as a trading body in 1963, as the result of The Restrictive Trade Practices Act 1956, and became a normal trade association to further the interests of its members and act as the mouthpiece of the industry. Any producer of sulphate of ammonia in the U.K. or in the Commonwealth can be a member. Membership represents over 90% of production in the U.K. and Northern Ireland. Affiliated to ABCM. 24/30 Gillingham Street, Victoria, London SW1.

British Sulphate of Copper Association (Export) Ltd

Formed 1924 to promote sales of copper sulphate, particularly for export. Only the few British producers are members. 50 Jermyn Street, London SW1.

The British Tanning Extract Manufacturers' Association

Formed about 1929 or earlier, with the normal functions of a trade association. Membership (6) is limited to manufacturers. The Association played an important part in World War II in the control of the raw materials and production of the industry. Sardinia House, 52 Lincoln's Inn Fields, Kingsway, London WC2.

The British Whiting Federation

Formed 1943 by amalgamation of the Northern and Southern Whiting Associations, to promote co-operation and to foster the trade of whiting manufacturers. Membership restricted to U.K. makers; the 15 firms represent over 90% of national output. The Federation promoted the formation, 1948, of the Whiting Research Council, in receipt of DSIR grant. The title was changed to Whiting and Industrial Powders Research Council 1962 on the setting up of the Industrial Powders Division. In October 1964, the Council amalgamated with the Chalk Lime and Allied Industries Research Association, to form Welwyn Hall Research Association. Available publications: *The Story of British Chalk Whiting* and *Whiting—Notes on origin, manufacture, properties and uses*. The Hall, Church Street, Welwyn, Herts.

The British Wood Preserving Association (BWPA)

A scientific association, founded 1930, to provide an impartial technical advisory service to promote the cause of wood preservation, which includes fire-proofing. Its articles are widely drawn and its activities are conducted on the broadest possible basis. There are seven categories of members, which are wide enough to include any organization or individual concerned with or interested in the production and/or use of wood preservatives. Membership over 450. It holds a yearly convention and publishes a wide range of papers, which are freely available, as is its advisory service. It is represented on international bodies such as the United Nations Food and Agriculture Organisation (FAO) working party on wood preservation. 6 Southampton Place, London WC1.

Coal Tar Research Association (CTRA)

Formed 1949 by the gas, coke oven and tar distilling industries with DSIR support. U.K. firms who treat coal tar products and plant makers are also eligible for membership as are foreign companies under certain conditions. Membership 60. Annual income about £140,000. Six-monthly review of coal tar technology and coal tar data book are publicly available. Administers the important Standardisation of Tar Products Tests Committee (STPTC) which has been active since 1927. Oxford Road, Gomersal, near Leeds.

Creosote Producers' Association Ltd

Incorporated 1936 as successor to the Creosote Producers' Export Association of many years' standing. The object is to provide co-operation between the members, who have to be creosote producers, to facilitate the bulk export of creosote oil, mainly to the U.S.A. Eight members. Managed by Tar Residuals Ltd, Plantation House, Mincing Lane, London EC3.

Cresylic Acid Refiners' Association

Formed 1945 by the U.K. refiners of cresylic acid to collect and disseminate statistical information. Membership 12. c/o Whinney, Smith & Whinney, Yorkshire House, Greek Street, Leeds 1.

The Federation of Gelatine & Glue Manufacturers Ltd

Formed 1921 as the Federation of Hide Gelatine and Glue Manufacturers Ltd, in succession to an un-incorporated organization already in existence for some time. In 1945, the Federation was enlarged by the inclusion of bone glue manufacturers and the present name adopted. The main purpose was to perform the usual functions of a trade association, and in particular to co-operate in regard to supplies of raw materials. From 1931 to 1959, there was a collective agreement with tanners for the purchase of raw materials but this has now been abandoned. Membership is limited to manufacturers, of whom there are 21, and to four merchants who collect raw materials. ABCM affiliate. Sardinia House, 52 Lincoln's Inn Fields, Kingsway, London WC2.

Fertiliser Manufacturers' Association (FMA) and
Superphosphate Manufacturers' Association (SMA)

Fertilizer manufacturers founded the Chemical Manure Manufacturers' Association 1875. Name changed to Fertiliser Manufacturers' Association 1904. Incorporated 1919. Present constitution dates from 1948, when the Superphosphate Manufacturers' Association was formed to look after the particular interests of superphosphate makers and to participate in the International Superphosphate Manufacturers' Association. The same staff operates the two organizations.

The FMA watches the interests of the fertilizer industry in general and particularly those of makers of compound fertilizers, in which form three-quarters of the plant nutrients reach the soil. It is not concerned with any form of trading or commercial policy. There are close working arrangements with the National Farmers' Union. Educational publicity scheme in operation. *Fertiliser Statistics* published annually and widely distributed. Restricted circulation for *Fertiliser Information.* FMA membership 51 (three in Eire). SMA membership 16 (three in Eire). SMA ABCM affiliate. 44 Russell Square, London WC1.

The Gelatine & Glue Research Association

Formed 1948, with DSIR support, as the British Gelatine and Glue Research Association, to provide co-operative research facilities in an old-established traditional industry. Name changed 1963. Ordinary and overseas membership limited to companies making gelatine and glue, or machinery or accessories for such manufacture. Users of gelatine or glue are associate members. Membership 55. Somewhat unique in that it has among its members makers and users in Australia, Canada, Eire, Europe, India, South Africa and the U.S.A. Receives grants from the Medical Research Council and the Agricultural Research Council. Income £30,000. Abstracts available to non-members. Headquarters: Sardinia House, 52 Lincoln's Inn Fields, London WC2.

Industrial Pest Control Association (IPCA)

Formed 1942 so that firms concerned with pest control in industrial premises, ships, food stores and homes, could co-operate and take common action when desirable. Maintenance of high standards for the products, methods and appliances a prominent feature. Organized First Pest Control Conference 1963; proceedings available. Associate membership open to overseas firms. Membership 53. Affiliated to ABCM. Cecil Chambers, 86 Strand, London WC2.

National Association of Charcoal Manufacturers

Formed 1942 to promote co-operation between charcoal manufacturers and to represent their interests *vis-à-vis* Government Departments and other organizations. Membership 20. c/o Shirley Aldred & Co. Ltd, Worksop, Notts.

The National Benzole and Allied Products Association (NBA)

Formed 1954 to continue and extend the activities of the National Benzole Association (NBA Ltd), created 1919. Main objects are to promote the interests of members, to organize technical activities, to standardize products, to publish technical information and to maintain contacts with Government Departments and other organizations. Membership, now 57, open to British firms, producing, refining and marketing benzole and allied products obtained from coal carbonization in the U.K. Range of publications on benzole, etc., generally available. It is British element of the International Conference of Benzole Producers for commercial and technical contacts. Affiliated to ABCM. 132–135 Sloane Street, London SW1.

The National Sulphuric Acid Association Ltd (NSAA)

Formed 1919, on recommendation of Ministry of Munitions Departmental Committee, to facilitate co-operation, particularly in the rationalization of the industry after the war-time over-production of sulphuric acid and to safeguard the supply of imported raw materials. Its objects were to provide (a) collective buying arrangements for sulphur, (b) agency shipping services for imported sulphur and pyrites, and (c) a statistical service; further, to help members and encourage new outlets for sulphuric acid. Membership open to any manufacturer of sulphuric acid. The 43 members represent 99% of U.K. acid production. In World War II, the association staff became the nucleus of the Sulphuric Acid Control of the Ministry of Supply. Affiliated to ABCM. Piccadilly House, 33/37 Regent Street, London SW1.

Paintmakers Association of Great Britain Ltd

Formed 1963 by the voluntary amalgamation of the National Paint Federation and the Society of British Paint Manufacturers Ltd, with the object of promoting and protecting the interests of the paint industry covering the manufacture of paint, varnish, lacquer and allied products. More than one class of membership. A full member has to be a manufacturer whose technical staff on research and routine testing represents not less than 5% of total employees. Other than overseas members, the roll of 168 groups and firms represents about 90% of U.K. production. There are seven local sections. There is an annual summer conference. The Association is a member of the European Committee of Paint and Printing Ink Manufacturers' Associations. Booklet, *More Than Meets the Eye*, on science, technology and careers in the industry is generally available. Prudential House, Wellesley Road, Croydon, Surrey.

Phenol Producers' Association

Formed 1927 with the broad objective of promoting the interests of U.K. producers of natural or synthetic phenol. In recent years its activities have been concerned with the collection and dissemination of statistics. Membership 15. c/o Whinney, Smith & Whinney, Yorkshire House, Greek Street, Leeds 1.

The Proprietary Articles Trade Association (PATA)

Formed 1896 to combat the then existing extreme price-cutting of pharmaceutical proprietaries, to ensure a fair remuneration to distributors and to eliminate the prevailing practice of substitution. Since the passing of the Restrictive Trade Practices Act 1956, there have been some changes in its objects. It continues to foster the interests of manufacturers, wholesalers and retailers who believe in price maintenance. Membership of some 500 pharmaceutical makers and wholesalers and 10,000 retail pharmacists. Premier House, 150 Southampton Row, London WC1.

Proprietary Association of Great Britain

Formed 1919 for the makers of proprietary medicines to provide for the normal trade association services, contacts with Government Departments and the general raising of standards. Membership, now 92, is open to makers, distributors and agents in the fields of proprietary medicines, proprietary foods and beverages, antiseptics, disinfectants, germicides and allied preparations, made or marketed in the U.K. and Northern Ireland. 623 Victoria House, Southampton Row, London WC1.

The Research Association of British Paint, Colour and Varnish Manufacturers

This organization, generally called the Paint Research Station, was formed, with DSIR sponsorship, in 1926 to serve not only the paint and varnish industry but also associates, including makers of pigments, synthetic resins, solvents, driers, printing inks, linoleum and vegetable oils. These are the full members. Those concerned with pre-treatment of metals and manufacture of plant and equipment can be associate members. Foreign membership is now permitted. For its subjects its library is the largest in the world. Its monthly *Review* contains 8000 technical abstracts yearly. Its 1964 budget was about £140,000. Membership covers 85% of paint production. Waldegrave Road, Teddington, Middx.

Rubber and Plastics Research Association of Great Britain

Formed October 1919, to provide research and associated services for the Rubber, Plastics and Allied Industries. Membership 340; open to firms in Great Britain and overseas. DSIR supported. Approximate annual income £250,000. Provides Rubber Abstracts for members and on payment to overseas non-members. Shawbury, Shrewsbury, Salop.

Scottish Association of Paint Manufacturers

Formed 1960 by paintmakers in Scotland, in order to promote a public relations campaign to further sales. Members must be active manufacturers and also members of The Paint and Oil Section of The Glasgow Chamber of Commerce. Membership 16. 48 West Regent Street, Glasgow C2.

The Society of British Printing Ink Manufacturers

Formed 1956 by amalgamation of two separate organizations with identical interests and objectives. Their main function is to protect and further the interests of the British printing ink industry. Membership open to U.K. makers of printing ink and accessories, not controlled by a printing firm. The 40 members represent 90% of U.K. printing ink production. Quarterly journal, *The British Ink Maker*, and *The Printing Ink Manual of Technology* generally available. Burley House, 5–11 Theobalds Road, London WC1.

The Society of British Soap Makers

Formed 1954, to promote the interests of the trade manufacturing or processing soap and/or other detergents, designed primarily for washing purposes, and to represent the industry *vis-à-vis* Government Departments and other organizations. Membership 63. 16 Northumberland Avenue, London WC2.

The Surface Coating Synthetic Resin Manufacturers' Association

Formed 1948 for unified action by synthetic resin manufacturers, particularly in negotiations for raw materials. Its activities now include the promotion of trade, the protection of members' interests generally, the dissemination of statistics and, where desirable, the conduct of scientific tests and investigations. Membership is open to all manufacturers of surface coating synthetic resins in Great Britain and Northern Ireland. Membership 19. Their *Surface Coating Resin Index* is purchasable by anyone. 3 The Grove, Ratton, Eastbourne, Sussex.

The United Kingdom Glycerine Producers' Association (UKGPA)

Originally formed after World War I for the orderly disposal of big stocks of Government war-purpose glycerine. Incorporated 1923. Its current objects are to represent the glycerine industry in negotiations with Government Departments and other organizations, and to promote the sale of glycerine. It is not a selling body. Producers of glycerine in the U.K. and Republic of Eire eligible for membership, at present 30. Operated rationing during World War II. Technical publicity bi-monthly bulletin *Glycerine News* and other publications freely available. 45 Portman Square, London W1.

Zinc Pigment Development Association

Formed 1943 with the objects of encouraging internal good relations, facilitating co-operation with Government Departments, promoting the use of zinc pigments (zinc oxide, zinc dust and lithopone) and interchanging information on their uses. Members (now only four) are required to have seven years' commercial production of zinc pigments. Association organizes large-scale exposure tests of pigments. All publications free to bona-fide enquirers. 34 Berkeley Square, London W1.

APPENDIX

SCIENTIFIC, TECHNICAL AND PROFESSIONAL ORGANIZATIONS OF SPECIAL INTEREST TO THE CHEMICAL INDUSTRY

Biochemical Society, The, 20 Park Crescent, London W1.

*British Association of Chemists, The, Hinchley House, 14 Harley Street London W1.

British Colour Council, 13 Portman Square, London W1.

British Pharmacopœia Commission, 44 Hallam Street, London W1.

British Weed Control Council, 95 Wigmore Street, London W1.

Chemical Council, 9/10 Savile Row, London W1.

*Chemical Society, The, Burlington House, Piccadilly, London W1.

European Federation of Chemical Engineering, London Secretariat, Institution of Chemical Engineers, 16 Belgrave Square, London SW1.

Faraday Society, The, 6 Gray's Inn Square, London WC1

Fertiliser Society, The, 44 Russell Square, London WC1.

Institute of Fuel, The, 18 Devonshire Street, Portland Place, London W1.

Institute of Packaging, The, Malcolm House, Empire Way, Wembley Park, Middx.

*Institute of Physics and The Physical Society, The, 47 Belgrave Square, London SW1.

*Institution of Chemical Engineers, The, 16 Belgrave Square, London SW1.

Lancastrian Frankland Society, c/o Nelsons Silk Ltd, Caton Road, Lancaster

National Society for Clean Air, Field House, Bream's Buildings, London EC4.

*Oil and Colour Chemists' Association, The, Wax Chandlers' Hall, Gresham Street, London EC2.

Pharmaceutical Society of Gt. Britain, The, 17 Bloomsbury Square, London WC1.

*Plastics Institute, The, 6 Mandeville Place, London W1.

*Royal Institute of Chemistry, The, 30 Russell Square, London WC1.

Royal Institution of Great Britain, The, 21 Albemarle Street, London W1.

Salters' Company, The, 36 Portland Place, London W1.

Salters' Institute of Industrial Chemistry, 36 Portland Place, London W1.

*Society for Analytical Chemistry, The, 14 Belgrave Square, London SW1.

*Society of Chemical Industry, The, 14 Belgrave Square, London SW1.

Society of Cosmetic Chemists of Great Britain, Lentheric, Lawrence Road, Tottenham, London N15.

*Society of Dyers and Colourists, PO Box 244, Dean House, 19 Piccadilly, Bradford 1.

*Society of Instrument Technology, The, 20 Peel Street, London W8.

Standardisation of Tar Products Test Committee, Oxford Road, Gomersal, nr. Leeds.

---

* These organizations have local sections.

Indexes

Part A: A History of the Modern British Chemical Industry 347

Part B: Companies of Importance in the British Chemical Industry 361

Part C: Trade Associations 377

Part A

Index to A History of the Modern British Chemical Industry

Abel, Frederick 54, 55
Academy of Sciences prize 21
Accum, Frederick 19
Acetate fibre, dyeing of 162
Acetylene 132, 144, 222, 230
Acetylene Illuminating Company 93
Acetyl-salicylic acid 12
Acriflavine 171
Acrilan 223
Acrylonitrile 222, 250
ACTH 178
Adrenaline 177
Advita Ltd 182
African Explosives & Chemical Industries Ltd 138
Agfa AG 70
Air Products (G.B.) Ltd 143
Albright, George 91
Albright & Wilson Ltd 32, 58, 92, 115, 120, 125, 132, 133, 134, 214, 215, 230, 255
Alcock (Peroxide) Ltd 139
Alderley Park, pharmaceutical factory at 171
Alizarin 66, 67, 68, 71
" Alizarin Crisis " 69
Alkali Act (1863) 35–6
Alkali Division (ICI) 121
Alkali waste 28
Alkali works, statistics of 36
" Alkathene " 251
Allen & Hanbury 177
" Alloprene " 235
Alsopps Brewery 67 n.
Alum 45–6, 155
Aluminium Chloride 120
Aluminium Company (Oldbury) 137
Aluminium sulphate 155
American Bakelite Company 116
American Colonies, effect of revolt of 20
6-Aminopenicillanic acid 175
Aminoplastics 198 ff.
Amlwch, production of sulphuric acid at 122
Ammonia 130–1
Ammonia–soda process 83 ff.
Ammonium nitrate 131
Ammonium sulphate 130
Amsterdam, soda and potash prices on market 20
Anaesthetics 179 ff.
Anhydrite, production of sulphuric acid from 123
Aniline 63, 64, 239
Annatto 64
Annual Abstract of Statistics 258
Anthracene 63
Anthraquinone, autoxidation of derivatives of 140
Antibiotics 173 ff.
Antimalarials 171–2
Antiseptic surgery 239
Antrycide 172
" Application factor " 8
Aquitania, S.S. 112
AR chemicals 7
Ardeer, explosives factory at 55
Ardil 212–13

Armstrong, Henry Edward 96
Armstrong, Whitworth & Company 106
Ashcroft, E. A. 90
Ashfield, Lord 103
Asiatic Petroleum Company 99
" Aspirin Age " 12, 252
Associated Ethyl Company 119, 147
Associated Lead Manufacturers Ltd 153–4
Associated Octel Ltd 255
Association of British Chemical Manufacturers 4, 111, 227
Astbury, Prof. W. T. 212
Atebrin (Mepacrine) 172
Athenas 21
Aylsworth, J. W. 197
Ayrshire, bauxitic clay mines in 155

Baekeland, L. H. 189, 196, 264
Bakelite 196
Bakelite Ltd 197
Bakelite Xylonite Ltd 193, 198, 210, 249
Ballistite 55
Banting, F. G. 177
Barger, G. 177
Barilla 16
Barium Chemicals Ltd 140
Barium peroxide 140
Barnard Castle, penicillin production at 174
Barry (S. Wales), manufacture of silicones at 215
BASF 71, 103, 253
Baskerville, John 196
Bayer AG 70, 170
Bayer, Dr. Otto 215
Beddoes, Dr. Thomas 74
Beecham Group 175
Beecham Laboratories 175–6
Beetle (trademark) 199
Beilstein's *Handbuch* 7
Belfast Lough, ICI works at 219, 224
Bell Brothers 85
Bell, Henry 50
Benzene 239
Benzole 70
Berger, Lewis 183
Bergius process 242–3
Berk & Company 183
Berlin airlift 244
Bernard, Claude 176
Best, C. H. 177
Bethell 63
Bevan, E. J. 79, 164, 189
Billingham Works 104, 106, 227, 243, 249
Biologically active chemicals 166 ff.
Bipyridyl 186
Bird, Thomas 88
" Birth pills " 179
Black ash 28
Black, Prof. Joseph 20
Bleaching powder 30, 31, 33
Bleaching Powder Association 42–3
Bleaching powder processes 119
Bleu de Lyons 66
Blinkhorn's chimney 33
" Blue John " 152
Boake, Roberts & Company 203
Bolton, works at 26
Boots Ltd 184
Bordeaux mixture 56, 184
Borden Chemical Company (U.S.) 217
Border Chemicals Ltd 249
Bosnische Elektrizitäts AG 230
Boussingault 59
Bowmans Chemicals Ltd 255
BP Chemicals 7
Bradford Dyers Association 270
Brin, Leon 59
Brins Oxygen Company 59, 142, 143
Britannia violet 66
British Aluminium Company 93, 155
British Association meetings
  Liverpool (1837) 49
  Manchester (1842) 63
  Birmingham (1849) 59
  Bristol (1898) 104
British Celanese Company 162, 248

British Cellulose & Chemical Manufacturing Company 80, 93
British Chrome & Chemicals Ltd 153
British Cyanides Company 62, 199
British Drug Houses Ltd 177, 178, 179
British Dyes Ltd 103
British Dyestuffs Corporation 103, 111, 156, 161, 162
British Dyewood Company 72
British Dynamite Company 55
British Enkalon 208
British Ethyl Corporation 147
British Geon 193, 194, 210
British Hydrocarbon Chemicals Ltd 185, 194, 237, 249
British Hydrocarbon Oil Act 243
British Industrial Plastics Ltd 62, 199
British Industrial Solvents Ltd 131, 236
British Nylon Spinners 207, 208
British Oxygen Company 60, 115, 142, 143, 222
British Petroleum Company 227, 237, 249
British Pharmacopoeia 7
British Resin Products 203, 225
British Titan Products Ltd 153
British Xylonite Company 78
Bromborough, manufacture of butyl alcohol at 237
Bromine 154–5
Brooke, Simpson & Spiller 68
Brotherton & Company 106, 115, 135
Broxil 176
Brunner, J. T. 85
Brunner, Roscoe 106
Brunner, Mond & Company 85, 89, 100, 105, 106, 111, 112, 121, 138, 144, 204
Bryant & May Ltd 92
BTR Industries 221
Buckland, William 49
Buna N 220
Buna S 220
Bunsen, R. W. 32
Burroughs Wellcome 172
Bush, W. J. & Company 255
Butadiene 94, 222, 251
Butakon A 222
Butakon S 222
Butane 244
Buxton Lime Firms 114
BX Plastics 203

Cabot Carbon Ltd 155
Cabot Corporation (U.S.) 155
Calcium Carbide 131–2, 144
Calder Hall, use of carbon dioxide at 146
Caledon Jade Green 161
Calgon 134
Calico Printers' Association 217, 219
Calor gas 244
Cantor Lectures 228
Capital in relation to labour 259
Carbide Industries Ltd 132, 222
Carbolan dyes 161
Carbon black 155
Carbon dioxide 145–6
Carbon disulphide 136, 229, 230
Carbon tetrachloride 230, 232
Carless, Capel & Leonard Ltd 242
Carothers, Wallace Hume 190, 191, 205, 218, 225
Carr, Thomas 25
Carrington, petrochemical plants at 249
Carvès, S. 239
Cassel Cyanide Company 138
Cassel Gold Extracting Company 61
Cassel, Henry Renner 61
Cassel Works 222–3
Castner cyanide process 138, 147
Castner, Hamilton Young 89, 137, 140, 146
Castner–Kellner Alkali Company 89, 101, 102, 141, 143, 230
Castner–Kellner Works
  Haber pilot plant at 106
  production of calcium carbide at 132
Castner–Solvay " long " cell 90

Categories of application of chemicals 8–11
Caustic soda 120
Celanese fibre 80
Celluloid 78
Cellulose & Chemical Company 94
Census of Production 258
Central Statistical Office 258
Chain, Ernst 173
Chambers, William 201
Chance Brothers 26
Chance & Hunt 114
Chance sulphur recovery process 41
Chaptal 24, 25
Chardonnet, de 79
Chemical engineering 95 ff.
Chemstrand Ltd 208, 223
Chibnall, Prof. A. C. 212
Chile nitre 48, 122
Chlorinated ethylene solvents 142
Chlorine industry 117
Chlorine liquefaction 102
Chlorine, use in manufacture of heavy organics 236
Chlorine in World War I 102
Chloro-benzene 229
Chloro-fluoro-methane derivatives 230, 255
Chloro-fluoro refrigerants 233
Chloro-hydrocarbons 227, 229
Chloro-methanes 232
Chloro-naphthalene waxes 235
Chloro-paraffin waxes 233
Chromium chemicals 153
Churchill, Winston 170
CIBA 71, 165, 194, 200, 217
Cibacron dyes 165
Civil War (American) 36
Clarence Salt Works 85
Claus & Company 70, 103
Clayton Aniline Company 71
Clegg, Samuel 62
Cliff, Joseph 47
Coal, heavy organic chemicals from 238–42
Coal, hydrogenation of 244
Coal tar distillation 62–3
Collison, J. 20
Conant 204
Consortium für Electrochemische Industrie 230
Consumption of chemicals 261
Cooper, William 56
Cooper, McDougall & Robertson Ltd 184
Copper oxychloride (fungicide) 184
Copper sulphate 56
Cordite 55
Corning Glass Works (N.Y.) 214
Cortisone 178
Couillet, ammonia–soda plant at 85
Courtaulds Ltd 137, 194, 207, 227
Courtelle 223
Crawford, J. W. C. 201
Creosote oil 63
Creslan 223
Cresols 241
Crookes, Sir William 104
Crosfield, Joseph & Sons 144, 149–50
Cross, C. F. 79, 164, 189, 218
Cruikshank 86
Cyanamid (G.B.) Ltd 182
Cyanamide process 105
Cyanides, manufacture of 60, 137–9
Cytamen 182

Dacron 219
Daily Herald, cartoon in 107
Dale, John 63
Dale & Company 67
Damard Company 196
Davis, George Edward 96, 97
Davy, Humphry 74, 86
DDT 185, 229
Deacon chlorine plants 101, 117
Deacon, Henry 35, 84, 85
Deacon–Hurter process 40, 42
Definitions of chemical industry 4
Degreasing of metals 231
Derby, Lord 35
" Derbyshire spar " 152
Derris 184, 185
Despretz 91
Detergent Age 252
Detergents 252–4

Detergents, foaming caused by 254
Deville–Castner aluminium process 148
Dewar, J. 55
Diaphragm cells 88
Dichloride of ethylene 231
Dichloroethylene 232
Dickson, J. T. 217
Diesbach 60
Diesbach, Henri de 163
Dimethyl phthalate 12
DIMP 12
Dingler's *Polytechnic Journal* 17
Dingley Tariff 44
Diphenylol propane 217
Distillers Company Ltd 115, 131, 143, 145, 193, 194, 224, 227, 236, 238, 249, 250
Distillers Company (Biochemicals) Ltd 182
" Dobane JN " 254
Domagk 170
Dorpat 74
Dow cell 88
Dow Chemical Company 147, 202, 214
Dow–Corning Corporation 214
Dow–Corning Silicones 214
Dow, Herbert 154
Downs, James Cloyd 90
Downs sodium cell 90
Dreyfus, Charles 70
Dreyfus, Henri and Camille 80
" Drikold " 146
Dumfries, manufacture of Ardil at 212
Duncan Flockhart & Company 75
Dundonald, Lord 20, 49
Dunlop Committee 168
Dunlop Company 221, 222
" Duplex " sodium perborate process 142
Du Pont Company 140, 191, 201, 206, 207, 211, 219, 225
" Duroprene " 235
Dyes and organic pigments 156 ff.
Dyestuffs (Import Regulation) Act 108, 111, 156, 167
Dynamite
Dynamos 86
Dynel 223

Eaglescliffe Chemical Company 153
Eastman Kodak 225
Edison, T. A. 197
Ehrlich, Paul 169
Electric Construction Company 91
Electric Reduction Company 92
Electrical energy, introduction of 86 ff.
Electrochemical Company (St. Helens) 101
Electroplating 61
Electrothermal processes 91 ff.
Elliot–Russell revolving furnace 40
Emmerich 173
Epichlorhydrin 217
Epoxy resins 216–17
Erinoid Ltd 80
Esparto grass, bleaching of 33
Esso Petroleum Company 222, 251
Ethyl alcohol 226, 236
Ethyl chloride 233
Ethyl–Dow Company 154
Ethylene 251
Ethylene dibromide 154
Ewins, Dr. Arthur James 170
Explosives in World War I 10
Exports 261
Extrudex Ltd 210

Falmouth Committee 244
Faraday, Michael 54, 75, 86
Farbenfabriken Bayer 225
Fardon 26
Fats, hydrogenation of 144
Fawcett, E. W. 205
Fawley 193
Ferguson, J. 180
Fertilizers
application in U.K. 10
early experiments 49–50
Fire-extinguishing fluids 255
Firms, size of 260–1

Fireproof Celluloid Syndicate 195, 196
Firestone 221
Fischer, Jesse 229
Fischer–Tropsch process 242
Fisons Ltd 115, 184
Fisons Overseas Ltd 263
Fleck, Lord 98
Fleming, Sir Alexander 173
Fletcher, Thomas 59
Florey, Lord Howard 173, 174
Fluon (PTFE) 211
Fluorine 9
Fluorine chemicals 150–2
Fluothane 180
Formica International Ltd 194
Forth Chemicals Ltd 194, 249
Fourcroy 20
Fourneau 170
Freeth, F. A. 101, 204
Frémy, Edmond 152, 253
Fresnel 83
Fry's soapworks 20
Fustic 64

Garbett, S. 16, 17, 20
Gas Council 239, 241
Gas, Light & Coke Company 52
Gay-Lussac 17, 253
Geigy Company 185
General Chemicals Group (ICI) 117, 120, 151
General Electric Company 214, 225
Geography of LeBlanc system 34–5
Gibbs diaphragm cell 101
Gibbs, E. A. 101 n.
Gibson, R. O. 205
Giessen Laboratory 50
Gilbert, Joseph Henry 50
Glass containers, use by pharmaceutical industry 167
Glass manufacture 16
Glauber's salt 74
Glaxo Company 174, 178, 182
Glover, John 17
Gold-mining, use of cyanide in 139
Goodlass Wall & Lead Industries 154
Goodrich, B. F. 209, 210
Goodrich Chemical Company 131
Goodyear 221
Gossage tower 30, 31
Gossage, William 26, 149
Government factories in World War I 99
Gräbe and Liebermann 66
Grace Chemicals 250
Gramoxone 186 n.
Grange Chemicals Ltd 249
Grangemouth 193
Greef, R. W. 154
Griesheim-Elektron 209
Griess, Peter 67, 158
Griffiths, Thomas and William 73
Grison 183
Growth rate of U.K. chemical industry 260
Guano 48
Guncotton 53
Gunpowder 53
Gunpowder engine 9
Gutta-percha (synthetic) 222

Haber ammonia process 81, 104, 106, 130
Haber, Fritz 103
Hall and Hérault aluminium processes 146
Hall, John & Sons 54
Hall, M. 140
Hancock, Thomas 77
Handbook of Chemical Engineering 97
Hargreaves–Bird cell 88, 101
Hargreaves, James 40, 88
Hargreaves–Robinson process 40
Harington 177
Hayle (Cornwall), manufacture of sulphuric acid at 122
Heavy inorganic chemicals, definition of 116
Hecogenin 178
Hedley, Thomas & Company 253
Hedon Chemicals Ltd 224
Héroult 91, 140
Hervey, Robert 47

Heworth Alkali Works 73
Hexachloroethane 232
Heylandt process 143
Hibitane 172
Hickson & Welch Ltd 255
Hill, William 18
Hitler, Adolf 220
Hofmann, A. W. 50, 63, 65, 188, 202
Hofmann's violet 66
Home, Francis 6, 49
Hooke, Robert 189
Hooker cell 88, 119
Hopkins, Gowland 182
Hormones 176 ff.
Houdry process 150
Hull Distillery Company Ltd 236
Humbolt 48
Hurter, Ferdinand 32, 41
Hutchinson, John 35
Huygens 9
Hydrazobenzene, autoxidation of 140
Hydrochloric acid 31, 33, 127–8
Hydrocyanic acid 61, 138, 147, 222
Hydrofluoric acid 151
Hydrogen peroxide 9, 60, 139

ICI 112, 113–15, 139, 147, 152, 154, 157, 165, 180, 184, 185, 193, 201, 204, 208, 209, 211, 212, 215, 231, 233, 235, 243, 250, 252, 255, 263
ICI–De Beers Company 138
ICI Fibres Division 219
ICI Fibres Ltd 219
ICI Pharmaceuticals Division 170
Ideal Manufacturing Company 197
IG Farbenindustrie 156, 164 n., 243
Imperial Automatic Light Company 93
Imperial College of Science and Technology 163
Imperial Smelting Corporation 140, 152, 255
Import Duties Act 111
Imports 262
Indigo 64, 71
Industrial gases 142 ff.
Industrial Revolution 9, 15, 16, 19, 27, 53, 68
Institution of Chemical Engineers 98
Insulin 177
International Synthetic Rubber Company 194, 222, 250, 251
Ionamine dyes 162
Islets of Langerhans 177
Isotactic polymers 224

Jealotts Hill 186
John, Hans 198
Johnson, Dr. Samuel 16 n.
Johnson, Matthey & Company 72, 106

Keir, James 19, 22
Kelp 16
Kemball, Bishop & Company 255
Kenfig, production of carbide at 131
Key Industry Duty 92
Kinetic Chemicals 211
King's Lynn, manufacture of butyl alcohol at 237
Kingzett, C. T. 37, 135
Kipping, Prof. F. S. 213
Kirkby, manufacture of polyphosphates at 134
Klatte and Rollet vinyl chloride process 209
Knietsch, Rudolph 102
Koller, Gustav 231
Kroll, W. J. 148
Kühne, Dr. 125
Kuhlmann–Ostwald nitric acid process 103
Kurtz, Andrew 73

Labour force 259
Lancashire Chemical Works Ltd 153
Langer, Otto 146
Laporte, Bernard 60, 139
Laporte Industries Ltd 139

Laporte Ltd 154
Laurent 235
Lawes Chemical Manure Company 52
Lawes, John Bennett 50, 52
Lead chromate 73
Lebach, Dr. H. 196
LeBlanc, Nicolas 21 ff.
LeBlanc process 24 ff., 30, 82
LeBlanc system 28
Leicester, Lovell & Company 217
Leigh 63
Leming's gas purification process 18
Leprince & Siveke 144
Leven (Fife), cyanide works at 94
Levinstein, Ivan 67, 69, 70, 83, 103
Lever Brothers 144
Leverkusen 125
Lewis Berger & Sons 73
Lewis, W. K. 97
Lewis, William 6
Liebig, Justus von 49, 74
Light-fastness, search for 159
Lilly, Eli 182
Lime burning 153
Lime-sulphur 183
Lindemann 216
Linnaeus 183 n.
Linstead, R. P. 163
"Lissapol N" 251, 253
Lister, Lord 239
Lithopone 73
Little, Arthur D. 96
"Liverpool prices" 37
London College of Hygiene and Tropical Medicine 173
London Orange 69
London Passenger Transport Board 210
Losh, William 20, 24, 25
Losh & Company 24, 25
Louis XVI 21
Low 173
Lucifer matches 58
Luft, Adolf 195
Lunar Society 19, 22
Lundström Brothers 59
Lunge, George 41
Luton, hydrogen peroxide production at 60
Lydd, testing explosive at 55
"Lysozyme" 173

MacArthur, John Stewart 61
Macintosh, Charles 30, 47, 49, 60, 62, 73
Madder 64, 67
Magenta 66
Magnesium, manufacture in U.K. 147
Malherbe 21
Manchester Brown 67
Manchester Technical School 96
Mansfield, Charles B. 63
Marbon Chemicals Ltd 250
Marchon Products Ltd 125, 134
Margraff 151
Mark and Wulff styrene process 202
Marseilles, soap industry at 23
Massachusetts Institute of Technology (MIT) 96, 181
Mathieson ammonia–soda process 86
Mauve, Perkin's 65, 159
Max Planck Institute 207
May & Baker Ltd 170
Maydown, production of carbide at 132
"M & B" 170
McGowan, Sir Harry 112, 113
McDougall Brothers 183
McDougall, William 184
McKechnie Brothers, manufacture of titanium by 148
McKinley Tariff 43, 44
MCPA 185, 186
Meister, Lucius & Bruning 71
Melamine 200
Meldola, R. 103
Mellon Institute 214
Mercury cells 88
Mergers 111 ff.
Merriam, L. P. 78
Mersey Chemical Works 135
Mersey White Lead Company 154

Métherie, J. C. de la 22, 23
Methylating Company Ltd 236
Methyl chloride 233
Methyl methacrylate 201
Methyl violet 66
Michelin Company 194
Michels, Prof. A. 205
Midland Bank 112
Midland Silicones Ltd 194, 215
Milan Polytechnic Institute 223
Miles, G. W. 79
Millardet 56
Miller & Company 51
Miller, John 52
Mills–Packard chamber 122, 126
Ministry of Housing & Local Government 254
Mitin FF 185
Moissan 93
"Monastral" pigments 164
Monceau, Duhamel du 19
Mond, Ludwig 41, 84, 85, 121, 146
Mond Nickel Company 147
Mond, Sir Alfred 107, 112, 113
Monsanto Company 194, 203, 223, 251
Morgan, Sir G. T. 228
Morris Motors 210
Morson, Thomas 75
Morton, Sir James 160
Mouldensite Ltd 197
Moulton, Lord 105
Murdoch, William 62
Murexide 65
Murgatroyd Salt & Chemical Company 119, 143
Murray, James 50
Muspratt, James 25, 26, 36, 49, 60, 73, 84
Muspratt, James Sheridan 50
Muspratt, Richard 50
Muspratt's chimney 33
Mustard gas 102

Naphtha 63
Naphthalene 63, 241, 251
National Benzole Association 242
National Health Service 168, 179
National Smelting Corporation 115
National Sulphuric Acid Association 124
Natta, Giulio 223
Natural gas, sulphur recovery from 126
Naugatuck Chemical Company 202
Naunton, W. J. S. 221
Nekal A 253
Neoprene 221, 222
Niagara Electrochemical Company 90
Nicholson, electrolysis of water by 86
Nicholson, William 24
Nieuwland, Father 221
Nitre cake 48
Nitre pots 99
Nitric acid 47–8, 129
Nitrogen, industrial uses of 143
Nitrogen Products Committee 104
Nitro-glycerine 55
Nobel, Alfred 54
Nobel Explosives 100
Nobel Industries 111, 112
Normann, Dr. W. 144
North British Rubber Company 221
Nylon 163, 187, 191, 207–8, 223

Octel Co Ltd, The Associated 122, 233
OECD 256
Oldbury Alkali Company 44, 62
Oleum 99
Onnes, Prof. Kamerlingh 205
Oppau, ammonia works at 105
Oppau Commissioners 107
Orbenin 176
Orlon 223
Orr, J. B. 73
Ortega y Gasset 12, 252
Ostelin 182
Ostromislensky 209
Ostwald nitric acid process 129
Output index 260

Overseas activities of U.K. chemical industry 262
Oxo-process 247, 248
Oxygen 59–60
Oxygen grid in Sheffield 143

Packard & Company 51
Packard, Edward 50
Paludrine 172
Pan Fibre 223
Paraffins 9
Paraquat 186
Parke, Davis & Company 168, 177
Parker, Mr. Justice 71
Parkes, Alexander 54, 77
Parkesine 77
Parkesine Company 78
Patent Law, German 68
Pathé Frères 80
Pattinson's White 73
Penbritin 176
Penicillin 173, 175
Penicillinase 175
Penicillium notatum 173, 174
Pentachloroethane 232
Pentothal 179
Perchloroethylene 232
Perkin & Sons 66, 67, 103
Perkin, William Henry 65
Perkin's Green 66
Peroxygen chemicals 139 ff.
"Persil" 149
Perspex 139, 201
Pest-control agents 183 ff.
Pest Control Ltd 184
Petersen tower process 126
Petrochemicals 245 ff.
"Petrol", origin of term 242
Petrol, synthetic 242
Petroleum as source of organic raw materials 192
Pfizer Ltd 250
Pharmaceutical industry 74–6, 167, 168
Pharmaceutical Society, formation of 75
Pharmaceuticals, synthetic 169 ff.
Pharmacy Acts 75
Phenol 63, 241
Phenol-formaldehyde 195 ff.
Philblack Ltd 155
Phillips, Dr. M. A. 170
Philips Petroleum Company 207
Phillips, Peregrine 17
Phosgene 102, 120
Phosphate rock 59
Phosphorus 58–9
Phosphorus chemicals 132–5
Phosphorus Company, the 91
Phosphorus sesquisulphide 92, 230
"Phossy jaw" 92
Phthalic anhydride 157
Phthalocyanine colours 164
Picric acid 55, 65, 229
Pigments 72–4
Pirelli Company 194, 221
Plant Protection Ltd 184, 186
Plasmoquin 172
Plastics, U.K. output of 1962 11
Platforming 247
Pneumatic Institute, the 74
Pointer, J. & Son 51
Polyacrylonitrile 222–3
Polycarbonates 225
Polyethylene 191, 203 ff., 206
Polyformaldehydes 225
Polypropylene 223–4
Polystyrene 202–3
Polytetrafluoroethylene (PTFE) 210–12
Polyurethanes 215–16
Polyvinyl acetate 94
Polyvinyl chloride 94, 191, 208–10
Portishead, manufacture of phosphorus at 134
Potassium chlorate 120
Potassium ferrocyanide 61
Potassium fertilizers 52
Price, Sir Keith 105
Priestley, Joseph 145
Process, definition of 5
Procion dyes 165
Productivity in U.K. chemical industry 259
Proguanil 172

Prontosil 170
Propane 244
Propathane 224
Protective tariffs, introduction of 108 ff.
Prussian Blue 73
Pyrethrum 184

Quassia, extract of 183
Quinol 63

Raistrick, Prof. 173
Rare gases 143
Raschen cyanide process 62, 136–7, 229
Raschig 198
Reactive dyes 164–5
Read Holliday & Sons 55, 63, 69, 103
Readman, Parker & Robinson's phosphorus process 91
Redman, Dr. L. V. 197
Redmanol Ltd 197
Refrigerants 255
  use of hydrofluoric acid in manufacture of 152
Reglone 186 n.
Regnault 208
Reppe, Dr. Walther 7
Research 263
"Rinso" 149
Ripper, Dr. 184
Robinson, Sir Robert 178, 205
Robinson, Thomas 40
Roebuck, Dr. John 16, 17, 20
Roessler & Hasslacher 141
Röhm & Hass 201
Roper, Robert 30
Roscoe, Prof. Sir Henry E. 42
Rose, F. L. 172
Rossiter, E. C. 199
Royal College of Chemistry 50, 67, 158
Rubber
  chlorination of 235
  natural 76
  synthetic 250
Runcorn
  manufacture of sodium peroxide at 141
  manufacture of vinyl chloride at 209
  soapery at 26

Safeguarding of Industries Act (1921) 110, 167
Safety Celluloid Company 80
Salt, electrolysis of 88, 90
Saltcake 28, 30
Salt tax, rebate of 21
Salvarsan 169, 170
Sanger, F. 177
Sankey, Mr. Justice 108
Saturn Gases Ltd 143
Scheele 58, 151
Schlosing–Rolland ammonia–soda process 85
Schönbein, C. F. 53, 54, 77
Schrötter, Anton 59
Scottish Cyanide Company 94
Schutze, Victor 80
Schutzenberger 79
Scottish Dyes Ltd 103, 160, 161
Secrosteron 179
Selly Oak, phosphorus factory at 58
Semet–Solway coking ovens 229
Shawinigan Chemicals 224
Shell Chemical Company 194, 203, 242, 248, 249, 253, 254
Shell Chemical Corporation (U.S.) 217
Shell Haven 254
Shipley, hydrogen peroxide plant at 60
Shirley Cotton Institute 184
Silica catalyst 150
Silicon chemicals 149–50
Silicones 213–15
  use-pattern of 215
Simon, E. 202
Simpson, J. Y. 75
Smith, John, alkali-maker, Great Lever 26
Smith, Sydney 27

Soap, hard, makers in U.K. 15
Sobrero, Ascanio 53
Soda (sodium carbonate) 30
Söderberg electrode 94
Sodium chlorate 120
Sodium cyanide 139
Sodium hyposulphite 135
Sodium (metal) 88, 147
Sodium perborate 139, 141
Sodium percarbonate 139, 142
Sodium peroxide 139
Sodium sulphate 135
Sodium sulphide 136
Sodium thiosulphate 135
Solvay & Company 89
Solvay process 42, 44, 81, 100, 121
Solway Chemicals Ltd 134
Solway Dyes Ltd 160
South Africa, gold-mining interests in 138
South Eastern Railway 53
Spence, Peter 47
Spence, Peter & Sons Ltd 155
Spill, Daniel 78
Spill, Daniel & Company 78
St. Rollox Works 25, 31
Standard Oil Company 207, 220, 247
Stanlow
  manufacture of polyurethanes at 217
  petrochemical plants at 249
Staphylococcus aureus 175
Statistics 257 ff.
Staudinger, Hermann 190, 202
Staveley Iron & Chemical Company 99, 143
Stearn, C. H. 78
Steigenberger, Louis 73
" Stergene " 253
Stilboestrol 178
Stormont Castle Government 223
Story, W. H. 196
Sturge, Edmund 58
Sturge, J. & E., Ltd 255
Styrene 190
Sulphanilamide 170
Sulphur, chlorides of 120
Sulphur dioxide 16
Sulphur Export Corporation 124
Sulphur, sources of 123, 124
Sulphuric acid 16 ff., 30, 99, 122ff., 123, 126
Superphosphate 51, 126–7
Swan, Joseph 195
Swan, J. W. 78
Swinburne, Sir James 195
Synthetic Ammonia & Nitrates Ltd 106, 107, 114
Syrolite Company 80

Tar
  British production (1870, 1880) 70
  distillation of 238 ff.
Tariff Act (American) 36
" Teepol " 253
TEL (tetraethyl lead) 12, 119, 147, 233
Telegraph Construction & Maintenance Company 206
Tennant & Company 51
Tennant, Charles 25, 27 n., 30, 36, 61
Tennant, Clow & Company 81
Tennant's " stalk " 33
Terephthalic acid 218, 219
" Terylene " 162, 163, 217–19, 223, 251
Tetrachloroethane 232
Tetryl 55
Thom, John 84
Thornhill, carbide plant at 93
Titanium 148
Titanium oxide 153–4
TNT (trinitro-toluene) 55, 99, 100
Tootal, Broadhurst Lee & Company 190
Topham, F. 78
Townsend's chimney 33
Trade Accounts of U.K. 258
Trade and Navigation Accounts 258
Trichloroethylene 180, 232
Trilene 180
Turnbull & Ramsay 84
Turner & Newall Group 200

Turpin 55
Tyne, LeBlanc factories on 37

Udhe (mercury) cells 119
Ugine (France) 250
Unilever 253
Union Carbide 251
Union Carbide Ltd 197
Union Carbide Corporation 193, 206
Unit operations 97
United Alkali Company 43 ff., 62, 89, 93, 95, 99, 101, 111, 112, 135, 229, 233
United Dyewood Corporation 72
United Sulphuric Acid Corporation 125
Uranium hexafluoride 151
Urea-formaldehyde resins 190, 198

Vauxhall Road (Liverpool) alkali works 26
Verel 223
Versailles, Treaty of 108
Vicara fibre 213
Vickers & Company 106
" Victane " 244
Vinyl acetate 224
Violet, Imperial 66
Viscose 79
Vitamin B4 182
Vitamin B12 182
Vitamins 181 ff.
Volumetric methods 6
Voss, Walter 183
Vulcanization 77

Waksman, Dr. Selman A. 174
Waldie 75
Wallsend-on-Tyne, sodium works at 138
Waltham Abbey 56
Ward, Joshua 16
Watt, James 20, 30
Weed-control agents 183 ff.
Weizman, C. 249
Welding, use of acetylene in 145
Weldon process 31 ff., 42, 101, 117
Weston Chemical Company 231
Westrosol 231
Whinfield, J. R. 217
Whitby, alum production at 47
Whitby, Prof. (Sir) Lionel 170
White, J. & J. Ltd 153
White lead 72, 153–4
Whitehaven, anhydrite at 125
Widnes 56, 186, 230, 235
Williams, Thomas & Dower 69
Willson, Thomas Leopold 92
Wilson, John Edward 58
Wilton Works 149, 219, 224, 249, 250
Withering, Dr. William 75
Woolwich 56
World War I, primary effects of 98 ff.

Xylonite Company 78

Yalding Manufacturing Company 183
Young, W. 239

Zefran 223
Ziegler, K. 207, 223, 249

Part B

Index of Companies of Importance in the British Chemical Industry

AB Insulin Ltd 274
A.B.M. Industrial Products Ltd 268
ACC (Chrome & Chemicals) Ltd, Canada 272
ACC (Fertilisers) Ltd 272
Abbey Chemicals Ltd 277
Abbott Alkaloidal Company 268
Abbott Laboratories Ltd 268
Acquitaine Fisons Ltd 329
Advita Ltd 324
Agfa 312
Agricare Products 313
Agricola Chemicals Ltd 308
Agricola Plant Protecting Chemicals Ltd 308
Agricultural Food Products Ltd 326
Air Products Ltd 268
Air Products (Great Britain) Ltd 269
Air Products and Chemicals Incorporated, U.S.A. 268
Albion Winding Co. Ltd 317
Albright & Wilson (Australia) Pty. Ltd 269, 270
Albright & Wilson Group (A. & W. Ltd) 269, 270, 273
Albright & Wilson (Ireland) Ltd 269, 270
Albright & Wilson (Mfg.) Ltd 269, 270
Albright & Wilson Match Phosphorus Co. Ltd 270
Albright & Wilson (Overseas Developments) Ltd 270
Alchemy Ltd 282
Alcock (Peroxide) Ltd 271
Alginate Industries Ltd 271
Alginate Industries (Ireland) Ltd 271
Alginate Industries (Scotland) Ltd 271
Allen Chlorophyll Co. Ltd 270
Allen, Frederick & Sons (Chemicals) Ltd 271
Allen & Hanburys Ltd 274, 294
Allen, Stafford, & Sons Ltd 269, 270
Alliance Feeds Ltd 272
Allied Chemical & Dye Corporation, U.S.A. 299
Allied Paints & Chemicals Ltd 273
Alumina Chemicals Ltd (Dublin) 271
Alumina Co. Ltd 271, 301, 325
Aluminium Company Ltd 299
Aluminium Sulphate Ltd 271, 301
Amalgamated Oxides (1939) Ltd 271
Amchem, U.S.A. 283
American Cyanamid Co. 287, 293
American Diversey Corporation 290
American Ethyl Gasoline Corporation 273
American Home Products Corporation 328
American Union Carbide Company 324
American Viscose Corporation 286
Ammonia Soda Co. 299
Anchor Chemical Co. Ltd 272
Anderson, James, & Co. (Colours) Ltd 293

Anglo-American Oil Co. 291
Anglo-American Plastics Ltd 285
Anglo-Dutch Margarine Union 323
Anglo-Iranian Oil Co. Ltd 274, 280
Anglo-Persian Oil Co. Ltd 274
Arabol Manufacturing Co. Ltd 307
Armour & Co. (U.S.) 318
Ashburton Chemical Works Ltd 293
Ashby, Morris, Ltd 271
Asprey, George, & Son Ltd 279
Aspro–Nicholas Group 296
Associated Adhesives Ltd 324
Associated British Cellulose Ltd 273
Associated British Maltsters Ltd 268
Associated Chemical Companies Ltd 270, 272
Associated Chemical Companies (Mfg.) Ltd 272
Associated Chemical Companies (Sales) Ltd 272
Associated Ethyl Company Ltd 273
Associated Fumigators Ltd 316
Associated Lead Manufacturers Ltd 295
Associated Octel Co. Ltd 273, 274
Associated Phosphate Manufacturers Ltd 305
Atlas Chemical Industries, U.S.A. 289
Ault & Wiborg Company, U.S.A. 273
Ault & Wiborg Ltd 273
Avon 302

BDH Group Ltd 273, 274, 294
BDH (Knights) Ltd 274
BDH (London Wholesale) Ltd 274
BDH (Middletons) Ltd 274
BDH (Nutritionals) Ltd 274
BDH (Woolley & Arnfield) Ltd 274
BIP Chemicals Ltd 323
BIP Engineering Ltd 323
BIP Gaydon Ltd 323
BIP Group 323
BIP Reinforced Products Ltd 323
BIP Tools Ltd 323
BNR Co. Ltd 283
BP California Ltd 274
BP Chemical Company Ltd 274, 275
BP Plastics 274, 275, 309
BTR Industries Ltd 302
BX Plastics Ltd 275, 289
Baird & Tatlock Group 299
Baird & Tatlock (London) Ltd 298
Bakelite Ltd 275, 324
Bakelite Xylonite Ltd (BXL) 275, 289, 290, 324, 329
Barium Chemicals Ltd 276, 301, 306
Barnoldby Crop Driers Ltd 312
Barron, Harveys & Co. 273
Barter Trading Corporation 288
Bayer Co. 282
Baywood Chemicals Ltd 282
Beadel, James, & Co. Ltd 315
Beck, Koller & Co. (England) Ltd 315
Becks, A. T., & Co. Ltd 295
Beecham Estates & Pills Ltd 276
Beecham Group Ltd 276
Beecham Research Laboratories Ltd 276
Beecham's Pills Ltd 276
Beetle Bond, Ltd 323
Beetle Products Co. Ltd 323
Bell's Asbestos & Engineering (Holdings) Ltd 283
Benbows Dog Mixture Co. 293
Benger Laboratories Ltd 292
Berger, Jenson & Nicholson Ltd 276
Berger, Lewis, & Sons Ltd 276, 318
Berger Traffic Markings Ltd 276
Berk Exothermic Ltd 277
Berk, F. W., & Co. Ltd 276, 278
Berk Leiner Ltd 277
Berk Ltd 276, 277
Berk Pharmaceuticals Ltd 277
Berk Spencer Acids Ltd 277, 278
Berlin Aniline Co. 272
Bexford Ltd 275
Bickford, Smith & Co. Ltd 300
Bierley Chemical Co. Ltd 305
Birmingham Chemical Co. Ltd 324
Block & Anderson Ltd 312
Blundell, Spence & Co. Ltd 277

Blundell–Permoglaze Ltd 277
Blythe Colour Works Ltd 303
Blythe, William, & Co. Ltd 277
Boake, Roberts A., & Co. Ltd 270
Boake, Roberts A. (Holding) Ltd 269, 270
Bohme, H. Th., Germany 298
Bolton, Thomas, & Sons Ltd 325
Boot, Jesse, & Co. Ltd 277
Booth & Co. (International) Ltd 280
Boots Pure Drug Co. Ltd 277
Borax Consolidated Ltd 278
Borax (Holdings) Ltd 277, 278
Borden Chemical Co., U.S.A. 306, 307
Border Chemicals Ltd 274, 280, 290, 301
Boroquimica Ltd 278
Bottomley, J. C., & Emerson Ltd 278
Bowmans Chemicals Ltd 278, 306
Bowmans (Warrington) Ltd 278
Bowring, C. T., & Co. (Fish Oils) Ltd 313
Boyd Stewart & Co. 291
Bradford Dyers Association Ltd 270
Bradley & Bliss Ltd 274
Bridge Chemicals Ltd 319
Bridge Colour Co. 329
Brifex Ltd 328
Briggs & Townsend Ltd 281
Brins Oxygen Co. Ltd 280
Bristol (Tar & Chemicals) Trading Ltd 279
Bristol & West Tar Distillers Ltd 278
British Alizarine Co. Ltd 300
British Alkaloids Ltd 313
British Aluminium Co. Ltd 271, 301
British Celanese Ltd 286, 290
British Cellophane Ltd 286
British Cellulose & Chemical Manufacturing Co. Ltd 286
British Chrome & Chemicals Ltd 272
British Cod Liver Oil Producers (Hull) Ltd 308
British Cod Liver Oils (Hull & Grimsby) Ltd 308
British Colloids Ltd 287
British Cyanide Co. 323
British DiaMalt Co. 268
British Disinfectant Co. 302
British Drug Houses Ltd 273, 274, 298
British Dyes Ltd 298, 300
British Dyestuffs Corporation Ltd 298, 299, 300
British Dyewood & Chemical Co. Ltd 279
British Dyewood Co. Ltd 279
British Dynamite Co. 300
British Electro Metallurgical Co. Ltd 275, 324
British Enka Ltd 286
British Ethyl Corporation Ltd 273
British Gelatine Co. Ltd 279, 280
British Geon Ltd 290
British Glues & Chemicals Ltd 279
British Hydrocarbon Chemicals Ltd (BHC) 274, 280, 288, 289, 290
British Industrial Plastics Ltd 323
British Industrial Solvents Ltd 288
British Match Corporation 312
British Nicotine Co. Ltd 280
British Nylon Spinners Ltd 286, 300
British Oxygen Chemicals Ltd 280
British Oxygen Co. Ltd 280
British Paints (Holdings) Ltd 281
British Petroleum Chemicals Ltd 280
British Petroleum Co. Ltd 273, 274, 275, 280, 288, 301, 309
British Resin Products Ltd 289, 290
British Tar Products Ltd 281
British Titan Products Co. Ltd (BTP) 281, 301
British Visqueen Ltd 301
British Xylonite Co. Ltd (BX) 275, 289, 329
Brock's Fireworks Ltd 281
Bromine Compounds Ltd, Israel 277
Brook, A. W., Ltd 306
Brook, Simpson & Spiller 297, 300
Brotherton & Co. Ltd 272, 304
Brown, James M., Ltd 321, 322
Brown & Sons 291
Brunner Mond & Co. Ltd 299
Bryant & May Ltd 270, 312

Bullough Securities Ltd 279
Burroughs, Wellcome & Co. 326, 327
Burt Boulton & Haywood Ltd 282, 300, 320
Burton, Wm., & Sons 305
Burton, Wm., & Sons (Bethnal Green) Ltd 305
Burts & Harvey Ltd 282, 320
Bush, Boake, Allen Ltd 269, 270
Bush Hellas Fruit Industries, S.A., Greece 269
Bush, W. J., & Co. Ltd 269, 270
Bush, W. J. & Co. (Nigeria) Ltd 269
Butler Chemicals Ltd 282
Butler, Wm., & Co. (Bristol) Ltd 278, 282
Butterley Co. Ltd 268, 269
Buxton Lime Firms Ltd 299

Cabot Carbon Ltd 282
Cabot Corporation, Boston 282
Calder & Mersey Extract Co. Ltd 292
Calfos Ltd 279
California Chemical Co. 274, 280, 289
Caltex 273, 274
Campbell, Rex, & Co. Ltd 283
Canadian Copper Co. 302
Cannon, B., & Co. Ltd 280
Carless, Capel & Leonard Ltd 282
Carnegie Bros. 325
Carnegie Chemicals (Welwyn) Ltd 325
Carnegie Organics Ltd 325
Carnegie Quinine Works Ltd 325
Carnegies of Welwyn Ltd 326
Carson, Walter, & Sons Ltd 283
Carson–Paripan Ltd 283
Carter, Gerald, Ltd 308
Cascelloid of Leicester 275
Casein Co. of America, Inc. 306
Cassel Cyanide Co. 299, 300
Cassel Gold Extracting Co. Ltd 299
Castner–Kellner Alkali Co. Ltd 299, 300
Catalin Corporation of America 283
Catalin Ltd 283
Cefoil Ltd 271
Celanese Corporation of America 281
Central Oil Refining Co. Ltd 328
Century Chemical Co. Ltd 283
Cephos Ltd 276
Chafer, J. W., Ltd 283
Chafer, J. W. (Scotland) Ltd 283
Champion Druce Ltd 295
Chance Bros. & Co. 299
Chance & Hunt 299, 300
Chapman, Messel & Co. 278
Charterhouse Group of Companies 317
Chattaway, W., Ltd 314
Chelsea Insecticides Ltd 316
Chemical Compounds Ltd 283
Chemical Supply Co. Ltd 283
Chemical Utilities Ltd 284
Chemstrand Ltd 284
Chevron Chemical Co. (California Chemical Co.) 289
Christopherson, Clifford, & Co. Ltd 269, 270
CIBA (ARL) Ltd 284
CIBA (Basle) 284
CIBA Chemicals Ltd 284
CIBA Clayton Ltd 284
CIBA Foundation 284
CIBA Laboratories Ltd 284
CIBA United Kingdom Ltd 284
Clark & Pinkerton 294
Claus & Co. Ltd 300
Claus & Rée 300
Clayton Aniline Co. Ltd 284, 293, 325
Cleckheaton Chemical Co. Ltd 305
Cleveland Product Co. 279
Coalite & Chemical Products Ltd 284
Coalite Oils & Chemicals Ltd 285
Coates Brothers & Company Ltd 285
Cocker Chemical Co. Ltd 315
Coke Oven Products Ltd 325
Cole, R. H., & Co. Ltd 309
Colgate–Palmolive Ltd 285

Collett, J. M., & Co. Ltd 268
Commercial Plastics Ltd 285
Commercial Plastics Industries Ltd 285
Commercial Solvents (G.B.) Ltd 288
Commercial Solvents Corporation (US) 288
Consolidated Zinc Corporation 322
Constance & Lintox 276
Cookson Lead & Antimony Co. Ltd 295
Cooper, McDougall & Robertson Ltd 301, 327
Cooper & Nephews 327
Cooper Technical Bureau 327
Co-operative Wholesale Society Ltd 285
Cory, Horace, & Co. Ltd 286
Cotopa Ltd 317
Coty (England) Ltd 313
Courtauld, Samuel, & Co. 286
Courtaulds Ltd 286, 290, 300, 325
Cowan Bros. (Stratford) Ltd 303
Cray Valley Products Ltd 285
Croda Ltd 287
Croda Organization Ltd 287
Croid Ltd 279, 280
Crookes Collosols Ltd 287
Crookes Laboratories Ltd 287, 313
Crosfield, Joseph, & Sons Ltd 323
Crowther & Graesser 296
Croydon Mouldrite Ltd 300
Cruickshank, R., Ltd 292
Cruickshank, R. (Cellulose) Ltd 307
Crystal-Brite Products Ltd 307
Cumbria Trading Co. Ltd 270
Cunningham, J. & J., Ltd 317
Cupola Mining & Milling Co. Ltd 305
Cuprinol Ltd 276
Curling, Geo., Wyman & Co. 273
Curtis's & Harvey Ltd 300
Cussons Group Ltd 287
Cussons (International) Ltd 287
Cussons, Sons & Co. Ltd 287
Cyanamid of Great Britain Ltd 287
Cyanamid Products Ltd 287
D. R. Syndicate Ltd 321
Dales Chemicals Ltd 306
Damard Lacquer Co. Ltd 275
Dampney, J., & Co. 281
Darton Manufacturing Co. Ltd 275
Dawson & Emerson 278
Dawson, James E. 304
Dawson, John, & Co. 279
De La Rue Group 293
Dean, F. J., Ltd 303
Decro Ltd 311
Demuth, Lewis, & Co. 308
Dennis, L., & Co. Ltd 288
Dental Manufacturing Co. Ltd 321
Deosan Ltd 290
Derby Chemical Colour Co. 300
Derbyshire Coalite Co. Ltd 285
Derbyshire Stone Group of Companies 299
Destree, SA Usines 315
Detarex Ltd 277
Dewy & Almy Ltd 295
Dickinson & Son 315
Diggory & Co. Ltd 304
Dinneford & Co. Ltd 276
Disinfestation Ltd 316
Dista Products Ltd 307
Distillers Co. Ltd 274, 275, 280, 286, 288, 301, 310, 324
Distillers Co. (Biochemicals) Ltd 290, 307
Distillers Plastic Services Ltd 290
Distrene Ltd 289, 290
Distugil (France) 290
Diversey (U.K.) Ltd 290
Domestos Ltd 324
Don Alum Ltd 271, 305
Doncaster Coalite Ltd 285
Dow Chemicals Ltd 301
Dow Chemicals, U.S.A. 283
Dow Chemie International AG, Zurich 322
Dow International 289, 290
Dow–Corning Corporation 270
Du Pont Company (United Kingdom) Ltd 290
du Pont de Nemours, E. I. & Co., U.S.A. 290, 299

Du-Bois Chemicals, U.S.A. 289
Dudleys (Redditch) Ltd 297
Dunlop Chemical Products Division 290
Dunlop Rubber Co. Ltd 290, 302
Duphar Ltd 313
Duramis Fuels Ltd 285
Durastic Ltd 295
Durham & Bonny Ltd 291
Durham Chemical Group Ltd 291
Durham Chemicals Ltd 291
Durham Raw Materials Ltd 291
Dutch State Mines 292
Duval, Claude, Ltd 274

EF Co. Ltd 286
Eaglescliffe Chemical Co. Ltd 272
Edinburgh Pharmaceutical Industries Ltd (EPI Group) 294
Edwardson, E. D., & Co. (Ware) Ltd 307
Elanco Products Ltd 307
Electric Reduction Co. of Canada Ltd 269, 270
Electro-Bleach & By-Products Co. 299
Electropol, Ltd 270
Eley Brothers Ltd 300
Ellison, Henry, Ltd 328
Ellison, Henry (Glasgow) Ltd 317
Eno Proprietaries Ltd 276
Erinoid Ltd 274, 309, 318
Esso Chemical Ltd 291
Esso Petroleum Company Ltd 291
Ethicon Ltd 311
Ethyl Export Corporation 273
Evans & Askin 302
Evans Medical Ltd 294
Evans Medical Supplies Ltd 294
Evans, Norman & Rais Ltd 268
Evans, Sons, Lescher & Webb Ltd 294
Eves, V. W., & Co. Ltd 292
Expanded Rubber Co. Ltd 275
Explosives Trades Ltd 300
Extrudex Ltd 275

FLT Ltd 292
Fairclough, Alfred, Ltd 279
Fairclough, J. H., Ltd 279
Fairlie, H. C., & Co. Ltd 272
Farmers Company Ltd 272
Farnell Carbons Ltd 292
Febrail Ltd 281
Federated Paints Ltd 291
Felling Chemical Works 322
Fergusson, Alexander, & Co. Ltd 295
Ferodo Ltd 323
Ferris & Co. Ltd 274
Fine Spinners & Doublers 286
Fireproof Celluloid Syndicate Ltd 275
Firestone 302
Fison, Packard and Prentice Ltd 292
Fisons Fertilizers Ltd 292, 325
Fisons Horticulture Ltd 292
Fisons Ltd 289, 292, 310
Fisons Overseas Ltd 292
Fisons Pest Control Ltd 292
Fisons Pharmaceuticals Ltd 292
Fixol & Stickphast Ltd 324
Flamingo Foam Ltd 309
Flockhart, Duncan, & Co. Ltd 294
Fluorspar Ltd 306
Food Industries Ltd 324
Forestal Industries (U.K.) Ltd 292
Forestal Land, Timber & Railway Co. Ltd 292
Formica International Ltd 293
Forth Chemicals Ltd 274, 280, 289, 290, 309
Foster, Blackett & James Ltd 295
Fricker's Metal & Chemical Co. Ltd 301
Frys Diecasting Ltd 295
Frys Metal Foundries Ltd 295
Fulford-Vitapointe 292
Fullers' Earth Union Ltd 305, 306
Fumigation Services Ltd 316

Gardinol Chemical Co. Ltd 298
Gas Light & Coke Co. 311
Gascoigne-Crowther Ltd 297
Gaskell, Deacon & Co. Ltd 299
Geddes, W. & A., Ltd 317
Geigy Colour Co. Ltd 293
Geigy (Holdings) Ltd 293
Geigy, J. R., SA 284

Geigy Pharmaceutical Co. Ltd 293
Geigy (U.K.) Ltd 293
Geka Trading Co. Ltd, Northern Nigeria 269
Gemec Ltd 324
Genatosan Ltd 292
General Industrial Paints Ltd 273
Genoxide Ltd 305
Gerard Brothers Ltd 287
Gerhardt, C. F., Ltd 293
Gerhardt-Penick Ltd 293
Gibbs, J. R., Ltd 274
Gibbs Pepsodent Ltd 324
Gibbs Proprietary Ltd 324
Givaudan & Co. Ltd 293
Givaudan, L., et Cie, SA, Switzerland 293
Glaxo Group 274, 329
Glaxo Group Ltd 293
Glaxo International Ltd 329
Glaxo Laboratories Ltd 293, 294
Glaxo Research Ltd 294
Glaxo–Allenburys (Export) Ltd 294, 329
Glebe Mines Ltd 305, 306
Glovers (Chemicals) Ltd 295
Gloy & Empire Adhesives Ltd 324
Glycerine Ltd 324
Golden Valley Colours Ltd 295
Golden Valley Ochre & Oxide Company (Wick) Ltd 295
Golden Valley Ochre and Oxide Co. (Yate) Ltd 295
Goodbody Ltd 270
Goodlass Wall & Co. Ltd 295
Goodlass Wall & Lead Industries Ltd 281, 295, 301
Goodrich, B. F. (U.S.A.) 290
Goodyear 302
Goulding, W. & H. M., Ltd 316
Grace, W. R., & Co. U.S.A. 295
Grace, W. R., Ltd 295
Graesser R., Ltd 296, 304
Graesser Salicylates Ltd 296
Graesser–Monsanto Ltd 296, 309
Graesser–Thomas, F. R., & Co. 296
Grange Chemicals Ltd 274, 280, 289, 290
Greef, R. W., & Co., Ltd 281, 301
Greenwich Leathercloth Co. 285
Greenwich Plastics Ltd 285
Gregory, Walter, & Co. Ltd 296
Griffiths Bros. & Co. London Ltd 296
Grove Chemical Co. Ltd 279
Guimet, Usines 315
Guinness, Arthur, Son & Co. Ltd 287
Gurr, Edward, Ltd 297
Gwalia Fertilisers (Briton Ferry) Ltd 306

Hadleigh Crowther Ltd 297
Haigh, John, & Co. Ltd 324
Haldane, Robert, & Co. Ltd 297
Halex Ltd 275
Hall, John, & Sons (Bristol & London) Ltd 276
Hampshire, F. W. 315
Hardie, J. & G., & Co. Ltd 272
Hardman & Co. 278
Hardman & Holden Ltd 278, 304
Hardy, Harry, Ltd 316
Harker Stagg Ltd 326
Harland, William, & Son Ltd 273
Harley, Thomas, & Co. Ltd 316
Harrison & Son (Hanley) Ltd 295
Harshaw Chemical Co, U.S.A. 297
Harshaw Chemicals Ltd 297
Harvey Pharmaceuticals 313
Hatrick, W. & R., Ltd 294
Healey Meters Ltd 297
Hearon, Squire & Francis Ltd 273
Hedon Chemicals Ltd 289
Hedon Monomers Ltd 286, 290
Hepworth, B., & Co. Ltd 297
Hercules Powder Co. 297
Hickson & Partners Ltd 298
Hickson & Welch Ltd 297, 298
Hickson & Welch (Holdings) Group 298
Hickson's Timber Impregnation Co. (GB) Ltd 298
Hill, Davy & Hodgkinsons Ltd 273
Hilton Hirst & Co. Ltd 316
Hinshelwood, Thos., & Co. Ltd 298

Hoffmann–La Roche, F., & Co., Basle 316
Holliday, L. B., & Co. Ltd 298
Holloways Pills Ltd 276
Holmes, G. B., Ltd 306
Holyoak Products Ltd 296
Honeywill & Stein Ltd 289
Honeywill-Atlas Ltd 289
Honeywill-DuBois Ltd 289
Hooker Chemical Corporation, Niagara 269
Hopkin & Williams Ltd 298, 322
Hornflowa Ltd 323
Horrocks, Thos., & Sons Ltd 304
Howard, E. R., Ltd 328
Howard & Sons 298, 306, 322
Howards of Ilford Ltd 306
Hoyt Metal Co. 277
Huffer & Smith Ltd 293
Hughes, F. A., & Co. Ltd 289, 290
Hughes & Lancaster 268, 269
Hull Chemicals Ltd 299
Hull Cod Liver Oils Ltd 308
Hull Fish Meal & Oil Co. Ltd 308
Hull & Liverpool Red Oxide Co. Ltd 329
Hunt Brothers 305
Hunt Brothers (Castleford) Ltd 305
Hunt, William, & Sons 299
Hunter–Penrose Littlejohn 303
Hutchinson, John, & Co. 299

ICI (Alkali) Ltd 273, 300
ICI (Dyestuffs) Ltd 300
ICI (Explosives) Ltd 300
ICI (Fertiliser and Synthetic Products) Ltd 300
ICI Fibres Ltd 300, 301
ICI (General Chemicals) Ltd 300
ICI (Hyde) Ltd 300, 301
ICI (Leathercloth) Ltd 300
ICI (Lime) Ltd 300
ICI (Metals) Ltd 300
ICI (Paints) Ltd 300
ICI (Plastics) Ltd 300
IG (Germany) 299, 312
Ibbetson & Co. 302
Ilford Ltd 275, 301
Imco Container Co. Ltd 325
Imperial Chemical Industries Ltd (ICI) 273, 274, 280, 281, 286, 291, 299, 300, 316, 317, 322, 325
Imperial Chemical Industries of Australia and New Zealand Ltd 270
Imperial Chemical (Pharmaceuticals) Ltd 300
Imperial Metal Industries Ltd 300, 301
Imperial Smelting Corporation Ltd (ISC) 276, 281, 301, 306, 322
Imperial Tobacco Co. (of Great Britain & Ireland) Ltd 280
Improved Liquid Glue Co. 279
Industrial Soaps Ltd 324
Insecta Laboratories Ltd 316
International Chemical Co. Ltd 328
International Nickel Ltd 302
International Nickel Co. of Canada Ltd 302
International Nickel Company (Mond) Ltd 302
International Printing Ink Corporation 273
International Protein Products Ltd 279
International Synthetic Rubber Co. Ltd 291, 302
International Technical Development Ltd 303
International Toxin Products 292
Iridon Ltd 285
Iron Jelloid Co. Ltd 276
Irving Yeast Vite Ltd 276
Izal Ltd 310
Izal (Overseas) Ltd 310

Jablo Plastics Industries Ltd 309
Jackson Polythene Ltd 275
Janda Chemicals Ltd 274
Jenson & Nicholson Ltd 276
Jeyes Group Ltd 302
Jeyes Sanitary Compounds Co. Ltd 302

Johnson & Johnson (Great Britain) Ltd 311
Johnson & Sons 303
Johnson & Sons (Assayers) Ltd 303
Johnson & Sons Manufacturing Chemists Ltd 303
Johnson & Sons Smelting Works Ltd 303
Johnson, Matthey & Co. Ltd 303
Johnson, W. W. & R., & Sons Ltd 295
Johnson's Ethical Plastics Ltd 311
Johnsons of Hendon (Holdings) Ltd 303
Johnsons-Hunter-Penrose Littlejohn Ltd 303
Jones Gas Process Co. Ltd 313
Judson, Daniel, & Son 296
Jungdahls, C., & Fabriks, Sweden 283

Kalle, Germany 312
Kaylene Chemicals Ltd 292
Kemball Bishop & Co 313
Kemet Products Ltd 324
Keystone Paint & Varnish Co. 276
Kleemann Group 309
Kleemann, O. & M., Ltd 274, 309
Kleestron Ltd 274, 309
Knight, John, Ltd 324
Knights Oil & Chemical Co. Ltd 274
Koch Laboratories Ltd 303
Koch-Light Laboratories Ltd 303
Kolok Manufacturing Co. Ltd 312
Kynoch Ltd 300

Lake & Cruickshank Ltd 304
Lambeth & Co. (Liverpool) Ltd 279
Lancashire Chemical Works Ltd 304
Lancashire Cotton Corporation 286
Lancashire Tar Distillers Ltd (LTD) 296, 304
Lankro Chemicals Ltd 304
Lansil Ltd 284
Laporte Acids Ltd 305, 306
Laporte, B., Ltd 305
Laporte Chemicals Ltd 276, 301, 305, 306
Laporte Industries Ltd 271, 305, 306
Laporte Minerals Ltd 305
Laporte Titanium Ltd 305, 306
Lawes Chemical Co. Ltd 306
Lawes Chemical Manufacture Co. Ltd 306
Learner, A., & Co. Ltd 273
Leas Cliff Products Ltd 313
Lederle Laboratories 287
Leech, Neal & Co. Ltd 300
Leicester, Lovell & Co. Ltd 306
Leiner, P., & Sons (Wales) Ltd 277
Leitch, John W., & Co. 298
Lennig Chemicals Ltd 307, 329
Leo Lines Ltd 270
Lever Brothers Ltd 323, 324
Levinstein Ltd 298, 300
Librex Lead Company Ltd 295
Light, L., & Co. Ltd 303
Lilly, Eli 290
Lilly, Eli, & Co. Ltd 288, 307
Lilly, Eli, International Corporation 307
Lilly Industries Ltd 307
Lilly Research Laboratories Ltd 307
Liversidge, Allen, Group of Companies 280
Lloyd Anphar 307
Lloyd Group 307
Lloyd, Howard, & Co. Ltd 307
Lloyd's Research 307
Lloyd-Hamol 307
Locke, Lancaster 295
Lomas Gelatine Works Ltd 279
London Lead Oxide Co. Ltd 295
London Tar & Chemical Co. Ltd 319
Lord, C., & Lord, S. A. 304
Lord, J. E. C., (Manchester) Ltd 304
Lovell & Christmas Ltd 306
Low Temperature Carbonisation Ltd 284

Lowe, Charles, & Co. (Manchester) Ltd 324
Lugsdale Chemical Co. 276

Macfarlan, J. F., & Co. Ltd 294
Macfarlan Smith Ltd 294
McKechnie Brothers Ltd 325
Macleans Ltd 276
Macpherson, Donald, & Co. Ltd 307
Magadi Soda Co., Kenya 299
Major & Co. 308
Malehurst Barytes Co. Ltd 305
Manchester Computer Centre Ltd 304
Manchester Creosote & Storage Co. Ltd 304
Manucol Products Ltd 271
Marchon Products Ltd 269, 270
Marfleet Refining Company Ltd 308
Marshall, William E., Ltd 272
Massey, Charles, & Son Ltd 279
May & Baker Ltd (M. & B.) 308
Mays Chemical Manure Co. 272
Mead Johnson & Co. of America 274
Mead Johnson Ltd 274
Mechema Ltd 308
Meggeson & Co. 270
Meggitt, S., & Sons Ltd 279
Meggitts (1917) Ltd 279
Menley & James Laboratories 319
Merck & Co., Inc., America 310
Merck Sharp & Dohme Ltd 310
Metallurgical Chemists Ltd 308
Metcalfe (Tar Distillers) Ltd 304
Methylating Co. Ltd 289
Metrotect Ltd 319
Michelin 302
Micronised Pigments Ltd 295
Middleton & Co. Ltd 274
Midland Silicones Ltd 269, 270
Midland Tar Distillers Ltd (MTD) 308, 316
Milne, Edward, & Co. 279
Minsal Ltd 327
Mirvale Chemical Co. Ltd 309
Mitcham Poultry Food Co. 279
Mobil 274
Mobil Chemicals 274
Mobil Chemicals Ltd 309
Mobil Chemical Sales Ltd 309
Mody & Co. Ltd 307
Monarch Chemical Co. 305
Mond Nickel Co. Ltd 302
Mondart Ltd 285
Monsanto Chemicals Ltd 274, 280, 291, 309
Monsanto Company of America 284, 296
Monsanto Europe S.A. 284
Morson, Thomas, & Son Ltd 310
Morton Chemical Co. (Chicago) 328
Morton Sundour Fabrics Ltd 300
Morton–Williams Ltd 328
Mouldensite Ltd 275
Mucklow & Co. 279
Mulford, H. K., & Co. Ltd 310
Murgatroyd's Mid-Cheshire Salt Works Co. 310
Murgatroyd's Salt & Chemical Co. Ltd 289, 290, 292, 310
Murgatroyd's Vacuum Salt Co. Ltd 310
Murphy Chemical Co. Ltd 294
Murray, O., & Co. Ltd 279, 280
Muspratt, James, & Sons Ltd 299

Nathan, Joseph, & Co. (Ltd) 294
National Carbon Company 324
National Coal Board (Coal Products Division) (NCB–CPD) 310
National Lead Co., U.S.A. 277
National Smelting Co. 301
National Titanium Pigments Ltd 305
Natural Chemicals Ltd (Phyllosan) 276
Nelson Acetate Ltd 286, 297
Nelson, James, Ltd 286
Nelson Silks Ltd 286

Nemlin Chemicals Ltd 304
Ness, Thomas, Ltd 310
Newalls Insulation Co. Ltd 322, 323
Newalls Insulation & Chemical Co. Ltd 322,
Newcastle-upon-Tyne Zinc Oxide Co. Ltd 291
Newton, Chambers & Co. Ltd 310
Nicholson, John, & Sons 305
Nicolson, W. B. (Scientific Instruments) Ltd 299
Nobel Chemical Finishes Ltd 300
Nobel Industries Ltd 299, 300
Nobel's Explosives Co. Ltd 300
North British Plastics Ltd 281
North British Rubber 302
North Thames Gas Board (NTGB) 311, 319
North Western Co-operative Tar Distillers Ltd 304
Noury & Van der Lande NV., Holland 311
Novadel Ltd 311
Nuodex Ltd 291
Nutfield Manufacturing Co. Ltd 311
Nutritional Consultants Ltd, Winnipeg 283

Oidas Metals Co. Ltd 295
Oldbury Alkali Co. Ltd 299
Oldbury Electro-Chemical Co., Niagara 269
Oldroyd, Wm., & Sons Ltd 279
Olin Mathieson Ltd 320
Oliver Wilkins 300
Onyx Chemical Corporation, U.S.A. 311
Optica United Kingdom Ltd 299
Organic Research Chemicals Ltd 303
Organon International Group 311
Organon Laboratories Ltd 311
Orobis, Ltd 289, 290
Oronite Chemical Co. 280
Orr's Zinc White Ltd 276, 301
Ortho Pharmaceuticals Ltd 311
Osmond Estates Ltd 312
Osmond & Sons Ltd 312
Osmonds Aerosols Ltd 312
Osmonds Pharmaceuticals Ltd 312
Ouseburn Trading Co. Ltd 291
Oxirane Ltd 304
Ozalid Company Ltd 312

PC Products (1001) Ltd 287
PR Chemicals Ltd 282
Pacific Borax 278
Pain, James, & Sons Ltd 312
Pains–Wessex Ltd 312
Palatine Gas Corporation Ltd 313
Pan Britannica Industries Ltd 321, 322
Paper Makers Chemicals Ltd 297
Parazone Co. Ltd 302
Paripan Ltd 283
Parke, Davis & Co. 312
Parsons, Thos., & Sons Ltd 308
Paterson, William, & Sons (Aberdeen) Ltd 294
Pearson, William, Ltd 312, 313
Pearson, William, (Holding) Co. Ltd 313
Pearson, William, (Hull) Co. Ltd 313
Pearson's Antiseptic Co. Ltd 313
Peboc Ltd 313
Pechiney SA, France 323
Penetone Co. Ltd 283
Penetone International Corporation of America 283
Penetone-Paripan 283
Penick, S. B., & Co. 293
Perkin & Sons 300
Permoglaze Ltd 277
Permutit Company Ltd 313
Pestcure Ltd 316
Petrochemicals Ltd 304, 318
Pfizer, Charles, & Co. 313
Pfizer Group 313, 329
Pfizer International 313
Pfizer Ltd 313
Phensic Ltd 276
Philblack Ltd 313
Philips Gloilampfabriken, NV 313

Philips-Duphar, Holland 287
Phillips Petroleum Co., U.S.A. 314
Phosferine (Ashton & Parsons) Ltd 276
Pictograph Ltd 303
Pictorial Machinery Ltd, Chemical Division 303
Pilkington, Wm., & Sons 299
Pinchin Johnson & Associates Ltd 286
Pinchin Johnson & Co. Ltd 286, 318
Pinkerton, Gibson & Co. Ltd 294
Pintus, R., & Co. Ltd 279
Pirelli 302
Pitch Plastic Polymers Ltd 310
Plant Protection Ltd 301
Plastic Containers Ltd 285
Plymouth Tar Distillers Ltd 278
Pointing, L. J., & Son Ltd 314
Polyglaze Ltd 309
Polythane Fibres Ltd 284
Portaccord Ltd 308
Potter, E. P., & Co. Ltd 272
Potter & Moore Ltd 269, 270
Powder Metallurgy Ltd 277
Premier Yeast Co. 313
Prentice Brothers 292
Price's (Bromborough) Ltd 324
Printar Industries Ltd 282
Proban Ltd 269, 270
Proctor & Gamble Ltd 314
Prodorite Ltd 314
Proprietary Perfumes Ltd 324
Pure Chemicals Ltd 301

Radiochemical Centre 314
Read, Holliday & Sons Ltd 298, 300
Reckitt & Colman Holdings Ltd 315, 326
Reckitt & Sons Ltd 315
Reckitt's (Colours) Ltd 315
Redmanol Ltd 275
Redwoods Chemical Works Ltd 278
Reed, R. F., Ltd 327
Reichhold Chemicals Ltd 315
Rentokil Group Ltd 316
Rentokil Laboratories Ltd 316
Research and Manufacturing Co. Ltd 283
Resinous Chemicals Ltd 281
Rexall Drug Co. Ltd 325
Rexall Drug and Chemical Co. (U.K.) Ltd 325
Rexolin Chemicals AB, Sweden 277
Rhone-Poulenc, France 308
Richardson Bros. Ltd 316
Richardsons Chemical & Manure Co. Ltd 301, 316
Richardsons Fertilisers Ltd 301, 316
Richmond Aerosols Ltd 287
Rid-Pest Ltd 316
Riker Laboratories Ltd 325, 326
Riley, John, & Sons Ltd 277
Rilite Ltd 285
Rio Tinto Dow of Canada 322
Rio Tinto-Zinc Corporation Ltd 281, 301, 322
Robinson Bros. Ltd 308, 316
Robinson, E. S. & A. 301
Robinson, James, & Co. Ltd 316
Robinson Plastics Films Ltd 301
Roche Products Ltd 316
Rohm and Haas, U.S.A. 283, 307, 327
Romney, House of 313
Ronsheim & Moore Ltd 298
Ronuk Ltd 310
Rose, J. L., Ltd 279
Rose, Sir W. A., & Co. Ltd 276
Rosenberg, J. & H., Group of Companies 293
Ross, James, & Co. (Lime Wharf) Ltd 317
Roussel Laboratories Ltd 323, 329
Roussel-Uclaf Ltd 329
Roussel-Uclaf SA, Paris 323
Rowe Bros. Ltd 295
Rowland, James, Ltd 274
Rowntree Organisation 321
Royal Dutch Shell Group of Companies 318
Ruberoid Co. Ltd 310
Runnymede Dispersions Ltd 283
Russell, G. C., Ltd 279

SAI Horticulture Ltd 317
SC Co. Ltd 305
St. Just, Theodore, & Co. Ltd 278
Sandoz Chemical Co. Ltd 317
Sandoz Ltd 284
Sandoz Products Ltd 317
Sandoz Products (Ireland) Ltd 317
Sandoz Trading & Shipping Co. Ltd 317
Saturn Industrial Gases Ltd 269
Saunders & Saunders Ltd 305
Schenectady Chemical Inc., New York 308
Schenectady Midland Ltd 308
Schermuly Limited 317
Schermuly Pistol Rocket Apparatus Ltd 317
Scientex Ltd 316
Scott, W. R., & Co. 279
Scott Bader & Co. Ltd 317
Scottish Agricultural Industries Ltd 301, 317, 322
Scottish Dyes Ltd 300
Scottish (Pentland) Fertilisers Ltd 317
Scottish Tar Distillers Ltd (STD) 317
Searle, G. D., & Co., Chicago 318
Searle, G. D., & Co. Ltd 318
Serta Ltd 302
Shankland, G. A., Ltd 279
Sharp & Dohme Ltd 310
Shawinigan Chemicals (Canada) 288, 289
Sheffield Chemical Co. Ltd 305
Shell Chemical Co. Ltd 304, 318
Shell Chemical Manufacturing Ltd 318
Shell Chemicals Ltd 318
Shell Company, 273, 274
Shellstar Ltd 318
Sherley, Prichard 276
Sherwoods Paints Ltd 307
Silica Gel Ltd 296
Simeon, C., & Co. Ltd 279, 280
Sinclair, Owen, & Co. Ltd 287
Sissons Brothers & Co. Ltd 315
Smith Kline & French Laboratories Ltd 319
Smith, Mark, Ltd 326
Smith & Nephew Associated Companies Ltd 318
Smith, Richard, Ltd 318
Smith, T. & H., Ltd 294
Smith, T. J., & Nephew 318
Smithie, W. C., & Co. Ltd 304
Society of Chemical Industry, Basle 284
Socony Mobil Oil Inc., U.S.A. 273, 309
Solway Chemicals Ltd 269, 270
Solway Dyes Ltd 300
Somerset Pharmaceuticals Ltd 296
Sondheimer, M., (Sondal Works) Ltd 307
Sorex (London) Ltd 326
South Eastern Gas Board (SEGB) 319
South Eastern Tar Distillers Ltd (SETD) 319
South Metropolitan Gas Company 319
South Western Gas Board 278, 282, 320
South Western Tar Distilleries Ltd (SWTD) 278, 319
Southern Gas Board 320
Spelthorne Metals Ltd 276
Spence, Peter, & Sons Ltd 281, 301, 305, 306
Spencer, Chapman & Co. 278
Spencer, Chapman & Messel Ltd 277, 278
Squibb, E. R., & Sons Ltd 320
Standard Essence Company Ltd 321
Standard Essence Co. (Ireland) Ltd 321
Standard Fireworks Ltd 320
Standard Oil Co., California 280
Standard Oil Co., New Jersey 291
Standard Soap Co. Ltd 279
Stanford Wylie & Fraser Ltd 273
Staveley Coal & Iron Co. Ltd 320
Staveley Iron & Chemical Co. Ltd 320
Stephenson, Robert, & Son Ltd 272
Stevenson & Howell Ltd 320
Storey, Joseph, & Co. Ltd 321

Strathclyde Paint Co. Ltd 291
Streetly Manufacturing Co. Ltd 323
Stretchables International Ltd 284
Sturge (Citric) Ltd 321
Sturge, John & E., Ltd 321
Styrene Co-Polymers Ltd 276, 318
Styrene Products Ltd 318
Sun Chemical Corporation, U.S.A. 328
Sunvi-Torrax Ltd 268
Sutcliffe, Speakman & Co. Ltd 321
Swallowfield Aerosols Ltd 296
Swan Chemical Co. Ltd 304
Synthetic Ammonia & Nitrates Ltd 299
Synthetic Chemicals Ltd 328
Synthite Ltd 289, 321, 322

Tangyes Ltd 269
Technical Products Ltd 318
Tees Bone Mill Ltd 279
Tees Refinery Co. 279
Tellam Products Ltd 324
Tennant, Charles, & Co. Ltd 321, 322
Tennant, Charles, & Co. of Carnoustie Ltd 322
Tennant, Charles, & Co. of St. Rollox Ltd 321
Tennant, Charles & Partners Ltd 299
Tennants Consolidated Ltd 321
Tennants (Lancs.) Ltd 321
Tennant's Textile Colours Ltd 321, 322
Thames Alum Ltd 271, 305
Thermalon Ltd 285
Thomas, J. L., & Co. Ltd 324
Thorium Ltd 301, 314, 322
Three Hands Products Ltd 302
Titan Co. 301
Titanium Intermediates Ltd 281
Todd, William R., & Son Ltd 273
Transparent Paper Ltd 325
Tupperware Manufacturing Co. Ltd 325
Turner Brothers Asbestos Co. Ltd 323
Turner & Newall Ltd 322, 323
Typke & King 291
Tyrer, Thomas, & Co. Ltd 269

Uclaf Ltd 323, 329
Uclaf SA, Paris 323
Ulster Fertilisers Ltd 301, 316
Ulster Manure Co. 301
Unigate Ltd 327
Unilever Ltd 285, 323, 324
Union Carbide Ltd 275, 291, 324
Union Carbide Corporation of U.S.A. 275, 289, 324
Union Carbide & Carbon Corporation, U.S.A. 299, 324
United Alkali Co. Ltd. 299, 300, 321
United Coke & Chemicals Co. Ltd 325
United Dairies Ltd 327
United Dyewood Corporation of New Jersey 279
United Steel Companies Ltd 325
United Sulphuric Acid Corporation Ltd 286, 301, 325
Universal Laboratories Ltd 313
Upjohn Ltd 325
Upsil Ltd 323

Vantorex Ltd 325, 326
Veno Drug Ltd 276
Vinatex Ltd 315
Vinyl Products Ltd 315
Visking Corporation 301
Vitamins Group 326
Vitamins Ltd 326
Vitax Ltd 326
Volker Lighting Corporation Ltd 322
Vulcan Chemical Co. Ltd 328

Waeco Ltd 312
Walker Chemical Co. Ltd 324
Walker, J. & T., Ltd 279
Walkers, Parker & Co. Ltd 295
Ward, Blenkinsop & Co. Ltd 326

Warrick Bros. 270
Washington Chemical Co. Ltd 322
Weardale Lead Co. Ltd 301
Weaver Refinery Co. Ltd 279
Weil, Joseph, & Sons Ltd 309
Wellcome Foundation Ltd 326, 327
Wembley Paint Company 273
Wessex Aircraft Engineering Co. 312
West Norfolk Fertilizers 292
Whey Products Ltd 327
Whiffen & Sons Ltd 292
White, A. J., Ltd 319
White, John & James, Ltd 272
Whitmoyer Laboratories Inc., U.S.A. 327
Whitmoyer-Reed Ltd 327, 329
Wiggin, Henry, & Co. Ltd 302
Wilkinson Group of Companies 307
Wilkinson, James, & Sons Ltd 305
Wilkinson, L. G., Ltd 307
Wilkinson Paints Ltd 308
Willesden Varnish Company 273
Williams Ansbacher Ltd 328
Williams (Hounslow) Ltd 327, 328
Williams (Hounslow) Yorkshire Ltd 328
Williamson & Corder Ltd 279
Williamson Morton & Co. 291
Wilson Blackadder & Co. Ltd 273
Wimsol Ltd 302
Witco Chemical Company Ltd 328
Witco Chemical Co. Inc., U.S.A. 328
Withins Paper Staining Co. Ltd 301
Wokingham Plastics Ltd 274, 309
Wood Preservation Ltd 316
Woodworm and Dry Rot Control Ltd 316
Woolley & Arnfield Ltd 274
Woolley, James, Sons & Co. Ltd 274
Wrigley, Fred H., Ltd 297
Wyeth, John, & Brother Ltd 328
Wyleys Ltd 270

YTD Plant Hire Ltd 328
Yeast Vite Ltd 276
Yorkshire Dyeware & Chemical Co. Ltd 328
Yorkshire Fumigation Services Ltd 316
Yorkshire Tar Corporation Ltd 328
Yorkshire Tar Distillers Ltd (YTD) 328
Young, B., & Co., Ltd 279
Young, Robert, & Co. Ltd 329
Youngs (Pharmaceuticals) Ltd 329

Zerolit Ltd 313
Zwanenberg Meat Co. 311

Part C
Index of Trade Associations

Acetylene Association, British 334
Adhesives, Association of Manufacturers of Vegetable 331
Adhesives Manufacturers' Association 331
Aerosol Manufacturers' Association, British 334
Agricultural Chemicals, Association of British Manufacturers of 332
Agricultural Products, Associated Manufacturers of Veterinary and 331
Ammonia (Sulphate of) Association and Committee 336
Ammonia (Sulphate of) Federation, British 336
Animal Medicine Makers' Association 331
Aromatic Compound Manufacturers' Association, British 334
Associated Manufacturers of Veterinary and Agricultural Products (AMVAP) 331
Association of British Chemical Manufacturers (ABCM) 330, 331, 332, 333, 334, 335, 336, 338, 339
Association of British Insecticide Manufacturers 332
Association of British Manufacturers of Agricultural Chemicals (ABMAC) 332
Association of British Organic and Compound Fertilisers Ltd 332
Association of British Organic Fertilisers 332
Association of the British Pharmaceutical Industry (ABPI) 332
Association of Chemical & Allied Employers (AC & AE) 330, 331, 332, 333, 334
Association of Drug & Fine Chemical Manufacturers 333
Association of Manufacturers of Vegetable Adhesives 331
Association of Tar Distillers (ATD) 333

Benzole and Allied Products Association, National 339
Biological Research Association, British Industrial 335
British Acetylene Association 334
British Aerosol Manufacturers' Association (BAMA) 334
British Aromatic Compound Manufacturers' Association 334
British Chemical & Dyestuffs Traders' Association Ltd (BCDTA) 334
British Chemical Industry Safety Council (BCISC) 331, 334
British Colour Makers' Association (BCMA) 335
British Disinfectant Manufacturers' Association (BDMA) 335
British Employers' Confederation (BEC) 330, 333
British Essence Manufacturers' Association 335
British Gelatine and Glue Research Association 338
British Industrial Biological Research Association (BIBRA) 335

British Institute of Management (BIM) 330
British Man-Made Fibres Federation 335
British National Committee on Surface Active Agents (BNC) 336
British Plastics Federation (BPF) 336
British Productivity Council (BPC) 330
British Rayon Federation 335
British Rayon and Synthetic Fibres Federation 335
British Road Tar Association (BRTA) 336
British Standards Institution (BSI) 330, 335
British Sulphate of Ammonia Federation Ltd (BSAF) 336
British Sulphate of Copper Association (Export) Ltd 336
British Tanning Extract Manufacturers' Association 337
British Whiting Federation 337
British Wood Preserving Association (BWPA) 337

Chalk Lime and Allied Industries Research Association 337
Charcoal Manufacturers, National Association of 339
Chemical (and Allied) Employers' Federation 333
Chemical & Dyestuffs Traders' Association, British 334
Chemical Industries Association Ltd (CIA) 330, 332
Chemical Manufacturers, Association of British 330, 331, 332, 333, 334, 335, 336, 338, 339
Chemical Manufacturers, Scottish Association of 333
Chemical Manure Manufacturers' Association 338
Coal Tar Research Association (CTRA) 337
Colour Makers' Association, British 335
Colour & Varnish (and Paint) Manufacturers, Research Association of British 340
Confederation of British Industry (CBI) 330
Copper (Sulphate of) Association (Export), British 336
Creosote Producers' (Export) Association 337
Cresylic Acid Refiners' Association 338

Disinfectant Manufacturers' Association, British 335
Drug Club 332
Drug & Fine Chemical Manufacturers, Association of 333
Drug Trade Association, Wholesale 332
Dyestuffs Traders' Association, British Chemical and 334

Essence Manufacturers' Association, British 335

Federation of British Industries (FBI) 330
Federation of (Hide) Gelatine and Glue Manufacturers Ltd 338
Fertiliser Manufacturers' Association (FMA) 338
Fertilisers, Association of British Organic and Compound 332

Gelatine and Glue Manufacturers, Federation of 338
Gelatine & Glue Research Association 338
Glue (and Gelatine) Research Association 338
Glue Manufacturers, Federation of Gelatine and 338
Glycerine Producers' Association U.K. 341

Industrial Pest Control Association (IPCA) 339
Insecticide Manufacturers, Association of British 332
Institute of Export 330

London Chamber of Horticulture 332

Man-Made Fibres Federation, British 335
Medicine Makers' Association, Animal 331

National Association of British Manufacturers (NABM) 330, 331
National Association of Charcoal Manufacturers 339
National Benzole (and Allied Products) Association (NBA) 339
National Paint Federation 335, 339
National Sulphuric Acid Association Ltd (NSAA) 339
Northern Whiting Association 337

Paint, Colour & Varnish Manufacturers, Research Association of British 340
Paint Manufacturers, Scottish Association of 341
Paint Research Station 340
Paintmakers Association of Great Britain Ltd 339
Pest Control Association, Industrial 339
Pharmaceutical Export Group 333
Pharmaceutical Industry, Association of the British 332
Phenol Producers' Association 340
Plastics Federation, British 336
Plastics (and Rubber) Research Association of Great Britain 341
Printing Ink Manufacturers, Society of British 341
Proprietary Articles Trade Association (PATA) 340
Proprietary Association of Great Britain 333, 340

Rayon (and Synthetic Fibres) Federation, British 335
Research Association of British Paint, Colour and Varnish Manufacturers 340
Resin, Synthetic (Surface Coating) Manufacturers' Association 341
Royal Society for the Prevention of Accidents (ROSPA) 330
Rubber and Plastics Research Association of Great Britain 341

Safety Council, British Chemical Industry 331, 334
Scientific & Industrial Research, Department of (DSIR) 331, 335, 336, 337, 338, 340, 341
Scottish Association of Chemical Manufacturers 333
Scottish Association of Paint Manufacturers 341
Scottish Wholesale Druggists Association 333
Soap Makers, Society of British 341
Society of British Paint Manufacturers Ltd 339
Society of British Printing Ink Manufacturers 341
Society of British Soap Makers 341
Southern Whiting Association 337
Standardisation of Tar Products Tests Committee (STPTC) 337
Sulphate of Ammonia Association 336
Sulphuric Acid Association, National 339
Superphosphate Manufacturers' Association (SMA) 338
Surface Active Agents, British National Committee 336

Surface Coating Synthetic Resin Manufacturers' Association 341
Synthetic Resin (Surface Coating) Manufacturers' Association 341

Tanning Extract Manufacturers' Association, British 337
Tar Association, British Road 336
Tar Distillers, Association of 333
Tar Products Tests Committee, Standardisation of 337
Tar Research Association, Coal 337

Ulster Wholesale Druggists Association 333
United Kingdom Glycerine Producers' Association (UKGPA) 341

Varnish (& Paint & Colour) Manufacturers, Research Association of British 340
Veterinary and Agricultural Products, Associated Manufacturers of 331

Wages Committee of Chemical Manufacturers 333
Welwyn Hall Research Association 337
Whiting Federation, British 337
Whiting and Industrial Powders Research Council 337
Whiting Research Council 337
Wholesale Drug Trade Association 332
Wholesale Druggists Association 333
Wholesale Sundries & Proprietaries Association 333
Wood Preserving Association, British 337

Zinc Pigment Development Association 342